Plant Fungal Diseases and Crop Protection

Plant Fungal Diseases and Crop Protection

Guest Editor

Ofir Degani

MDPI

Basel • Beijing • Wuhan • Barcelona • Belgrade • Novi Sad • Cluj • Manchester

Guest Editor
Ofir Degani
Faculty of Sciences and Technology
Tel-Hai Academic College
Upper Galilee
Israel

Editorial Office
MDPI AG
Grosspeteranlage 5
4052 Basel, Switzerland

This is a reprint of the Special Issue, published open access by the journal *Journal of Fungi* (ISSN 2309-608X), freely accessible at: www.mdpi.com/journal/jof/special_issues/0O7RW5OI67.

For citation purposes, cite each article independently as indicated on the article page online and as indicated below:

Lastname, A.A.; Lastname, B.B. Article Title. *Journal Name* **Year**, *Volume Number*, Page Range.

ISBN 978-3-7258-3994-0 (Hbk)
ISBN 978-3-7258-3993-3 (PDF)
https://doi.org/10.3390/books978-3-7258-3993-3

Contents

About the Editor

Ofir Degani

Dr. Ofir Degani completed his Ph.D. at the Technion-Israel Institute of Technology (Haifa, Israel), specializing in genetic engineering approaches in phytopathology. He carried out his post-doctoral studies at the Migal-Galilee Research Institute (Israel). Currently, Dr. Degani serves as the research group director of the Phytopathology Laboratory at this institute and a senior staff member at Tel-Hai Academic College (Israel). His research integrates molecular, biochemical, and microbiological tools to advance the understanding and management of fungal diseases in plants. His group focuses on the following research areas:

1. Developing biological, chemical, and agrotechnical strategies to control major plant diseases such as maize late wilt and stalk rot, cotton charcoal rot, and onion basal rot.

2. Engineering and utilizing the plant microbiome to create eco-friendly solutions for crop protection.

3. Investigating pathogen–pathogen interactions and their implications in plant pathology to enhance integrated disease management practices.

Preface

Fungal pathogens constitute the most diverse and widespread group of plant disease agents, posing a major threat to global agriculture. Infecting leaves, roots, and seeds, these pathogens cause substantial yield losses and threaten food security. Their complex infection mechanisms and adaptability make disease control an ongoing scientific challenge, underscoring the urgent need for innovative and sustainable strategies.

Chemical fungicides have traditionally formed the foundation of plant protection. Although highly effective, particularly during severe outbreaks, their prolonged use raises concerns regarding environmental safety, human health, and the development of fungicide-resistant strains. These challenges have prompted a shift toward alternative or integrated approaches that align with sustainability objectives. Among them, combined chemical–biological (mixed-method) strategies offer promising, environmentally conscious solutions.

The deployment of resistant plant cultivars is widely recognized as a sustainable method for managing fungal diseases. However, the emergence of aggressive pathogen strains that are capable of breaching genetic resistance highlights the need for complementary strategies. In recent years, biological control agents (BCAs) such as *Trichoderma*, *Bacillus*, and *Pseudomonas* species have gained significant attention. These beneficial microbes not only suppress pathogens but also promote plant resilience through symbiotic interactions.

This Special Issue compiles recent advances in fungal disease management, featuring fourteen contributions—original research and reviews—spanning a wide array of crops and topics. These include disease control in staple crops like soybean, potato, and wheat, as well as specialized hosts such as fruit trees, mushrooms, and woody oil plants. The Special Issue also explores the influence of friendly or non-pathogenic fungi on disease development and highlights innovative tools such as microbiome engineering, nanotechnology, AI-driven diagnostics, and molecular techniques for pathogen detection and study.

Together, these studies reflect global efforts to enhance sustainable agriculture in the face of the rising challenges posed by climate change, natural habitat reduction, and increasing food demands. I extend my deepest thanks to all authors and reviewers for their valuable contributions. Their work not only expands our understanding of plant–fungal interactions but also points toward practical, field-ready solutions for protecting crops.

This collection is intended to inform and inspire researchers, agronomists, farmers, and policymakers to advance resilient, eco-friendly agricultural systems.

Ofir Degani
Guest Editor

Journal of *Fungi*

Editorial

Plant Fungal Diseases and Crop Protection

Ofir Degani [1,2]

[1] Plant Sciences Department, MIGAL—Galilee Research Institute, Tarshish 2, Kiryat Shmona 1101600, Israel; d-ofir@migal.org.il or ofird@telhai.ac.il; Tel.: +972-54-678-0114
[2] Faculty of Sciences, Tel-Hai College, Upper Galilee, Tel-Hai 1220800, Israel

Received: 26 March 2025
Accepted: 31 March 2025
Published: 1 April 2025

Citation: Degani, O. Plant Fungal Diseases and Crop Protection. *J. Fungi* **2025**, *11*, 274. https://doi.org/10.3390/jof11040274

Fungi represent the largest group of plant pathogens, infecting their hosts via leaves, seeds, and roots. These pathogens cause significant crop damage through diverse mechanisms, leading to reduced global production and an ongoing demand for innovative control strategies [1,2]. Effective management of fungal diseases in crops depends upon a comprehensive understanding of the fungi involved, identifying susceptible growth stages, and recognizing environmental factors that influence disease progression [3].

Chemical pesticides have long been fundamental in crop protection strategies against phytopathogens [4]. Nevertheless, growing public concerns regarding synthetic chemical usage [5] and the emergence of fungicide-resistant fungal strains [6] necessitate the exploration of alternative, environmentally friendly approaches [7,8]. Despite these concerns, traditional chemical methods have proven highly effective, especially in severe disease outbreaks, benefiting from well-established production and supply chains, and are familiar to farmers and agricultural field guides. Therefore, integrating biological and chemical approaches (mixed-method management) has been proposed as a transitional strategy from conventional chemical interventions toward greener technologies [9–11]. Such combinations may also enhance the robustness and stability of protective microorganisms under varying environmental conditions, ultimately benefiting agricultural practices.

At present, there is an increasing global scientific emphasis on developing sustainable crop production methods capable of meeting growing demands without environmental harm. Utilizing resistant cultivars is a promising method due to its low cost and compatibility with various agricultural practices [12]. However, phytopathologists have noted the emergence of highly aggressive pathogen strains capable of overcoming cultivar resistance, thus compromising their effectiveness [13,14].

Recently, biological control agents (BCAs) have gained considerable attention as potential tools for managing fungal diseases [15–17]. Specific BCAs, such as *Trichoderma* spp., *Bacillus* spp., *Pseudomonas fluorescens*, *Beauveria bassiana*, and *Gliocladium* spp., offer numerous benefits for crop protection. Some BCAs function as beneficial endophytes, living symbiotically within plants [18–20].

Exploring innovative green technologies to enhance fungal pathogen control is a critical priority, and the full potential of these advancements is only beginning to unfold [21]. Cutting-edge technologies, including microbiome engineering, nanotechnology, artificial intelligence (AI), genome editing, RNA interference, and functional peptides, could facilitate precise and sustainable management strategies for fungal diseases in agriculture [22–24]. Microbiome-based solutions and nanoparticles enable targeted pathogen suppression with minimal environmental impacts, while AI and machine learning significantly enhance disease detection accuracy—such as analyzing remote-sensing data for field monitoring and treatment optimization [25,26]. Furthermore, advancements in molecular diagnostics, such as endpoint PCR and real-time PCR, have revolutionized accurate pathogen identification, enabling early detection and improved disease management strategies [27,28].

This Special Issue comprises two review articles and eleven original research papers, highlighting recent innovations and scientific advances in plant fungal diseases and their agricultural implications. It also includes discussions of non-pathogenic fungi associated with plants that influence disease onset and severity. The broad range of topics covered underscores the complex challenges we face in modern agriculture, including significant diseases affecting various hosts—from fruit trees and woody oil plants to staple crops, such as soybean, potato, wheat, strawberries, and even edible mushrooms, such as morels (*Morchella* spp.) and white button mushrooms (*Agaricus bisporus*). Furthermore, it explores innovative biocontrol methods and novel weed management approaches by mycoherbicide.

I extend my sincere appreciation to all contributing authors and reviewers whose valuable efforts were crucial for the successful publication of this Special Issue. Given the complexity and dynamic nature of ensuring global crop health, food security, and sustainable agriculture under global changes, including climate change and industrialization, there are significant challenges ahead in developing and optimizing effective fungal control technologies. The promising outcomes of ongoing research should encourage stakeholders and industry partners to actively translate these innovative management strategies into practical agricultural applications.

Conflicts of Interest: The author declares no conflict of interest.

References

1. Mapuranga, J.; Zhang, N.; Zhang, L.; Chang, J.; Yang, W. Infection strategies and pathogenicity of biotrophic plant fungal pathogens. *Front. Microbiol.* **2022**, *13*, 799396.
2. Rodriguez-Moreno, L.; Ebert, M.K.; Bolton, M.D.; Thomma, B.P.H.J. Tools of the crook- infection strategies of fungal plant pathogens. *Plant J.* **2018**, *93*, 664–674.
3. Doehlemann, G.; Ökmen, B.; Zhu, W.; Sharon, A. Plant pathogenic fungi. *Microbiol. Spectr.* **2017**, *5*. [CrossRef]
4. Zhou, W.; Li, M.; Achal, V. A comprehensive review on environmental and human health impacts of chemical pesticide usage. *Emerg. Contam.* **2025**, *11*, 100410.
5. Shekhar, C.; Khosya, R.; Thakur, K.; Mahajan, D.; Kumar, R.; Kumar, S.; Sharma, A.K. A systematic review of pesticide exposure, associated risks, and long-term human health impacts. *Toxicol. Rep.* **2024**, *13*, 101840.
6. Castroagudín, V.L.; Ceresini, P.C.; de Oliveira, S.C.; Reges, J.T.; Maciel, J.L.; Bonato, A.L.; Dorigan, A.F.; McDonald, B.A. Resistance to QOI fungicides is widespread in Brazilian populations of the wheat blast pathogen *Magnaporthe oryzae*. *Phytopathology* **2015**, *105*, 284–294.
7. Madlhophe, S.; Ogugua, U.V.; Makhubu, F.N.; Figlan, S. Use of biological control agents for managing fungal pathogens in Solanaceae crops: Progress and future perspectives—A review. *Discov. Appl. Sci.* **2025**, *7*, 83.
8. Thambugala, K.M.; Daranagama, D.A.; Phillips, A.J.L.; Kannangara, S.D.; Promputtha, I. Fungi vs. Fungi in biocontrol: An overview of fungal antagonists applied against fungal plant pathogens. *Front. Cell. Infect. Microbiol.* **2020**, *10*, 604923.
9. Gordani, A.; Hijazi, B.; Dimant, E.; Degani, O. Integrated biological and chemical control against the maize late wilt agent *Magnaporthiopsis maydis*. *Soil Syst.* **2023**, *7*, 1. [CrossRef]
10. Ons, L.; Bylemans, D.; Thevissen, K.; Cammue, B.P.A. Combining biocontrol agents with chemical fungicides for integrated plant fungal disease control. *Microorganisms* **2020**, *8*, 1930. [CrossRef]
11. Ruano-Rosa, D.; Arjona-Girona, I.; López-Herrera, C. Integrated control of avocado white root rot combining low concentrations of fluazinam and *Trichoderma* spp. *Crop Prot.* **2018**, *112*, 363–370. [CrossRef]
12. Mundt, C.C. Durable resistance: A key to sustainable management of pathogens and pests. *Infect. Genet. Evol.* **2014**, *27*, 446–455.
13. Shofman, G.; Degani, O. Mixed fungal strains challenge host resistance: Insights into *Magnaporthiopsis maydis* pathogenicity in maize. *Front. Microbiol.* **2025**, *16*, 1520237. [CrossRef]
14. Holmes, G.J.; Ojiambo, P.S.; Hausbeck, M.K.; Quesada-Ocampo, L.; Keinath, A.P. Resurgence of Cucurbit Downy Mildew in the United States: A Watershed Event for Research and Extension. *Plant Dis.* **2015**, *99*, 428–441.
15. Ghazy, N.; El-Nahrawy, S. Siderophore production by *Bacillus subtilis* mf497446 and *Pseudomonas koreensis* mg209738 and their efficacy in controlling *Cephalosporium maydis* in maize plant. *Arch. Microbiol.* **2020**, *203*, 1195–1209.

16. Pandit, M.A.; Kumar, J.; Gulati, S.; Bhandari, N.; Mehta, P.; Katyal, R.; Rawat, C.D.; Mishra, V.; Kaur, J. Major biological control strategies for plant pathogens. *Pathogens* **2022**, *11*, 273. [CrossRef]
17. Elnahal, A.S.; El-Saadony, M.T.; Saad, A.M.; Desoky, E.-S.M.; El-Tahan, A.M.; Rady, M.M.; AbuQamar, S.F.; El-Tarabily, K.A. The use of microbial inoculants for biological control, plant growth promotion, and sustainable agriculture: A review. *Eur. J. Plant Pathol.* **2022**, *162*, 759–792.
18. Degani, O.; Danielle, R.; Dor, S. The microflora of maize grains as a biological barrier against the late wilt causal agent, *Magnaporthiopsis maydis*. *Agronomy* **2021**, *11*, 965. [CrossRef]
19. Deng, Y.; Chen, H.; Li, C.; Xu, J.; Qi, Q.; Xu, Y.; Zhu, Y.; Zheng, J.; Peng, D.; Ruan, L. Endophyte *Bacillus subtilis* evade plant defense by producing lantibiotic subtilomycin to mask self-produced flagellin. *Commun. Biol.* **2019**, *2*, 368.
20. Gond, S.K.; Bergen, M.S.; Torres, M.S.; White, J.F., Jr. Endophytic *Bacillus* spp. Produce antifungal lipopeptides and induce host defence gene expression in maize. *Microbiol. Res.* **2015**, *172*, 79–87. [CrossRef]
21. Gómez-Lama Cabanás, C.; Mercado-Blanco, J. Groundbreaking technologies and the biocontrol of fungal vascular plant pathogens. *J. Fungi* **2025**, *11*, 77. [CrossRef] [PubMed]
22. Nizamani, M.M.; Hughes, A.C.; Zhang, H.-L.; Wang, Y. Revolutionizing agriculture with nanotechnology: Innovative approaches in fungal disease management and plant health monitoring. *Sci. Total Environ.* **2024**, *928*, 172473. [CrossRef] [PubMed]
23. Islam, M.S.; Ahmed, M.R.; Noman, M.; Zhang, Z.; Wang, J.; Lu, Z.; Cai, Y.; Ahmed, T.; Li, B.; Wang, Y.; et al. Integrating RNA interference and nanotechnology: A transformative approach in plant protection. *Plants* **2025**, *14*, 977. [CrossRef]
24. Montesinos, E. Functional peptides for plant disease control. *Annu. Rev. Phytopathol.* **2023**, *61*, 301–324.
25. Chen, A.; Jacob, M.; Shoshani, G.; Charter, M. Using computer vision, image analysis and uavs for the automatic recognition and counting of common cranes (*Grus grus*). *J. Environ. Manag.* **2023**, *328*, 116948. [CrossRef]
26. Yang, C. Remote sensing and precision agriculture technologies for crop disease detection and management with a practical application example. *Engineering* **2020**, *6*, 528–532. [CrossRef]
27. Hariharan, G.; Prasannath, K. Recent advances in molecular diagnostics of fungal plant pathogens: A mini review. *Front. Cell. Infect. Microbiol.* **2021**, *10*, 600234.
28. Degani, O.; Becher, P.; Gordani, A. Real-time PCR early detection of Trichoderma treatments efficiency against cotton charcoal rot disease. *J. Nat. Pestic. Res.* **2023**, *4*, 100027. [CrossRef]

Journal of *Fungi*

MDPI

Article

Population Structure of *Phytophthora infestans* in Israel Changes Frequently Due to the Import of Asymptomatic Late Blight-Infected Potato Seed Tubers from Europe

Yigal Cohen

Faculty of Life Sciences, Bar Ilan University, Ramat Gan 5290002, Israel; ycohen@biu.ac.il or yigal.cohen1@gmail.com

Abstract: Late blight, caused by the oomycete *Phytophthora infestans*, is a devastating disease of potato worldwide. In Israel, potatoes are grown twice a year, in autumn and spring, with late blight causing extensive damage in both seasons. While tuber seeds for the autumn planting are produced locally, seed tubers for the spring planting are imported from Europe due to dormancy of local tubers. Here, we demonstrate that seed tubers imported from Europe for the spring season carry asymptomatic infection with EU genotypes of *P. infestans*, which alters the population structure of the pathogen each spring. The proportion of imported tubers carrying asymptomatic infections ranged between 1.2 and 3.75%, varying by year and cultivar. Asymptomatic tubers produced late blight-infected sprouts about one month after planting. The sporangia produced on these sprouts served as primary inoculum, causing intensive foliage attacks on neighboring plants. When sprout-infected plants were uprooted and the mother tuber was washed, sliced, and placed in moistened dishes at 18 °C, profuse sporulation of *P. infestans* developed on the slices' surfaces within 1–2 days. The dominant genotype of *P. infestans* in the autumn season in Israel is 23A1, but genotypes in the following spring season changed to include 13A2 or 36A2. Surprisingly, genotype 43A1, which might be resistant to CAA and OSBPI fungicides and appeared in Europe in 2022, emerged in Israel in spring 2024. The immigrating genotypes do not persist in the country, allowing 23A1 to regain predominance in the following autumn. Long-term monitoring data suggest that the population structure of *P. infestans* changes yearly but temporarily due to the import of new genotypes from Europe.

Keywords: CAA fungicides; disease control; EuroBlight; OSBPI fungicides; mandipropamid; oomycetes; oxathiapiprolin; tomato

Citation: Cohen, Y. Population Structure of *Phytophthora infestans* in Israel Changes Frequently Due to the Import of Asymptomatic Late Blight-Infected Potato Seed Tubers from Europe. *J. Fungi* **2024**, *10*, 549. https://doi.org/10.3390/jof10080549

Academic Editor: Ofir Degani

Received: 14 July 2024
Revised: 1 August 2024
Accepted: 3 August 2024
Published: 4 August 2024

1. Introduction

Potatoes are grown in Israel twice a year, in autumn (planting in September–October, harvest in December–January) and spring (planting in December–January, harvest in April–May). Tuber seeds for the autumn season are produced locally in the previous spring season, while tuber seeds for the spring planting are imported yearly from Europe due to the dormancy of local tubers. Late blight, caused by *Phytophthora infestans* (Mont.) DeBary, is a devastating disease in both seasons, attacking from November until May.

The population structure of *P. infestans* in Israel changes frequently, mainly in the spring season. The reasons responsible for these changes are poorly understood. The objective of this study was to provide adequate explanations for why and how these changes in the population structure take place.

Monitoring the population structure of *P. infestans* in Israel began in 1983 when A2 mating type metalaxyl-resistant (R) isolates emigrated from Europe to the country [1]. Between 1983 and 1991, clonal populations of exclusively A2 mating types were found, with a steady increase in the frequency of R isolates over the years. From 1993 until 2012, the A1 mating type dominated and coexisted with the A2 mating type, suggesting that sexual recombination most likely occurred [2,3], allowing for isolates partially resistant to metalaxyl (intermediate, I) to appear. The frequency of R isolates drastically declined from 1993 to 2012 [4]. Thus, *P. infestans* underwent three major genetic changes during the period 1983–2000 [5]. The A2 R isolates were more competitive than A1 S (metalaxyl-sensitive) isolates on potato foliage [6], whereas the A1 S isolates were more competitive in tubers under storage [7,8].

From 2004 onward, monitoring of the potato population of *P. infestans* in Israel was accompanied by SSR genotype analysis. The present study describes the population structure of *P. infestans* in Israel over the last 21 years (2004–2024). It shows that the population structure significantly changed each spring but did not persist into the next autumn season. This study also proves that imported seed tubers from the EU are responsible for the changes in the spring. The entrance of new genotypes into the country in the spring occurred via asymptomatic seed tubers carrying latent infections of *P. infestans*.

2. Materials and Methods

2.1. Latent Infection in Imported Tubers—2016

Certified potato seed tubers cv Nicola were imported from Holland (producer unknown; supplied by Yaham Ltd., Magen, Israel). Five hundred tubers were first inspected visually for the absence of late blight symptoms and thereafter sown in soil on 6 December 2016 inside a low net house (No 1, 7 × 3 × 100 m) covered with 50 mesh white screen, located on campus. The soil had not been exposed to potato or tomato cultivation in the past 10 years. The plants were drip-irrigated twice a week with water containing 0.5% NPK fertilizer.

2.2. Latent Infection in Imported Tubers—2017

Two hundred seed tubers of cv Nicola and 200 seed tubers of cv Mondial (both imported from Europe (unknown producer; supplied by Yaham Ltd., Magen, Israel) were first inspected visually, to verify that there were no late blight symptoms, and thereafter sown on 5 January 2017 in soil in net house No 9 on campus. The soil had not been exposed to potato or tomato cultivation in the past 10 years. Another 200 seed tubers of cv Mondial were sown after similar inspection in a high net house (No 1, 16 × 7 × 25 m) in 120 L (0.5 × 0.2 × 1.2 m) polystyrene containers filled with a pasteurized peat:perlite (10:1) mixture, with 5 tubers per container. The plants were drip-irrigated twice a week with water containing 0.5% NPK fertilizer.

2.3. Latent Infection in Imported Tubers—2024

Certified potato seed tubers of 8 cultivars were imported from Europe (supplied by Yaham Ltd., Magen, Israel). Table 1 provides the names of the cultivars and their European producers. Eighty tubers of each cultivar were first inspected visually, to verify that there were no late blight symptoms, and thereafter sown on 1 January 2024 in 120 L (05; ×, 02; ×, 12; m) polystyrene containers filled with a pasteurized peat:perlite (10:1) mixture, with 5 tubers per container. The containers were placed inside a high net house (No 7, 7 × 5 × 50 m) located on campus. The plants were drip-irrigated twice a week with water containing 0.5% NPK fertilizer. Weather data were retrieved from a local meteorological station.

Table 1. List of the potato cultivars used in this study, including their origin and producer.

Cultivar	Producer	Country of Origin
Celtiane	Germicopa	France
Gelly	Eroplant	Germany
Mozart	HZPC	Holland
Orchestra	Meijer Potato	Holland
Rosana	Germicopa	Holland
Sifra	HZPC	Holland/France
Soprano	Meijer Potato	Holland
VR 808	Saltire Seeds	Scotland

2.4. Collection of Field Isolates 2004–2024

About 1120 isolates of *P. infestans* were collected from infected potato fields in the western Negev, Israel, during the autumn and spring seasons of 2004–2024 (30–40 isolates per season). Samples of 5–10 infected leaves were collected from an infected field, placed in a moistened plastic bag in a cooler, and shipped to the laboratory within a few hours. Infected leaves were placed on moist filter paper in 14 cm Petri dishes and incubated at 18 °C in the dark for ~15 h to allow for sporulation of the pathogen. The produced sporangia were harvested into ice-cold distilled water and used for propagation and DNA extraction. To propagate an isolate, sporangia were drop-inoculated onto detached leaflets of potato (Sifra) or tomato (Roter Gnom) and placed for a week in a growth chamber at 18 °C for sporulation as described before [9].

2.5. Extraction of DNA from Sporangia of Phytophthora Infestans

DNA was extracted as described for *Pseudoperonospora cubensis* [10]. Briefly, a sample of about 1×10^6 sporangia was macerated in 1.5 mL micro-tubes using disposable pellet pestle grinders. Maceration continued after adding 0.6 mL CTAB (hexadecyltrimethyl-ammonium bromide) buffer [1.4 M NaCl, 20 mM EDTA, 100 mM TRIS-Cl, 2% (w/v) CTAB pH 8.0], and the sample was incubated at 60 °C for 45 min. The sample was then extracted with 0.6 mL chloroform/isoamyl alcohol (24:1) and centrifuged at 12,000× g for 5 min. The aqueous phase was transferred to a 1.5 mL tube where the DNA was precipitated with an equal volume of cold (−20 °C) isopropanol. DNA concentration was determined with a ND-1000 spectrophotometer (NanoDrop, Waltham, MA, USA).

2.6. Genotype Identification

The DNA samples were shipped to Dr. David Cooke (The James Hutton Institute, Invergowrie, Dundee DD2 5DA, UK). Genotyping of the samples and reference isolates was performed as described by Li et al. [11] using a 12-plex PCR technique employing multiplexing of twelve SSR markers. The markers used in this study were D13, G11, Pi4B, Pi04, Pi63, Pi70, PinfSSR2, PinfSSR3, PinfSSR4, PinfSSR6, PinfSSR8, and PinfSSR11 [12].

2.7. Data Analysis

The percentage of leaf area infected with late blight in the 2024 experiment was visually assessed in each of the 8 cultivars from 41 days after planting until 76 days after planting. Tukey's HSD (honestly significant difference) test was performed to detect significant differences at =0.05 level between the mean late blight-infected leaf area of the 8 potato cultivars at 76 days after planting.

3. Results

3.1. Latent Infection 2016

Imported tubers were sown in soil on 6 December 2016. On 11 January 2017 (36 days after planting), 6 out of 500 (1.2%) emerging plants showed late blight symptoms on sprouts. The infected plants were scattered randomly in the net house (Figure 1A). Symptoms were visible on the leaves and/or stem (Figure 1B). No late blight symptoms were visible on the skin of the mother seed tubers when the symptomatic plants were uprooted from the soil, whereas symptoms were seen on the below-ground emerging stems (Figure 1C). The six infected plants were uprooted, their mother tubers were washed, surface sterilized, sliced, and the slices placed on a dry filter paper in dishes at 18 °C in the dark for 48 h. Profuse sporulation of *P. infestans* was observed on the slices of all mother tubers (Figure 1D). No such sporulation occurred in mother tubers taken from neighboring healthy plants (n = 10). SSR analysis of sporangia collected from the sporulating slices revealed genotype 13A2. Virulence analysis revealed a complex race 123456791011 of *P. infestans*, intermediately resistant to metalaxyl. The largest number of isolates belonging to genotype 13A2 were collected during spring 2016 (see below).

Figure 1. Potato seed tubers (*cv* Nicola, imported from Holland) carrying asymptomatic infection with *Phytophthora infestans* developed late blight symptoms upon germination. Tubers were sown on 6 December 2016. Symptoms were seen on 11 January 2017, 36 days after planting. (**A**) The appearance of 500 plants 5 weeks after sowing. (**B**) Emerging sprouts showing late blight symptoms. (**C**) Below-ground stems showing necrotic symptoms of late blight, while mother tuber appears healthy. (**D**) Sporulation of *Phytophthora infestans* on tuber slices (arrows) that were taken from the mother tuber shown in (**C**).

3.2. Latent Infection 2017

The imported tubers were sown on 5 January 2017. On 15 February 2017 (41 days after planting), 4 out 400 (1%) Nicola plants and 1 out of 200 (0.5%) Mondial plants showed late blight symptoms on sprouts (Figure 2A–D). Slices taken from the mother tubers of infected plants showed heavy sporulation of *P. infestans* (Figure 2E,F).

Figure 2. Potato seed tubers imported from Europe carrying asymptomatic infection with *Phytophthora infestans* developed late blight symptoms upon germination. Tubers were sown on 5 January 2017. Symptoms were detected on 15 February 2017, 41 days after sowing. (**A**) Nicola plants in net house 1. (**B**) Mondial plants in net house 9. (**C**) symptoms (arrow) of late blight at ground level. (**D**) symptoms of late blight on sprout apex (arrows). (**E**) sporulation of *Phytophthora infestans* on a tuber slice of cv Nicola. (**F**) Sporulation of *Phytophthora infestans* on a tuber slice of cv Mondial.

3.3. Latent Infection 2024

Eighty seed tubers of each of the eight cultivars (Table 1) were sown on 1 January 2024 in 128 containers filled with pasteurized soil in a net house at BIU Farm. The meteorological conditions during the growing season were conducive to late blight development (Figure 3). On 7 February 2024, 37 days after sowing, infected sprouts were visible in two Rosana plants and three VR-808 plants (Figure 4A–F). Upon uprooting the plants from the soil and washing, the mother tubers showed no symptoms of late blight. However, when the tubers were sliced and placed in moistened dishes, profuse sporulation of *P. infestans* was observed on the slices' surfaces (Figure 4F,G).

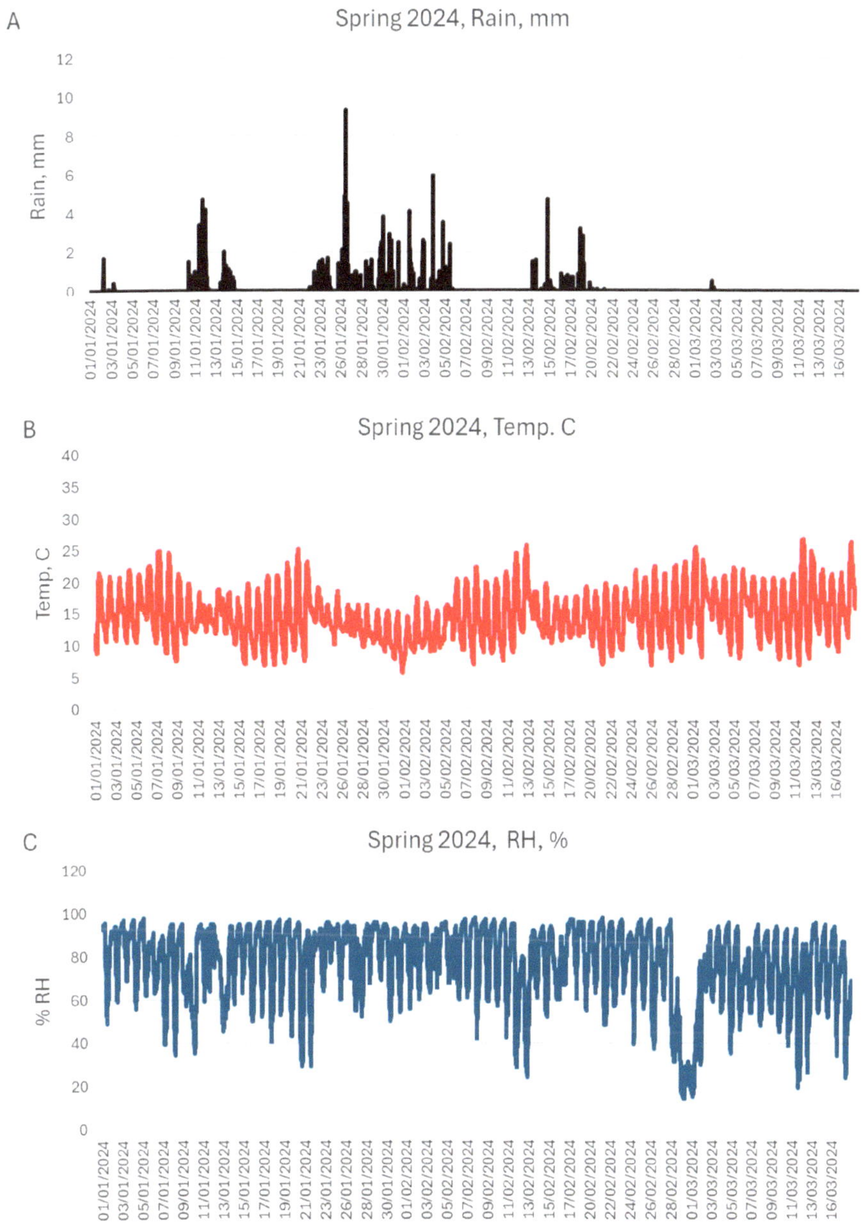

Figure 3. Meteorological conditions prevailing in Spring 2024 at BIU Farm during the epidemics of late blight caused by *Phytophthora infestans* in eight cultivars of potato whose seeds were imported from Europe. (**A**) rain (total = 370 mm). (**B**) air temperature (mean = 14.6 °C; min = 5.8 °C; max = 26.6 °C). (**C**) % RH (mean = 76.2%; min = 14%; max = 98%).

Figure 4. Potato seed tubers (*cv* Rosana and VR 808) carrying asymptomatic infection with *Phytophthora infestans* developed late blight symptoms upon germination. Imported tubers were sown on 1 January 2024 and sprout symptoms were observed on 7 February 2024, 36 days after sowing. (**A**) Net house with germinating potato plants at 36 days after planting. (**B–D**) Late blight symptoms on a germinating plant cultivar Rosana with no external symptoms on tubers. (**E**) An infected plant of cv VR 808 at 40 days after planting. Note the two healthy plants alongside. (**F**,**G**) Sporulation of *P. infestans* on surface of tuber slices cut from symptomless tubers of VR-808 and Rosana, respectively.

On 15 February 2024, 45 days after planting, the disease appeared across the net house in all cultivars, making it impossible to determine whether the source of infection was aerial or from tubers. Leaf samples were collected from each cultivar separately, and the sporangia produced thereafter in moistened Petri dishes were subjected to several tests and SSR analysis. A similar collection of infected leaves was performed at 76 days after planting. The traits of sporangia collected from the tubers slices and leaves are shown in Table 2. The isolates retrieved from the seed tubers belonged to the 13A2 genotype. The isolates collected from the leaves at 45 days after planting belonged to genotypes 23A1, 13A2 or 43A1, depending on the cultivar, whereas those collected at 76 days after planting all belonged to 13A2, confirming the highly competitive virulence of this genotype. The retrieved 13A2 isolates were all resistant to mefenoxam and infective to 10 out 11 potato differential lines (race 12345678910). They were all incompatible with tomatoes (Figure 5). When inoculated with a low sporangial dose (10 sporangia/droplet), they sporulated on potato leaves but produced no symptoms on tomato leaves (Figure 5A). When isolates 23A1 or 13 A2 were spray-inoculated at a high sporangial dose (5000 sporangia/mL), the former produced heavy sporulation on fruits (Figure 5B) and leaves (Figure 5C) of tomato, whereas the latter produced symptoms with no sporulation on fruits or leaves of tomato.

Table 2. Genotypes of *Phytophthora infestans* retrieved from seed tubers (at 38 days after planting) or leaves (at 45 and 88 days after planting) of seven cultivars of potato. Seed tubers were imported from Europe (see Table 1) and sown on 1 January 2024 in pasteurized soil in a net house at BIU Farm. (No data are available for cultivar Celtiane). nt = not tested. Each genotype is marked with a specific background color.

		Genotype	
Cultivar	**38 Days Seed Tubers**	**45 Days Leaves**	**88 Days Leaves**
VR 808	EU_13_A2	EU_23_A1	EU_13_A2
Rosana	EU_13_A2	EU_13_A2	EU_13_A2
Gelly	nt	EU_13_A2	EU_13_A2
Orchestra	nt	EU_13_A2	EU_13_A2
Soprano	nt	EU_43_A1	EU_13_A2
Mozart	nt	EU_13_A2	EU_13_A2
Sifra	nt	EU_23_A1	EU_13_A2

Figure 5. Compatibility to potato and tomato of genotypes 23A1 and 13A2 retrieved from potato. (**A**) In detached tomato and potato leaves. (**B**,**C**) In tomato fruits. (**D**,**E**) In tomato leaves. Note heavy sporulation of 23A1 on tomato fruits and leaves as against hypersensitive response to 13A2 with no sporulation.

Large differences were observed in the progress of late blight on the foliage of the eight potato cultivars in the net house (Figure 6A). The most susceptible cultivars were VR 808, Rosana and Celtiane, while the least susceptible ones were Gelly, Mozart, and Soprano (Figure 6B).

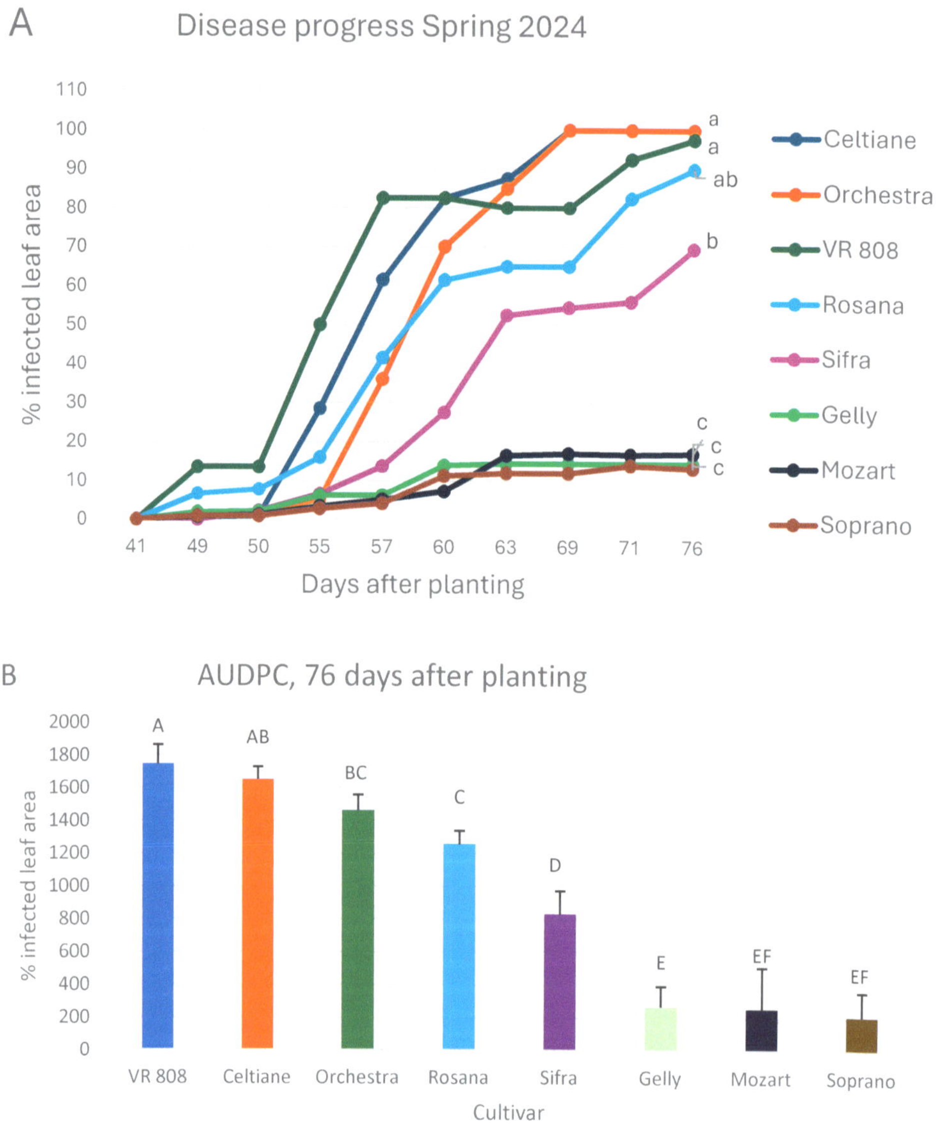

Figure 6. Progress of late blight on foliage of eight cultivars of imported potato cultivars. (**A**) Disease progresses in each cultivar during a 76-day period after planting. (**B**) Area under disease progress curves (derived from data in (**A**)). Different letters on curves or columns indicate a significant difference between cultivars at $\alpha = 0.05$ (Tukey's HDS).

3.4. Population Structure 2004–2024

Five genotypes of *P. infestans* were detected in potato fields in Israel during the past 21 years: 23A1, US7-like, 13A2, 36A2, and 43A1. The yearly fluctuation in their frequency is shown in Figure 7. The total number of isolates belonging to these genotypes was 857, 60, 115, 84, and 1, respectively.

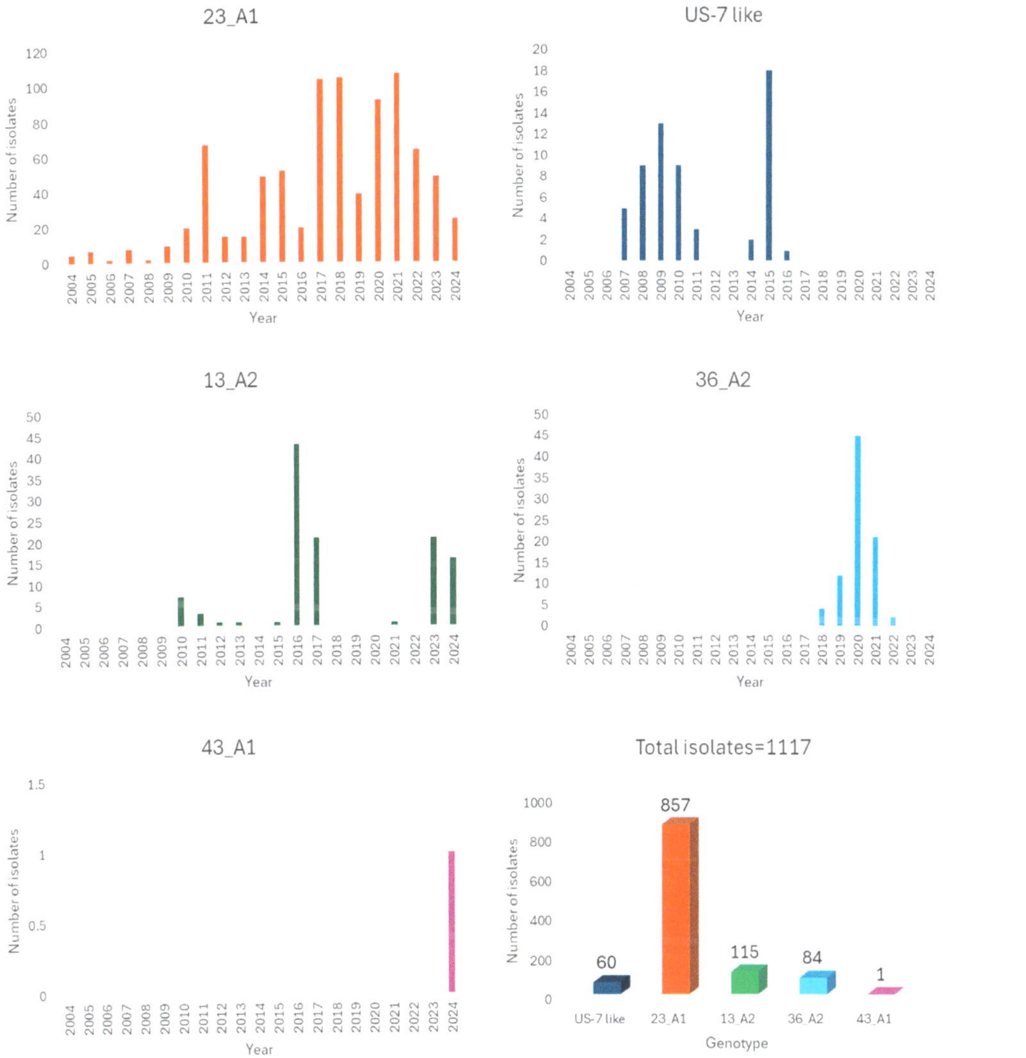

Figure 7. Annual frequency of genotypes of *Phytophthora infestans* in potato crops in Israel during the period 2004–2024.

Genotype 23A1 appeared in 2004, US7-like in 2007, 13A2 in 2010, 36A2 in 2018, and 43A1 appeared in spring 2024. They were detected in the country during 21, 8, 10, 5, and 1 year(s), respectively. Their order of appearance and decline in Israel may reflect the order of their appearance and decline in Europe (EuroBlight) (See Table 3).

Table 3. Population structure of *Phytophthora infestans* in potato crops in Israel for the 17-year period of 2007–2024. Figures with a background color represent the number of isolates that were collected in the spring season. Each genotype has a different background color.

Year	Season	Genotype and Number of Isolates				
		23_A1	US-7 like	13_A2	36_A2	43_A1
2007	Autumn	8	5	0	0	0
2008	Spring	0	4	0	0	0
2008	Autumn	2	5	0	0	0
2009	Spring	1	11	0	0	0
2009	Autumn	9	2	0	0	0
2010	Spring	8	9	7	0	0
2010	Autumn	12	0	0	0	0
2011	Spring	11	3	3	0	0
2011	Autumn	56	0	0	0	0
2012	Spring	5	0	1	0	0
2012	Autumn	10	0	0	0	0
2013	Spring	10	0	1	0	0
2013	Autumn	5	0	0	0	0
2014	Spring	20	2	0	0	0
2014	Autumn	29	0	0	0	0
2015	Spring	26	6	0	0	0
2015	Autumn	26	12	1	0	0
2016	Spring	14	1	36	0	0
2016	Autumn	6	0	7	0	0
2017	Spring	63	0	21	0	0
2017	Autumn	41	0	0	0	0
2018	Spring	45	0	0	4	0
2018	Autumn	60	0	0	0	0
2019	Spring	18	0	0	3	0
2019	Autumn	21	0	0	9	0
2020	Spring	46	0	0	44	0
2020	Autumn	46	0	0	1	0
2021	Spring	61	0	0	21	0
2021	Autumn	46	0	1	0	0
2022	Spring	29	0	0	2	0
2022	Autumn	35	0	0	0	0
2023	Spring	4	0	21	0	0
2023	Autumn	45	0	0	0	0
2024	Spring	25	0	16	0	1

Table 3 shows the seasonal shift (autumn vs. spring) in the population structure during the last 17 years (2007–2024). (No such data are available for 2004–2006.) The data confirm that the detection of genotypes 13A2 and 36A2 mostly occurred in the spring season, probably harbored inside infected seed tubers imported from Europe. Thus, out of 116 isolates that belonged to genotype 13A2, 108 (93.1%) were detected in the spring seasons. Similarly, out of 84 isolates that belonged to genotype 36A2, 74 isolates (88.1%) were detected in the spring. Genotype 43A1 showed up in spring 2024, soon after its dramatic appearance in Europe with resistance to CAA and OSBPI fungicides. Out of 60 isolates that belonged to US7-like, 36 isolates (60%) appeared in the spring. In contrast, genotype 23A1 was more frequent in the autumn season (55% of the isolates) as compared to the spring season (45% of the isolates), suggesting a local over-summering.

4. Discussion

Global trade of potato tubers facilitates the worldwide transportation of *Phytophthora infestans*. This has been occurring since the 1840s when *P. infestans* first migrated northward from Peru to Colombia, Mexico, the USA, and then to Europe and Asia [13]. From Europe, the pathogen has spread to other countries, including India [14,15], Egypt [12], north Africa, and South Africa [16]. Detailed routes of migration in Europe, the USA, Asia, and Latin America are available from the internet sites EuroBlight, USABlight, AsiaBlight, and Tizón Latino, respectively.

Reports in EuroBlight confirm that all five genotypes that were detected in Israel were previously detected in Europe. Genotype 23A1 has been prevalent since 2004 on tomatoes in south Europe. In 2023, it comprised most of the population in Italy. Genotype 13A2 was first detected in Europe, in Holland, and Germany, in 2004. It reached Britain in 2005 and displaced the population by 2006 [17]. It dominated the population for a decade and then declined to 9.3% in 2019, 7.6% in 2020 and 4.9% in 2021. Genotype 36A2 appeared in Europe in 2015. It rose from 20.8% in 2020 to 36.7% in 2021. In 2023, it comprised 46, 64 and 76% of the samples in Belgium, England, and France, respectively. Comprising 37% of the 2023 samples, genotype 36A2 was the most frequently sampled genotype, suggesting it remains fitter than other clones. The frequency of genotype 43A1 increased from 17% in 2022 to 23% of the population in 2023. It has extended its range to ten European countries [18,19].

In this paper, we show that *P. infestans* is continuously migrating from Europe to Israel via potato tuber seeds. These migrations occur through asymptomatic infected tubers, which are undetectable by buyers during field inspections before harvest and by growers at the time of sowing. Asymptomatic infected tubers only manifest the disease on germinating sprouts about one month after sowing in soil.

The asymptomatic infection of potato tubers with *P. infestans* remains a biological mystery. It is unclear how and when the pathogen reaches the tubers, whether mycelia/sporangia occupy the tuber buds (eyes) or lenticels, or whether the mycelium has penetrated the parenchyma. Additionally, it is unknown why external rot does not develop, nor is it known what the mechanism that keeps the pathogen silent during tuber dormancy until sprouting. This behavior might mimic *Uncinula necator*, the powdery mildew fungal agent of grapevine, which overwinters in buds of the twigs and develops heavily mildewed "flag" shoots in the spring, thus serving as a source of primary inoculum [20].

Using artificial inoculation of tuber buds, Johnson and Cummings [21] observed that latent infection of potato seed tubers and the production of viable sporangia of *P. infestans* occurred after long-term cold storage at around 4 °C. Latent infection in tubers was asymptomatic, without discoloration or necrosis in external or internal tissues at the end of storage. However, symptoms and sporangia developed on asymptomatic tubers placed at temperatures of 15 °C and above. They concluded that *P. infestans* could survive asymptomatically in potato seed tubers for extended periods at around 4 °C. Latent infection of seed tubers poses a challenge for late blight management as visual inspection will not reveal latently infected tubers, and tubers with low infection severity may be overlooked. In another report, Johnson [22] showed that mist periods favored the expression of *P. infestans* in infected shoots emerging from late blight-infected seed tubers. Moist conditions that favor the emergence of infected shoots also favor sporulation and repeated infections in the field.

Unlike these reports, our potato seed tubers showed no symptoms after three weeks of incubation at 20–25 °C. In our 2016, 2017 and 2024 trials, a small proportion (1–3.75%) of asymptomatic seed tubers developed infected plants in the field. Heavy rains were associated with the emergence of infected sprouts from the asymptomatic late blight-infected seed tubers. Of the eight cultivars tested in 2024, Rosana and VR 808 showed the emergence of infected sprouts. In all trials, mother tubers of the infected plants produced heavy sporulation of *P. infestans* upon slicing. SSR analysis revealed the arrival of genotypes 13A2 and 43A1 from Europe. This is the first report on the occurrence of 43A1 in Israel.

The genotype 23A1 has predominated in the Israeli population for the past 21 years, comprising about 76.7% of all isolates. It has replaced genotype US7-like, and its long persistence is likely related to its compatibility with tomatoes, which are abundantly grown during the summer (June–November) in areas adjacent to potato-growing regions in the country. No experimental data are available to prove its competitive fitness on potato with other genotypes of *P. infestans* present in the Israeli population. This genotype is also common in north Africa and South Africa [16]. In Egypt, SSR genotyping of 152 isolates collected from potato and tomato (and one from eggplant) during 2013–2018 showed that, in contrast to 2010–2012 [23], when the proportion of the 13A2 lineage was 35%, all later isolates belonged to the 23A1 lineage [12]. The widespread distribution of the 23A1 clonal lineage on both potato and tomato crops in Egypt could be attributed to its adaptation to Egyptian climatic conditions [12]. Some isolates of the 23A1 lineage might have become tolerant to warmer Egyptian agro-ecological conditions [23].

Genotype 43A1 arrived in Israel in spring 2024 in infected asymptomatic seed tubers produced in Europe in summer 2023. This genotype has recently increased in the EU population. Recent reports by EuroBlight indicated a transition to relatively new genotypes such as EU36, EU37, EU41, EU43, and EU46, with 68% of the samples being genotypes unknown nine years ago. Conversely, the older genotypes, such as EU8A1, EU6A1, and EU13A2, only represented 17% of the population compared to 68% in 2014. Genotype EU43 increased from 2% of the sampled European population in 2021 to 17% in 2022, and to 23% in 2023. Marked increases in the frequency of EU43 were observed in The Netherlands (from 42% to 55%), Germany (from 7% to 50%), and Belgium (from 12% to 35%) from 2022 to 2023. The sampled range of EU43 also widened to include France, Ireland, and Spain in 2023, in addition to the seven other countries where it was previously reported. EU43 has been reported to be resistant to the CAA and OSBPI groups of fungicides [18,19,24,25]. Industry reports show that there are resistant and sensitive isolates to both fungicide groups in genotype EU43 [19]. It is highly likely that the reported resistances account for the dominance of the EU43 lineage in the intensively sampled regions of The Netherlands, Belgium, and northern Germany in 2023. According to EuroBlight, these later changes in *P. infestans* populations may directly influence the efficacy of plant protection products, the deployment of resistant cultivars, and the performance of disease warning systems.

Our unpublished data suggest that the EU43A1 isolate that has arrived in Israel in spring 2024 is sensitive to 0.3 ppm ai oxathiapiprolin.

In conclusion, the potato industry in Israel is continuously threatened with new late blight genotypes immigrating to the country from Europe via asymptomatic seed tubers. Disease management should include strict visual inspection of the newly sown potato fields in the spring season as soon as sprouts emerge from soil.

Funding: This study was supported by The Plant Council of Israel Not Grant Number 002223.

Institutional Review Board Statement: Not applicable.

Informed Consent Statement: Not applicable.

Data Availability Statement: The data that support the findings of this study are available from the corresponding author upon reasonable request.

Acknowledgments: The authors acknowledge M. Galperin, Uri Zig, and the Plant Council of Israel for their valuable support, and David Cooke and staff at the James Hutton Institute, Dundee, UK for genotyping the samples.

Conflicts of Interest: The authors declare no conflicts of interest.

References

1. Cohen, Y.; Reuveni, M. Occurrence of metalaxyl-resistant isolates of *Phytophthora infestans* in potato fields in Israel. *Phytopathology* **1983**, *73*, 925–927. [CrossRef]
2. Cohen, Y.; Farkash, S.; Reshit, Z.; Baider, A. Oospore production of *Phytophthora infestans* in potato and tomato leaves. *Phytopathology* **1997**, *87*, 191–196. [CrossRef] [PubMed]
3. Cohen, Y.; Farkash, S.; Baider, A.; Shaw, D.S. Sprinkling irrigation enhances production of oospores of *Phytophthora infestans* in field-grown crops of potato. *Phytopathology* **2000**, *90*, 1105–1111. [CrossRef] [PubMed]
4. Dietrich, C.; Hermann, D.; McKenzie, Y.; Cohen, Y.; Gisi, U. Phenylamides: Market Trends and Resistance Situation 35 Years After First Product Introduction. In *Fungicide Resistance in North America*; Stevenson, K.L., McGrath, M.T., Wyenandt, C.A., Eds.; APS Press: Saint Paul, MN, USA, 2019; 411p.
5. Cohen, Y. Populations of *Phytophthora infestans* in Israel underwent three major genetic changes during 1983–2000. *Phytopathology* **2002**, *92*, 300–307. [CrossRef] [PubMed]

6. Kadish, D.; Cohen, Y. Population dynamics of metalaxyl-sensitive and metalaxyl-resistant isolates of *Phytophthora infestans* in untreated potato crops. *Plant Pathol.* **1989**, *38*, 271–276. [CrossRef]
7. Kadish, D.; Cohen, Y. Overseasoning of metalaxyl-sensitive and metalaxyl-resistant isolates of *Phytophthora infestans* in potato tubers. *Phytopathology* **1992**, *82*, 887–889. [CrossRef]
8. Gisi, U.; Cohen, Y. Resistance to phenylamide fungicides: A case study with *Phytophthora infestans* involving mating type and race structure. *Annu. Rev. Phytopathol.* **1996**, *34*, 549–572. [CrossRef] [PubMed]
9. Cohen, Y.; Weitman, M. Mobility of oxathiapiprolin in and between tomato plants. *Pest. Manag. Sci.* **2023**, *79*, 1102–1112. [CrossRef]
10. Cohen, Y.; Rubin, A.E.; Galperin, M.; Ploch, S.; Runge, F.; Thines, M. Seed Transmission of Pseudoperonospora cubensis. *PLoS ONE* **2014**, *9*, e109766. [CrossRef]
11. Li, Y.; Cooke, D.E.; Jacobsen, E.; van der Lee, T. Efficient multiplex simple sequence repeat genotyping of the oomycete plant pathogen *Phytophthora infestans*. *J. Microbiol. Methods* **2013**, *92*, 316–322. [CrossRef]
12. El-Ganainy, S.M.; Ismail, A.M.; Soliman, M.S.; Ahmed, Y.; Sattar, M.N.; Chellappan, B.V.; Cooke, D.E.L. Population dynamics of *Phytophthora infestans* in Egypt reveals clonal dominance of 23_A1 and displacement of 13_A2 clonal lineage. *J. Fungi* **2023**, *9*, 349. [CrossRef] [PubMed]
13. Patarroyo, C.; Lucca, F.; Dupas, S.; Restrepo, S. Reconstructing the global migration history of *Phytophthora infestans* towards Colombia. *Phytopathology*, 2024; *online ahead of print*. [CrossRef]
14. Chowdappa, P.; Nirmal Kumar, B.J.; Madhura, S.; Mohan Kumar, S.P.; Myers, K.L.; Fry, W.E.; Cooke, D.E.L. Severe outbreaks of late blight on potato and tomato in South India caused by recent changes in the *Phytophthora infestans* population. *Plant Pathol.* **2015**, *64*, 191–199. [CrossRef]
15. Dey, T.; Dwivedi, S.K.; Datta, S.; Cooke, D.E.L.; Roy, S.G. Understanding the temporal dynamics of invasive late blight populations in India for improved management practices. *Phytopathology* **2024**. [CrossRef]
16. McLeod, A.; De Villiers, D.; Sullivan, L.; Coertze, S.; Cooke, D.E.L. First report of *Phytophthora infestans* lineage EU23 causing potato and tomato late blight in South Africa. *Plant Dis* **2023**, *108*, 231. [CrossRef] [PubMed]
17. Cooke, D.E.L.; Cano, L.M.; Raffaele, S.; Bain, R.A.; Cooke, L.R.; Etherington, G.J.; Deahl, K.L.; Farrer, R.A.; Gilroy, E.M.; Goss, E.M.; et al. Genome Analyses of an Aggressive and Invasive Lineage of the Irish Potato Famine Pathogen. *PLoS Pathog.* **2012**, *8*, e1002940. [CrossRef] [PubMed]
18. Carboxylic Acid Amides (CAA) Working Group Annual Meeting Season 2022 on January 17th, 2023 Protocol of the discussions and recommendations of the CAA Working Group of the Fungicide Resistance Action Committee (FRAC). Available online: https://www.frac.info/docs/default-source/working-groups/caa-fungicides/caa-wg/minutes-of-the-2023-caa-meeting-recommendations-for-2023.pdf (accessed on 1 August 2024).
19. Minutes of the FRAC OSBPI Working Group Meeting 7 February 2024, Corteva lab, Eschbach, Germany. Participants -Corteva Jean-Luc Genet (Chair) Przemek Szubstarski Mamadou Mboup Syngenta Stefano Torriani David Ranner Irina Metaeva Bayer Jürgen Derpmann Andreas Mehl Christian Zupanc. Available online: https://www.frac.info/docs/default-source/working-groups/osbpi-wg/minutes-of-the-2024-osbpi-wg-meeting-recommendations-for-2024---7-jan-24.pdf (accessed on 1 August 2024).
20. Cortesi, P.; Ottaviani, M.-P.; Milgroom, M.G. Spatial and genetic analysis of a flag shoot subpopulation of *Erysiphe necator* in Italy. *Phytopathology* **2004**, *94*, 544–550. [CrossRef]
21. Johnson, D.A. Transmission of *Phytophthora infestans* from infected potato seed tubers to emerged shoots. *Plant Dis.* **2010**, *94*, 18–23. [CrossRef] [PubMed]
22. Johnson, D.A.; Cummings, T.F. Latent infection of potato seed tubers by *Phytophthora infestans* during long-term cold storage. *Plant Dis.* **2009**, *93*, 940–946. [CrossRef]
23. El-Ganainy, S.M.; Iqbal, Z.; Awad, H.M.; Sattar, M.N.; Tohamy, A.M.; Abbas, A.O.; Squires, J.; Cooke, D.E. Genotypic and phenotypic structure of the population of *Phytophthora infestans* in Egypt revealed the presence of European genotypes. *J. Fungi* **2022**, *8*, 468. [CrossRef]
24. Kaur, A.; Doyle, D.; Cooke, D.E.L.; Mullins, E.; Kildea, S. First report of the *Phytophthora infestans* EU_43_A1 clonal lineage and associated PiCesA3 mutation G1105S in Ireland. *New Dis. Rep.* **2024**, *49*. [CrossRef]
25. Abuley, I.K.; Lynott, J.S.; Hansen, J.G.; Cooke, D.E.L.; Lees, A.K. The EU43 genotype of *Phytophthora infestans* displays resistance to mandipropamid. *Plant Pathol.* **2023**, *72*, 1171–1344. [CrossRef]

Journal of *Fungi*

MDPI

Review

RNAi-Based Approaches to Control Mycotoxin Producers: Challenges and Perspectives

Alexander A. Stakheev [1,*], Michael Taliansky [1], Natalia O. Kalinina [1,2] and Sergey K. Zavriev [1]

1 M.M. Shemyakin and Yu.A. Ovchinnikov Institute of Bioorganic Chemistry, 117997 Moscow, Russia
2 A.N. Belozersky Institute of Physico-Chemical Biology, Lomonosov Moscow State University, 119992 Moscow, Russia
* Correspondence: stakheev.aa@gmail.com

Abstract: Mycotoxin contamination of food and feed is a worldwide problem that needs to be addressed with highly efficient and biologically safe techniques. RNA interference (RNAi) is a natural mechanism playing an important role in different processes in eukaryotes, including the regulation of gene expression, maintenance of genome stability, protection against viruses and others. Recently, RNAi-based techniques have been widely applied for the purposes of food safety and management of plant diseases, including those caused by mycotoxin-producing fungi. In this review, we summarize the current state-of-the-art RNAi-based approaches for reducing the aggressiveness of key toxigenic fungal pathogens and mycotoxin contamination of grain and its products. The ways of improving RNAi efficiency for plant protection and future perspectives of this technique, including progress in methods of double-stranded RNA production and its delivery to the target cells, are also discussed.

Keywords: RNA interference; double-stranded RNAs; plant protection; mycotoxins; *Fusarium*; *Aspergillus*; HIGS; SIGS

1. Introduction

Fungal diseases are among the most significant factors of agricultural losses and food security threats worldwide [1,2]. Climate change along with the development of transport and trade lead to the emergence of new pathogenic species as well as expansion of the area of existing ones. Besides direct impacts on quality and quantity of yields, fungi of genus *Fusarium*, *Aspergillus*, *Penicillium* and *Alternaria* are able to produce a wide range of toxic secondary metabolites known as mycotoxins. The ability to synthesize toxic secondary metabolites was acquired by different fungal taxa during evolution [3] and became an important tool for occupying novel ecological niches [4]. Nowadays, the number of known mycotoxins is estimated as 300–400 [5], but the most widely distributed and important among them are aflatoxins, trichothecenes, fumonisins and zearalenone. These compounds are highly toxic to mammals, possessing carcinogenic, teratogenic, hepatotoxic and immunosuppressive properties, inhibiting protein biosynthesis and causing mitochondrial dysfunction [6–11]. Moreover, trichothecenes, especially deoxynivalenol (DON) and toxins of the recently described NX group [12], act as a virulence factor in host plants, facilitating tissue colonization and inhibiting plant immune response [13,14].

The toxic effect of mycotoxins, as well as direct economic impacts caused by their producers, makes it necessary to develop efficient methods to control toxigenic fungi in agricultural plants. Traditional approaches, including crop rotation or developing resistant cultivars, are time-consuming and unable to prevent mycotoxin accumulation completely. The application of fungicides has been the most widely applied technique to control mycotoxigenic fungi [15]. However, this approach has several disadvantages, including possible induction of mycotoxin synthesis [16,17], negative effects on human and animal health [18] and impact on biodiversity [19]. The development of pathogens' resistance is another serious challenge, reducing the efficiency of this approach for combating toxigenic

Citation: Stakheev, A.A.; Taliansky, M.; Kalinina, N.O.; Zavriev, S.K. RNAi-Based Approaches to Control Mycotoxin Producers: Challenges and Perspectives. *J. Fungi* **2024**, *10*, 682. https://doi.org/10.3390/jof10100682

Academic Editor: Ofir Degani

Received: 26 August 2024
Revised: 26 September 2024
Accepted: 27 September 2024
Published: 29 September 2024

fungi [20,21]. The use of mycotoxin biosynthesis inhibitors of natural origin has been a promising alternative to fungicides, mainly in terms of biological safety [22–25], but it is often difficult to isolate an individual active compound from a total extract and to obtain it in quantitates appropriate for massive use.

Recently, attention has been paid to methods based on the control of gene expression via RNA-interference (RNAi). After the first observation in petunia plants [26], RNAi was described in many other organisms of different taxa. Generally, RNAi represents a conserved mechanism, the central role in which is played by small RNAs (sRNAs), able to regulate gene expression at the post-transcriptional level (post-transcriptional gene silencing, PTGS) [27]. RNAi is involved in various cellular processes, such as maintenance of genome stability, adaptation to stressful conditions, chromatin modification, protection against viruses and alien nucleic acids, development and pathogenesis [28–34].

Romano and Machino [35] first demonstrated that transformation of the fungus *Neurospora crassa* with different portions of two carotenogenic genes (*al-1* and *al-3*) resulted in phenotypic changes. Later, Fire et al. [36] showed that injection of specific exogenous dsRNA into the nematode *Caenorhabditis elegans* led to gene-specific silencing. These discoveries have opened up an opportunity of using RNAi as an efficient biotechnological tool. Particularly, it has been used for analyses of plant and fungal gene functions, as well as for improving traits of industrially, agronomically and medically important fungal strains [37–39]. Since the 2000s, RNAi-based approaches have been applied for solving different problems of plant protection, including management of mycotoxins and their producers. In this review, we discuss what has been undertaken in the field of application of RNAi to control mycotoxigenic fungi and what should be conducted to bring this approach to widescale use in agricultural practice.

2. RNAi Principles and Its Current Applications

2.1. Mechanism of RNAi and Key Components of the Process

The main players of RNAi processes in fungi, as well as in other eukaryotes, are proteins of Dicer and Argonaute families, as well as RNA-dependent RNA-polymerases (RdRps) [40–42]. In brief, the RNAi pathway includes two steps. In the first step, the trigger RNA (either dsRNA or hairpin RNA) is processed into small interfering RNAs (siRNAs), a class of sRNAs, typically 21–25 base pairs in length containing 3′ dinucleotide overhangs. This process is catalyzed by Dicer or Dicer-like (DCLs) enzymes, members of the RNase III family. The siRNAs are then amplified by RdRPs. siRNAs produced by Dicer cleavage and those synthesized by RdRPs are called primary and secondary siRNAs, respectively [43]. In the second step, siRNAs are loaded into the Argonaute (AGO) protein to form the RNA-induced silencing complex (RISC). Then, one of the siRNA strands (passenger) is removed from the RISC and another (guide) one directs RISC to the mRNA target to cleave it by the AGO protein [44–47].

Despite the fact that RNAi is a highly conserved mechanism, fungal species differ by the number of RNAi machinery components they possess. Several species, such as *Saccharomyces cerevisiae* and *Ustilago maydis*, do not have DCLs and AGOs since they were lost during evolution [48–50]. However, the latter species is able to produce sRNAs from tRNAs and 5.8 S ribosomal RNA [51]. Dicer-independent mechanisms of RNAi were also identified in some other fungi [28,52]. *F. graminearum*, one of the most widespread and devastating toxigenic species, possesses two DCLs, two AGOs and five RdRps. According to Chen et al. [45], only *Fg*AGO1 and *Fg*DCL2 are necessary for silencing. Later, Zeng et al. [53] and Gaffar et al. [54] demonstrated that *Fg*AGO2 and *Fg*DCL1 participate in sexual ascosporogenesis and pathogenic processes, while *Fg*AGO1 and *Fg*DCL2 regulate asexual conidia formation and germination. *F. graminearum* RdRps were shown to be responsible for conidiation [54], but their role in the production of secondary siRNAs has been unclear. According to Song et al. [55], *F. asiaticum*, a member of the *Fusarium graminearum* species complex (FGSC), lacks the ability to produce secondary siRNAs and this results in short-time silencing of a target gene. This inability can be considered as a serious potential obstacle for applying RNAi to manage fungal pathogens in the field.

2.2. Cross-Kingdom RNAi and dsRNA Uptake

It was recently described that sRNAs can bi-directionally move between plants and pathogenic organisms, being an important element of an arms race between them. This phenomenon is called 'cross-kingdom RNA interference' (ckRNAi) [56–58]. Pathogens send sRNAs to the host cells to suppress plant immunity, while plants use this mechanism to inhibit pest virulence [59–63]. Interestingly, sRNAs produced by a fungal pathogen often affect target gene(s) expression via the host plant's RNAi machinery. For instance, Weiberg et al. [59] demonstrated the presence of *Botrytis cinerea*-produced sRNAs associated with AGO1 of *Arabidopsis thaliana*. sRNAs of a vascular pathogen *F. oxysporum* f.sp. *lycopersici* bind to tomato Argonaute 4a (SlyAGO4a) [64]. Toxigenic *F. graminearum* was shown to send sRNAs, targeting the resistance-related *CEBiP* gene, into the cells of common wheat [65]. Mechanisms of ckRNAi are far from being fully understood; however, some of them have been elucidated. Plants can translocate sRNAs by cell-to-cell mechanism via the plasmodesmata (short-range transfer) or via the vascular system, predominantly phloem (long-range movement) [66,67]. Recent studies found that plants synthesize extracellular vesicles (EVs) to carry RNA molecules and other effectors directed against pathogenic fungi [62,68,69]. In turn, many fungal species release EVs to deliver RNAs into host plant cells [70]. Surprisingly, apart from sRNAs, full-length mRNAs can also be transported into target cells and translated [71]. On the other hand, several reports suggest that EVs play a small role or no role in sRNAs movement, suggesting the existence of other mechanisms of secretion, movement and maintaining the stability of sRNAs [72–74].

The ability to uptake dsRNA varies among fungal species. Several fungi, such as *Botrytis cinerea*, *Sclerotinia sclerotiorum*, *Rhizoctonia solani*, *Aspergillus niger* and *Verticillium dahliae*, as well as some *Fusarium* spp., demonstrated a high rate of dsRNA uptake [75–77]. At the same time, *Trichoderma virens* and the oomycete *Phytophthora infestans* had limited uptake; *Colletotrichum gloeosporioides* and *Zymoseptoria tritici* showed no uptake at all [75,78]. It is believed that uptake efficiency is related to fungal lifestyle and that necrotrophic fungi are more susceptible to dsRNA application. Mechanisms of uptake are still unknown for most fungal species; however, it was demonstrated that clathrin-mediated endocytosis, a well-known eukaryotic mechanism, is involved in dsRNA uptake in *S. sclerotiorum* [79]. In this study, live cell imaging experiments showed that hyphal tips of younger, rapidly growing hyphae of *S. sclerotiorum* are responsible for most active uptake processes. Several proteins, including vacuolar ATPase (VATPase), the heavy chain of the clathrin complex (CHC), clathrin adaptor protein 2 complex (AP2), ADP ribosylation factor-like 1 protein (Arf72A) and amphiphysin (Amph) were shown to be essential for successful dsRNA uptake. At the same time, specific receptors responsible for dsRNA binding have yet to be identified.

2.3. HIGS and SIGS

The very first studies on the use of RNAi to control mycotoxin production were carried out by integration of siRNAs-encoding sequences into the genomes of toxigenic strains [80,81] followed by in vitro and in planta tests of mycotoxin accumulation inhibition. These studies demonstrated the principal possibility of the use of RNAi to inhibit mycotoxin biosynthesis and fungal growth. In addition, they allowed uncovering functions of some genes and their opportunity to be used as targets for RNAi. At the same time, further development of RNAi-based techniques as tools for pathogen management in the field has led to the emergence of such approaches, as host-induced gene silencing (HIGS) and spray-induced gene silencing (SIGS).

In HIGS, the natural mechanism of ckRNAi is used for plant protection against pathogenic organisms by expressing exogenous dsRNAs homologous to target pathogen's gene(s). A simplified scheme of HIGS is shown in Figure 1. To perform HIGS, a dsRNA or a hairpin RNA-encoding construct that targets a specific pathogen's gene is transformed into the host plant [82]. The first successful application of HIGS against toxigenic fungus was described in 2010 for tobacco plants, expressing a hairpin β-glucuronidase (*GUS*) gene-targeted dsRNA to inhibit *F. verticillioides* [83]. Since then, a number of studies have been carried out demonstrating the applicability of this approach to inhibit different toxigenic species and corresponding toxins. HIGS has been successfully used to protect wheat [84,85], maize [86–91], barley [92,93], Arabidopsis [92], peanut [94] and groundnut [95] (summarized in Table 1). However, this technique has several shortcomings that make it necessary to search for improvements and possible alternatives. Transgenic approaches are time-consuming and technically sophisticated. The application of HIGS requires transformation of the corresponding host plant, though there are still no transformation protocols for many crops. The key methods used for plant transformation are microprojectille bombardment, *Agrobacterium* infiltration and transgenesis, but all these techniques have specific limitations and drawbacks [96]. It is anticipated that the introduction of the CRISPR/Cas9 system will facilitate plant genome engineering [97,98], but to date several challenges remain. A principal obstacle of the wider use of HIGS is that the legislation of GMO is country-specific and there are many countries where these organisms are prohibited to develop and use even for scientific purposes [99]. Another method technically similar to HIGS is VIGS (virus-induced gene silencing) which represents a virus vector-based transient synthesis of dsRNA in a host plant [100]. The most often-used virus vector for VIGS is barley stripe mosaic virus (BSMV) [101]. BSMV-based VIGS was successfully used to protect wheat against toxigenic *F. culmorum* [85,102] and several non-toxigenic fungi, such as *Blumeria graminis* and *Puccinia striiformis* [82,101,103]. Obvious restrictions of this approach are methodological difficulties of efficient plant infection and possible entry of the virus into the environment. As an alternative to the use of BSMV, Zhang et al. [104] described the application of *F. graminearum* gemytripvirus (FgGMTV1)-based vector to convert an *F. graminearum* strain into hypovirulent ones by silencing two pathogenicity-related genes *TRI101* and *FgPP1*, and therefore, to protect wheat against infection. The authors showed that the obtained hypovirulent strains can be used as biocontrol agents themselves, inhibiting fungal infection and decreasing DON content after wheat spikes being infected by these strains' mycelia together with conidial suspension of virulent *F. graminearum* strain PH-1. Mycovirus-based VIGS seems to be promising to manage mycotoxin producers, including under field conditions. However, mycoviruses is a relatively poor-studied group of viruses; thus, corresponding vectors able to mediate efficient silencing have yet to be developed.

SIGS is based on the application of specific dsRNA (s) directly to the pathogen or plant surface, generally using a foliar spray. A simplified scheme of SIGS is shown in Figure 1. Koch et al. [76] described the first successful use of SIGS to control mycotoxigenic fungi. A long-coding dsRNA (CYP3dsRNA), targeting three cytochrome P450 lanosterol C-14α-demethylases of *F. graminearum*, was applied at Arabidopsis and barley leaves by spraying. Six days after the inoculation, the development of brownish lesions on dsRNA-treated leaves was significantly lower than in control samples and the expression levels of three target genes were reduced in the range of 48% to 58%. Moreover, the authors demonstrated that silencing and inhibition of fungal infection were observed in distant (non-treated) parts of barley leaves, indicating dsRNA translocation within the plant as well as the fact that fungal, not the plant's, RNAi machinery plays the key role in the silencing process. Later, the movement of sprayed dsRNA from leaves to stems and roots of barley during three days after treatment was confirmed by Biedenkopf et al. [105]. The direct comparison of SIGS and HIGS methods, performed by Koch et al. [93] using the aforementioned *CYP51*-based system demonstrated that the growth inhibition of *F. graminearum* on barley leaves was significantly higher for SIGS in comparison to HIGS (in lab conditions). For instance, infected leaf areas were reduced by 7% (HIGS) and 80% (SIGS) when CYP51A-dsRNA is applied. At the same time, there is a lack of studies, comparing HIGS vs. SIGS efficiencies both in vitro and *in planta*. By now, SIGS has proven to be effective in protecting wheat [77,102,106–108] and barley [109] against toxigenic *Fusarium*. Therefore, SIGS-based approaches have great prospects, but there are several serious limitations, hindering the possibility of massive use of these techniques. These limitations are summarized in Figure 2 and discussed below.

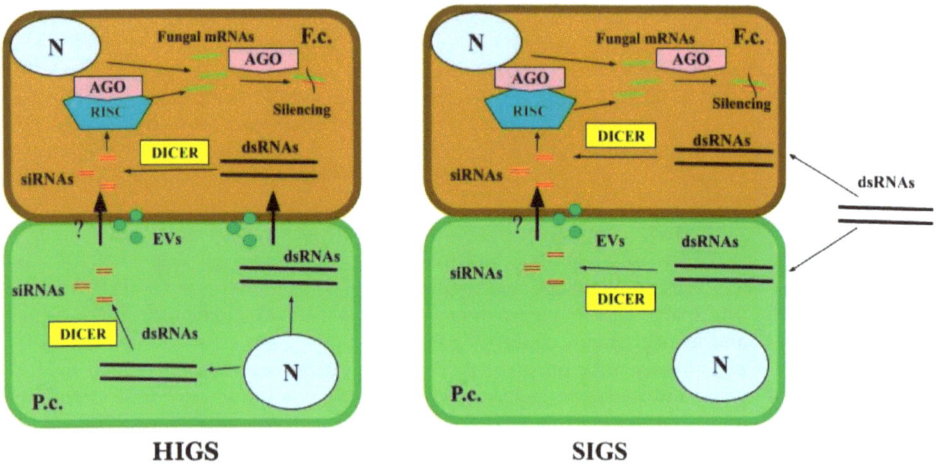

Figure 1. A schematic representation of HIGS (**left**) and SIGS (**right**) mechanisms. HIGS: a cell of a transgenic plant generates sequence-specific dsRNAs to target corresponding fungal genes. These dsRNAs are both transferred to fungal cells by extracellular vesicles or other mechanisms and cleaved into siRNAs by plant DCL proteins. dsRNAs transferred to fungal cells also cleaved into siRNAs by fungal DCL proteins. siRNAs are loaded into the AGO protein to form the RNA-induced silencing complex (RISC). Then, the one of the siRNA (passenger) strand is removed from the RISC and another (guide) one directs RISC to the mRNA target to cleave it. SIGS: The dsRNAs are sprayed onto the plant surface. One part of dsRNAs are taken by the fungus and then produced into siRNAs by the fungal RNAi machinery. Another part of dsRNAs is taken by the plant cells and produced into siRNAs by plant RNAi machinery and then transferred into fungal cells. The following processes are analogous to those of HIGS. P.c.—plant cell; F.c.—fungal cell; N—nucleus; EVs—extracellular vesicles.

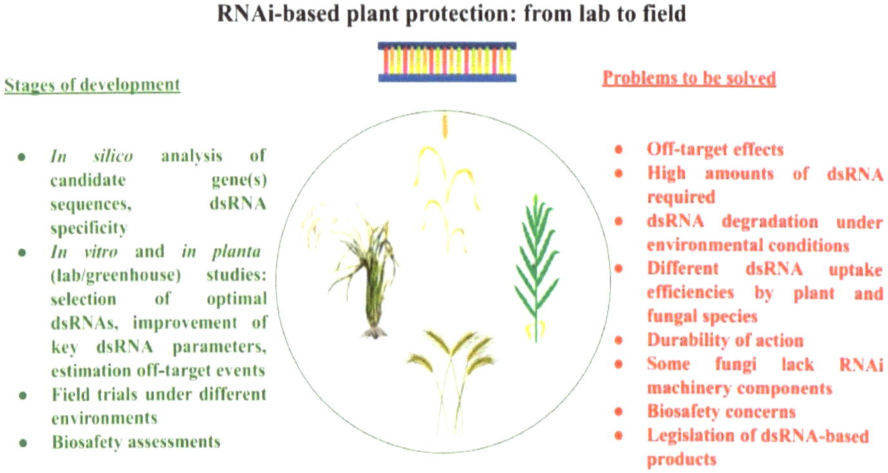

Figure 2. Key stages of experimental adaptation of RNAi-based products to field applications and main problems to be solved on the way.

Table 1. Summary of HIGS and SIGS assays to control toxigenic fungal species described to date.

Fungal Species	Method(s)	Host Plant(s)	Target Gene(s)	Effect on Mycotoxin Production	Reference
F. graminearum	SIGS	Barley	*CYP51A* + *CYP51B* + *CYP51C*	-	[76]
F. asiaticum	SIGS	Wheat	*Myo5* (myosin 5)	-	[77]
F. verticillioides	HIGS	Tobacco	*GUS* (β-glucuronidase)	-	[83]
F. graminearum	HIGS	Wheat	*Chs3b* (chitin synthase 3b)	78–85% reduction (DON)	[84]
F. culmorum	HIGS, VIGS	Wheat	*FcFg1* (secreted lipase); *FcFmk1* (MAP-kinase); *FcGls1* (β-1,3-Glucan synthase); *FcChsV* (chitin synthase V)	-	[85]
A. flavus	HIGS	Maize	*Amy1* (alpha-amylase)	Drastically reduced (AFB1, AFB2)	[86]
A. flavus	HIGS	Maize	*AFLR*	14-fold reduction (AFB1, AFB2)	[87]
A. flavus	HIGS	Maize	*Alk* (alkaline protease)	84–87% reduction (aflatoxin)	[88]
A. flavus	HIGS	Maize	*AFLM* (versicolorin dehydrogenase)	54.2–95.3% reduction (AFB1)	[89]
A. flavus	HIGS	Maize	*p2c* (polygalacturonase)	30.3–93.7% reduction (aflatoxin)	[90]
A. flavus, A. parasiticus	HIGS	Maize	*AFLC*	100% reduction (AFB1)	[91]
F. graminearum	HIGS	Arabidopsis, barley	*CYP51A* + *CYP51B* + *CYP51C* (sterol 14α-demethylase; single product)	-	[92]
F. graminearum	HIGS, SIGS	Arabidopsis, barley	*CYP51A*; *CYP51B*; *CYP51C*	-	[93]
A. flavus	HIGS	Peanut	*AFLS* (transcriptional regulator) + *AFLR* (transcriptional regulator) + *AFLC* (polyketide synthase) + *Pes1* (nonribosomal peptide synthetase) + *AFLep* (efflux pump)	100% reduction (AFB1, AFB2)	[94]
A. flavus	HIGS	Groundnut	*NsdC* (Cys_2His_2 zinc finger transcriptional regulator) + *VeA* (development and secondary metabolism regulator) + *AFLR* + *AFLM*	>1000-fold reduction (AFB1)	[95]
F. culmorum	SIGS, VIGS	Wheat	*TRI5* (trichodiene synthase)	53–85% reduction (DON, in vitro)	[102]
F. asiaticum	SIGS	Wheat	*β2-tubulin*	-	[106]
F. graminearum	SIGS	Wheat	*TRI6* (transcriptional regulator)	72% reduction (DON)	[107]
F. graminearum	SIGS	Wheat	Chs7 (chitin synthase 7); Gls (glucan synthase); Pkc (protein kinase C)	25–50% reduction (DON)	[108]
F. graminearum	SIGS	Barley	*AGO* (ARGONAUTE); *DCL* (DICER)	-	[109]

-: mycotoxin content was not estimated; AFB1/2: aflatoxins B1/B2.

3. The Ways to Improve RNAi Efficiency and Facilitate Its Transition to the Field-Scale Application

3.1. Selection of Target Gene(s)

The selection of a target gene is crucial for the successful application of RNAi-based disease control. An ideal RNAi gene target must be essential for survival and pathogenicity and should not have functional redundancy [110]. Generally, the potential target genes for mycotoxin producers can be divided into the following groups: (i) genes involved in mycotoxin biosynthesis or regulating it; (ii) genes encoding global regulation factors; and (iii) genes encoding fungal virulence factors. The first group includes the genes of the trichothecene biosynthetic cluster in *Fusarium* spp., such as *TRI5* and *TRI6* [102,107], or different genes of the aflatoxin biosynthetic cluster in *Aspergillus* spp. [87,89,91,94]. These genes are specific for these mycotoxin producers; therefore, their use should reduce the risk of unintended silencing of other targets (off-target effects, OTEs) [111]. On the other hand, these genes can be more polymorphic even at the intraspecific level than housekeeping genes [112] and this fact can potentially reduce the efficiency of RNAi by a higher number of mismatches. Moreover, a fungal isolate may be able to produce different types of mycotoxins, thus silencing of a specific cluster gene does not inhibit the production of other toxic metabolites. A possible alternative is to use genes encoding global transcription factors that regulate different fungal processes. An example of such a gene is *Vel*, responsible for growth, hyphal development, conidia formation and secondary metabolite production in *Fusarium*. This gene, as well as its homolog *veA*, was successfully used to control non-toxigenic *F. oxysporum* f. sp. *cubense* [113] and *A. flavus* [95], respectively. Potential target genes related to fungal virulence include those encoding enzymes and other factors that interfere with plant immunity and can therefore facilitate the infection process. Among these are *Amy1*, *Alk* and *p2c* genes applied to protect maize against *A. flavus* [86,88,90] and chitin synthases that proved effective against *F. graminearum* and *F. culmorum* infection in wheat [84,85,108]. Several studies have shown the great potential of using genes targeted by the main fungicides, such as azoles [76,77,92,93,106]. The genes encoding key components of fungal RNAi machinery (AGOs and DCLs) have also been effective as targets for silencing [109]. In some of the aforementioned studies, the authors silenced multiple genes at once. This approach looks promising in terms of efficiency and reliability, although it can be limited because of higher costs and off-target risks. In further studies, it would be interesting to expand the range of regulatory genes; firstly, all those encoding global regulators, participating in both developmental processes and mycotoxin production. In addition, some groups of genes involved in toxin synthesis, including those that participate in zearalenone biosynthesis in *Fusarium* spp., have not been tested. Finally, little is known regarding the real efficiency of different targets under field conditions.

3.2. Design and Optimization of dsRNAs

One of the most challenging tasks that any RNAi-based study faces is potential OTEs. An example of difficulties caused by the off-target effect is described by Masanga et al. [87]. In this study, host-induced silencing of gene encoding transcriptional factor *AFLR* in *A. flavus* led to simultaneous lowering of aflatoxin content (14-fold), stunting and reduced kernel placement in the transgenic maize. This observation was attributed to the silencing of unintended genes in transformed plants by *AFLR*-targeted siRNAs. A genome-wide bioinformatics analysis is still the best tool to find possible off-target sequences. The ability of the widely used Basic Local Alignment Search Tool (BLAST, [114]) algorithm to predict local alignments of short sequences is limited [115], but some alternative bioinformatics tools, such as dsCheck [116], ERNAi [117], GESS [118], si-RNA finder [119], pssRNAit [120] and SeedMatchR [121] were developed. The remaining challenge is that a great number of plant, fungal and bacterial genomes are still to be sequenced. Nevertheless, several recommendations to minimize off-target effects can be proposed. For instance, long GC-rich fragments as well as those containing $\geq$16 nt-long sequences that are complementary to any non-target mRNA should be avoided [122,123]. According to Chen et al. 2024 [124], the critical sequence identity of off-target effects is approximately 80%. Another possible problem is that dsRNAs produced from different regions of the same gene often demonstrate different silencing efficiencies. For instance, Johnson et al. [125] showed that targeting different segments of *FUM1* and *FUM8* genes of *F. verticillioides* led to different rates of fumonisin B1 synthesis inhibition. Similar results were obtained by Song et al. [77] for five different regions of *F. asiaticum MYO5* gene. Baldwin et al. [126] demonstrated that targeting both long (~600 nt) and short (250–300 nt) regions of *TRI6* gene of *F. graminearum* resulted in repeatable patterns of the produced siRNAs. Unequal distribution of siRNAs derived from fragments of five aflatoxin biosynthetic genes of *A. flavus* was later shown by Power et al. [127]. Moreover, the authors elucidated that most of siRNAs possess uracil as the 5′ terminal base and cytosine as the 3′ terminal base. These observations can be explained by the specificities of DCL processing, AGO binding and amplification of dsRNA by RdRps [128–130]. Today, there are no algorithms and software for accurately predicting profiles of siRNAs, derived from a certain long dsRNA. Therefore, in vitro and in vivo testing of dsRNAs, targeting different segments of the same gene(s) followed by a selection of the most efficient candidate is still an unavoidable stage of any RNAi experiment.

Selecting an optimal concentration of dsRNA for SIGS is another important task, influencing the silencing efficiency and potential off-target effects. It is believed that the concentration should be determined for an individual pathogen and that the dsRNA dose that proved effective in vitro might not show the same efficiency in open air. To date, there is a lack of studies comparing the effectiveness of different dsRNA concentrations to protect plants against mycotoxin producers. Tretiakova et al. [102] demonstrated that the strongest effect on DON accumulation was obtained using 0.96 μg of dsRNA complementary to the *TRI5* gene of *F. culmorum* as well as that higher concentrations were not more efficient. A dsRNA length is another important factor of designing an optimal assay. As in the case of optimal dsRNA concentrations, it seems to be adopted for a specific pathogen and genes to be silenced. Several studies [76,126] demonstrated that longer dsRNAs (500–600 nt) are more efficient than shorter variants. On the other hand, Höfle et al. [131] demonstrated that SIGS-based efficiencies decreased from 80% for 200–500 nt constructs to 65% for 800 nt constructs and to 50% for >1500 nt dsRNA constructs. Moreover, Tretiakova et al. [102] used a short 161 nt dsRNA to successfully silence *TRI5* gene expression. As mentioned above, the silencing efficiency of dsRNAs is determined by their nucleotide sequences rather than molecules' length *per se*. In theory, shorter molecules that are efficiently processed by DCL should be more promising for RNAi than longer ones giving low siRNAs yield. On the other hand, longer dsRNAs may contain different regions of high silencing efficiencies, thus increasing overall efficiency of the process.

3.3. dsRNA Production: Increasing Quantities and Lowering Costs

In laboratory conditions, dsRNA is often synthesized using in vitro transcription kits, such as MEGA script®RNAi kit (Thermo Fisher Scientific, Waltham, MA, USA) and others. However, for field trials, this approach would be unacceptable due to the high costs of production (~USD 700/mg). Chemical synthesis allows large quantitates of dsRNA to be obtained but it is also very expensive with the cost increasing in accordance with the length of molecule [132]. A possible alternative is using large-scale in vivo production techniques based on genetically engineered bacteria, such as *Escherichia coli* [133]. During the last decades, the yields of dsRNA obtained by *E. coli* have significantly increased [134] mainly due to the modification of growing and nutrition conditions and development of new expressing vectors and host strains. Another principal parameter is what extraction and purification methods are used. Recent studies [135,136] showed that using cheap and highly efficient protocols for dsRNA extraction could drastically increase the total dsRNA yield and reduce its cost. Verdonckt and Broeck [137] applied a well-known system of *E. coli* HT115 strain transformed with the L4440 expressing vector to optimize the extraction and purification protocol and to show that extracted dsRNAs are not cytotoxic. The estimated price of dsRNA, obtained using optimized protocol that includes heating pre-treatment and

LiCl precipitation is 43.2 EUR/mg. Apart from *E. coli*, other microorganisms can be applied to produce high quantities of dsRNA. For instance, Niehl et al. [138] developed a stable and accurate in vivo dsRNA replication system based on *Pseudomonas syringae* modified with components of phi6 bacteriophage replication complex. The risks of using bacteria to produce dsRNA for plant protection are their potential release to the environment and impact on animal and human health. As an alternative, dsRNA can be produced in other organisms, which are considered safe for human health, such as *Yarrowia lipolytica* [139] or *Saccharomyces cerevisiae* [140]. In vivo-produced dsRNAs have been successfully used for plant protection against viral [141], fungal [142] and insect [143] pathogens. Another promising way is the use of cell-free systems, able to produce dsRNA, characterized by high yield and extremely low cost (as little as USD 0.5/g, [144]).

3.4. Improving Host Plants' dsRNA Uptake

An obvious problem that must be addressed for successful SIGS application in the field is dsRNA uptake by a host plant. Several factors, including dsRNA stability on the plant surface, density and behavior of stomata and specific features of waxy cuticle may significantly influence the uptake efficiency [145]. Among the approaches applied to facilitate the uptake process are using surfactants, high-pressure spraying, abrasion and stomata flooding [145–147]. In theory, after overcoming the cuticle barrier, a dsRNA molecule should bypass the apoplast–symplast system, entering the plant cell to be processed by plant RNAi machinery. However, in the case of toxigenic fungi such as *F. graminearum*, pathogens mainly uptake unprocessed dsRNAs and digest them by its own RNAi machinery [76]. Therefore, the development of methods improving dsRNA movement via plant vasculature is needed. One of these approaches, based on the use of carrier molecules and different types of formulations, is considered in detail in the next section.

3.5. dsRNA Formulation: Improving Delivery Efficiency and Overcoming Degradation

In the majority of studies performed so far, dsRNAs are delivered to a mycotoxigenic fungus in a naked, unformulated variant, characterized by a restricted lifetime and efficiency, especially in planta. One of the most promising approaches to solve these problems is using nanoparticles as dsRNA carriers. An example of the successful application of carriers to protect plants against mycotoxin producers is the work of Power et al. [127]. The authors used 0.6 μm gold particles to coat both dsRNA and dsDNA and for their delivery to *A. flavus*-infected peanut plants by particle bombardment. This approach has proven effective, but it is technically sophisticated and hard to be widely applied. On the other hand, nanocarriers have been actively used in RNAi-based plant protection against different pathogens. Positively charged layered double hydroxide (LDH) nanosheets, also called 'BioClay', were among the first nanomaterials successfully used in plant pathology. Mitter et al. [148] applied LDH to protect cowpea and tobacco against the cucumber mosaic virus (CMV) and pepper mild mottle virus (PMMoV). The authors declared that LDH-formulated dsRNA provided up to 20-days protection of plants against viruses, compared to 5 days of protection using naked dsRNA. Mosa and Youssef [149] used LDH nanosheets to protect dsRNAs targeted to essential genes of *F. oxysporum* f. sp. *radicis-lycopersici* causing crown and root rots in tomatoes and showed that this technology provided plant protection for at least 60 days. Chen et al. [150] declared that using dsRNAs loaded on LDH nanosheet significantly improved the efficiency of management maize and tobacco infections caused by *Rhizoctonia solani*. Another material, widely used to protect dsRNA and to deliver it to target cells is chitosan (poly β-1,4-D-glucosamine, [151]). Chitosan is a biocompatible material, characterized by low toxicity, inexpensive production and a wide range of modifications [152]. Chitosan-based nanoparticles have been used to protect plants against insects, such as cotton bollworm (*Helicoverpa armigera*, [153]) and stem borer (*Chilo supressalis*, [154]). In a recent study [155], six different types of nanoparticles—chitosan (CS), polyethyleneimine (PEI), protamine, carbon quantum dot (CQD), polyamidoamine (PAMAM) and chitosan/star polycation complex (CSC)—were used as carriers for dsRNAs targeting two genes related to the virulence of *Rhizoctonia solani*. All the nanoparticles could assemble with dsRNA, but CSC also enhanced the stability of the molecule, reduced pathogen infection on rice and prolonged the protection time up to 20 days. Based on the result, the authors declared that SIGS-based CSC-dsRNA delivery systems have the potential to be commercialized and used for plant fungal disease management in the field.

Qiao et al. [156] proposed to use artificial nanovesicles for dsRNA delivery to combat *Botrytis cinerea* on tomato and grape fruit leaves. It was shown that AVs application extended the time of dsRNA-based protection to 10 days on tomato and grape fruits and to 21 days on grape leaves. Another recent study demonstrated that *E. coli*-derived anucleated minicells can be applied for dsRNA production, encapsulation and delivery [157]. In this work, the authors obtained an *E. coli* mutant producing a large number of minicells with compromised RNase-III activity and transformed it with vectors expressing chitin synthase class III (Chs3a, Chs3b) and DICER-like proteins (DCL1 and DCL2) genes of *Botryotinia fuckeliana*. The production of minicell-encapsulated dsRNA in bioreactors was about 100 mg/L. Topical spray application of dsRNA packaged in minicells on strawberry 1 h before inoculation of *B. fuckeliana* led to complete inhibition of fungal growth and disease development (while naked dsRNA was applied, a minimal fungal growth was observed). It was shown that the use of minicell-encapsulated dsRNAs prolonged the strawberry protection against *B. fuckeliana* up to 12 days. Niño-Sánchez et al. [158] proposed to use non-pathogenic bacteria as a system to deliver RNAi to fungal cells as an alternative to HIGS and SIGS assays. In the study, the RNAse III null-mutant strain of *E. coli* HT115 was transformed with two plasmid vectors able to express dsRNA against the *AFLC* gene involved in aflatoxin synthesis in the *A. flavus* and *BcSaS1* genes, responsible for virulence in *Botrytis cinerea*. The authors demonstrated that under in vitro conditions, exposure to both living bacteria or whole-cell lysates induced silencing of the target genes, leading to inhibiting aflatoxins production and mycelial growth, respectively.

Generally, different prospective types of potential nanocarriers are thoroughly discussed in several reviews [159–161]. The use of nanoparticle-based approaches significantly improve dsRNA stability, enhance its uptake by plant and target organisms, and prolong the time of protection. At the same time, it should be noted that several types of nanoparticles may be linked to higher toxicity in plants, animals and soil microflora. Moreover, they potentially negatively impact human health: for example, inhaling nanoparticles may result in lung inflammation and heart problems, and their ability to overcome the blood–brain barrier may lead to delivering toxic compounds to the brain and central nervous system [162,163]. Therefore, extensive studies on the safety of nanoparticles and estimation of corresponding risks are of great importance.

3.6. Nanoparticle-Free Approaches for RNAi-Based Systems

Recently, techniques of increasing the stability of dsRNA and its delivery efficiency avoiding the use of nanoparticles are being developed. Such an innovative approach was proposed by Singewar and Fladung [164] and called 'mycorrhiza-mediated dsRNA delivery method'. In this approach, stressed plants are inoculated with mycorrhiza cultures carrying dsRNA (obtained through the direct uptake of dsRNA or by transformation) which targets a specific pathogen's gene. In the mycorrhizal system, the nutritional exchange takes place between fungal and plant root cell walls. The fungus is able to alter the plant's cell wall and cause beneficial invasion, helping to transfer dsRNA to the plant where the molecule is further transported to the stem and leaves through the vascular system. Therefore, in theory, this approach can provide a double benefit, including traditional advantages of mycorrhiza together with dsRNA-based silencing of pathogens. Another promising technique is chemical modifications of dsRNA [165,166]. The therapeutic meaning of chemically modified short RNAs is well known in medicine [167], but their potential to be used in agricultural systems is still to be elucidated. In one of the first studies, Howard et al. [168] demonstrated that phosphorothionate- and 2′-fluoro-modified long dsRNAs were able to induce RNAi in *Drosophila* cell lines, live green stink bug (*Nezara viridula*) nymphs and live western corn rootworm (*Diabrotica virgifera virgifera*) larvae with higher efficacy compared to unmodified dsRNAs. In live insects, treatment by the modified molecules resulted in mortality in the latter two species. In addition, the modified dsRNAs were stable to environmental nucleases, including those present in soil and stink bug saliva. These results indicate the possibility of using chemically modified long dsRNAs for RNAi-based plant protection. Further studies on the different pathogens, including fungi, with the use of different modifications are required to develop RNAi-based control protocols.

4. Conclusions and Perspectives

Since mycotoxins represent one of the most significant risks for agriculture and food safety, methods of reliable, eco-friendly and cheap control of mycotoxigenic fungi are of high importance. RNAi-based techniques have been extensively used for different scientific and applied purposes. Both HIGS and SIGS have proven effective to protect agricultural plants against producers of key mycotoxins and have the great potential to replace traditional fungicides. However, the ultimate goal, undoubtedly, should be the implementation of this approach in agricultural practice and the development of RNAi-based biopesticides. To meet this challenge, the following issues should be addressed: (i) expansion of the list of fungal species and target genes tested; (ii) development and improvement of algorithms for rational dsRNA design and identification of potential off-targets; (iii) elucidating the optimal dsRNA parameters (concentration, molecule length) for in vitro and *in planta* applications; (iv) improvement of techniques for cost-effective and large-scale dsRNA production; (v) developing the methods of efficient dsRNA protection and targeted delivery to a pathogen. Besides solving these practical tasks, further basic research of fungal RNAi machinery, mechanisms of RNA uptake, transport in plants and between host and pathogen and factors influencing these processes are needed. Little is known about the persistence, durability and side effects of dsRNAs on host plants. In addition, implementation of RNAi-based approaches in agricultural practice will require consideration of the biosafety of new dsRNA pesticides. Finally, the current legislations should be adopted to approve the RNAi-based methods and products. Solving these problems will pave the way for these products to successful field trials and entering the market.

Author Contributions: Conceptualization, A.A.S. and M.T.; writing—A.A.S.; editing—M.T., N.O.K. and S.K.Z. All authors have read and agreed to the published version of the manuscript.

Funding: Russian Science Foundation, grant № 23-74-30003, supported this work.

Institutional Review Board Statement: Not applicable.

Informed Consent Statement: Not applicable.

Data Availability Statement: Data are contained within the article.

Conflicts of Interest: The authors declare no conflict of interest.

References

1. Almeida, F.; Rodriguez, M.L.; Coelho, C. The still underestimated problem of fungal diseases worldwide. *Front. Microbiol.* **2019**, *10*, 214. [CrossRef]
2. Fones, H.N.; Bebber, D.P.; Chaloner, T.M.; Kay, W.T.; Steinberg, G.; Gurr, S.J. Threats to global food security from emerging fungal and oomycete crop pathogens. *Nat. Food* **2020**, *1*, 332–342. [CrossRef] [PubMed]
3. O'Donnell, K.; Rooney, A.P.; Proctor, R.H.; Brown, D.W.; McCormick, S.P.; Ward, T.J.; Frandsen, R.J.N.; Lysøe, E.; Rehner, S.A.; Aoki, T.; et al. Phylogenetic analyses of *RPB1* and *RPB2* support a middle Cretaceous origin for a clade comprising all agriculturally and medically important fusaria. *Fungal Genet. Biol.* **2013**, *52*, 20–31. [CrossRef] [PubMed]
4. Ciegler, A. Do mycotoxins function in ecological processes? *J. Food Saf.* **1983**, *5*, 23–30. [CrossRef]
5. Latham, R.L.; Boyle, J.T.; Barbano, A.; Loveman, W.G.; Brown, N.A. Diverse mycotoxin threats to safe food and feed cereals. *Essays Biochem.* **2023**, *67*, 797–809. [CrossRef] [PubMed]
6. Bin-Umer, M.A.; McLaughlin, J.; Basu, D.; McCormick, S.; Tumer, N.E. Trichothecene mycotoxins inhibit mitochondrial translation—Implication for the mechanism of toxicity. *Toxins* **2011**, *3*, 1484–1501. [CrossRef] [PubMed]

7. Cundliffe, E.; Cannon, M.; Davies, J. Mechanism of inhibition of eukaryotic protein synthesis by trichothecene fungal toxins. *Proc. Natl. Acad. Sci. USA* **1974**, *71*, 30–34. [CrossRef] [PubMed]
8. Escrivá, L.; Font, G.; Manyes, L. In vivo toxicity studies of *Fusarium* mycotoxins in the last decade: A review. *Food Chem Toxicol.* **2015**, *78*, 185–206. [CrossRef] [PubMed]
9. Kumar, V.V. Aflatoxins: Properties, toxicity and detoxification. *Nutr. Food Sci.* **2018**, *6*, 555696. [CrossRef]
10. Pickova, D.; Ostry, V.; Toman, J.; Malir, F. Aflatoxins: History, significant milestones, recent data on their toxicity and ways to mitigation. *Toxins* **2021**, *13*, 399. [CrossRef] [PubMed]
11. Ropejko, K.; Twarużek, M. Zearalenone and its metabolites—General overview, occurrence, and toxicity. *Toxins* **2021**, *13*, 35. [CrossRef] [PubMed]
12. Hao, G.; McCormick, S.; Tiley, H.; Gutiérrez, S.; Yulfo-Soto, G.; Vaughan, M.M.; Ward, T.J. NX trichothecenes are required for *Fusarium graminearum* infection of wheat. *Mol. Plant-Microbe Interact.* **2023**, *36*, 294–304. [CrossRef] [PubMed]
13. Bai, G.H.; Desjardins, A.E.; Plattner, R.D. Deoxynivalenol-nonproducing *Fusarium graminearum* causes initial infection, but does not cause disease spread in wheat spikes. *Mycopathologia* **2002**, *153*, 91–98. [CrossRef] [PubMed]
14. Langevin, F.; Eudes, F.; Comeau, A. Effect of trichothecenes produced by *Fusarium graminearum* during *Fusarium* head blight development in six cereal species. *Eur. J. Plant Pathol.* **2004**, *110*, 735–746. [CrossRef]
15. Haidukowski, M.; Pascale, M.; Perrone, G.; Pancaldi, D.; Campagna, C.; Visconti, A. Effect of fungicides on the development of *Fusarium* head blight, yield and deoxynivalenol accumulation in wheat inoculated under field conditions with *Fusarium graminearum* and *Fusarium culmorum*. *J. Sci. Food Agric.* **2005**, *85*, 191–198. [CrossRef]
16. Cendoya, E.; Nichea, M.J.; del Pilar Monge, M.; Zachetti, V.G.L.; Chiacchiera, S.M.; Ramirez, M.L. Effect of fungicides commonly used for Fusarium head blight management on growth and fumonisin production by *Fusarium proliferatum*. *Rev. Argent. Microbiol.* **2021**, *53*, 64–74. [CrossRef] [PubMed]
17. Ramirez, M.; Chulze, S.; Magan, N. Impact of environmental factors on growth and deoxynivalenol production by *Fusarium graminearum* isolates from Argentinian wheat. *Crop Prot.* **2004**, *23*, 117–125. [CrossRef]
18. Tao, H.; Bao, Z.; Jin, C.; Miao, W.; Fu, Z.; Jin, Y. Toxic effects and mechanisms of three commonly used fungicides on the human colon adenocarcinoma cell line Caco-2. *Environ. Pollut.* **2020**, *263*, 114660. [CrossRef]
19. Cordero-Bueso, G.; Arroyo, T.; Valero, E. A long term field study of the effect of fungicides penconazole and sulfur on yeasts in the vineyard. *Int. J. Food Microbiol.* **2014**, *189*, 189–194. [CrossRef] [PubMed]
20. de Chaves, M.A.; Reginatto, P.; da Costa, B.S.; de Paschoal, R.I.; Teixeira, M.L.; Feuntefria, A.A. Fungicide resistance in *Fusarium graminearum* species complex. *Curr. Microbiol.* **2022**, *79*, 62. [CrossRef]
21. Fraaije, B.; Atkins, S.; Hanley, S.; Macdonald, A.; Lucas, J. The multi-fungicide resistance status of *Aspergillus fumigatus* populations in arable soils and the wider European environment. *Front. Microbiol.* **2020**, *11*, 599233. [CrossRef] [PubMed]
22. Abbas, A.; Wright, C.W.; El-Sawi, N.; Yli-Mattila, T.; Malinen, A.M. A methanolic extract of *Zanthoxylum bungeanum* modulates secondary metabolism regulator genes in *Aspergillus flavus* and shuts down aflatoxin production. *Sci. Rep.* **2022**, *12*, 5995. [CrossRef] [PubMed]
23. Dzhavakhiya, V.G.; Voinova, T.M.; Popletaeva, S.B.; Statsyuk, N.V.; Limantseva, L.A.; Shcherbakova, L.A. Effect of various compounds blocking the colony pigmentation on the aflatoxin B1 production by *Aspergillus flavus*. *Toxins* **2016**, *8*, 313. [CrossRef]
24. Liang, D.; Xing, F.; Selvaraj, J.N.; Liu, X.; Wang, L.; Hua, H.; Zhou, L.; Zhao, Y.; Wang, Y.; Liu, Y. Inhibitory effect of cinnamaldehyde, citral, and eugenol on aflatoxin B1 biosynthesis in *Aspergillus flavus*. *J. Food Sci.* **2015**, *80*, M2917–M2924. [CrossRef] [PubMed]
25. Stakheev, A.A.; Erokhin, D.V.; Meleschuk, E.A.; Mikityuk, O.D.; Statsyuk, N.V. Effect of compactin on the mycotoxin production and expression of related biosynthetic and regulatory genes in toxigenic *Fusarium culmorum*. *Microorganisms* **2022**, *10*, 1347. [CrossRef]
26. Napoli, C.; Lemieux, C.; Jorgensen, R. Introduction of a chimeric chalcone synthase gene into petunia results in reversible co-suppression of homologous genes in trans. *Plant Cell* **1990**, *2*, 279–289. [CrossRef]
27. Elbashir, S.M.; Harborth, J.; Lendeckel, W.; Yalcin, A.; Weber, K.; Tuschl, T. Duplexes of 21-nucleotide RNAs mediate RNA interference in cultured mammalian cells. *Nature* **2001**, *411*, 494–498. [CrossRef]
28. Billmyre, R.B.; Calo, S.; Feretzaki, M.; Wang, X.; Heitman, J. RNAi function, diversity, and loss in the Fungal Kingdom. *Chromosome Res.* **2013**, *21*, 561–572. [CrossRef] [PubMed]
29. Chang, Z.; Billmyre, R.B.; Lee, S.C.; Heitman, J. Broad antifungal resistance mediated by RNAi-dependent epimutation in the basal human fungal pathogen *Mucor circinelloides*. *PLoS Genet.* **2019**, *15*, e1007957. [CrossRef]
30. Covey, S.N.; Al-Kaff, N.S.; Lángara, A.; Turner, D.S. Plants combat infection by gene silencing. *Nature* **1997**, *385*, 781–782. [CrossRef]
31. Son, H.; Park, A.R.; Lim, J.Y.; Shin, C.; Lee, Y.-W. Genome-wide exonic small interference RNA-mediated gene silencing regulates sexual reproduction in the homothallic fungus *Fusarium graminearum*. *PLoS Genet.* **2017**, *13*, e1006595. [CrossRef]
32. Wang, X.-B.; Wu, Q.; Ito, T.; Cillo, F.; Li, W.-X.; Chen, X.; Yu, J.-L.; Ding, S.-W. RNAi-mediated viral immunity requires amplification of virus-derived siRNAs in *Arabidopsis thaliana*. *Proc. Natl. Acad. Sci. USA* **2010**, *107*, 484–489. [CrossRef]
33. Yu, R.; Wang, X.; Moazed, D. Epigenetic inheritance mediated by coupling of RNAi and histone H3K9 methylation. *Nature* **2018**, *558*, 615–619. [CrossRef]

34. Lax, C.; Tahiri, G.; Patiño-Medina, J.A.; Cánovas-Márquez, J.T.; Pérez-Ruiz, J.A.; Osorio-Concepción, M.; Navarro, E.; Calo, S. The evolutionary significance of RNAi in the fungal kingdom. *Int. J. Mol. Sci.* **2020**, *21*, 9348. [CrossRef]
35. Romano, N.; Macino, G. Quelling: Transient inactivation of gene expression in *Neurospora crassa* by transformation with homologous sequences. *Mol. Microbiol.* **1992**, *6*, 3343–3353. [CrossRef]
36. Fire, A.; Xu, S.; Montgomery, M.K.; Kostas, S.A.; Driver, S.E.; Mello, C.C. Potent and specific genetic interference by double-stranded RNA in *Caenorhabditis elegans*. *Nature* **1998**, *391*, 806–811. [CrossRef]
37. Desmukh, R.; Purohit, H.J. siRNA mediated gene silencing in *Fusarium* sp. HKF15 for overproduction of bikaverin. *Bioresour. Technol.* **2014**, *157*, 368–371. [CrossRef]
38. Ma, C.; Liu, J.; Tang, J.; Sun, Y.; Jiang, X.; Zhang, T.; Feng, Y.; Liu, Q.; Wang, L. Current genetic approaches to investigate gene functions in *Trichoderma reesei*. *Microb. Cell Factories* **2023**, *22*, 97. [CrossRef]
39. Salazar-Cerezo, S.; de Vries, R.P.; Garrigues, S. Strategies for the development of industrial fungal producing strains. *J. Fungi* **2023**, *9*, 834. [CrossRef]
40. Hutvagner, G.; Simard, M.J. Argonaute proteins: Key players in RNA silencing. *Nat. Rev. Cell Biol.* **2008**, *9*, 22–32. [CrossRef]
41. Makeyev, E.V.; Bamford, D.H. Cellular RNA-dependent RNA polymerase involved in posttranscriptional gene silencing has two distinct activity modes. *Mol. Cell* **2002**, *10*, 1417–1427. [CrossRef]
42. Tijsterman, M.; Plasterk, R.H.A. Dicers at RISC: The mechanism of RNAi. *Cell* **2004**, *117*, 1–3. [CrossRef] [PubMed]
43. Pak, J.; Fire, A. Distinct populations of primary and secondary effectors during RNAi in *C. elegans*. *Science* **2007**, *315*, 241–244. [CrossRef] [PubMed]
44. Baulcombe, D. RNA silencing in plants. *Nature* **2004**, *431*, 356–363. [CrossRef]
45. Chen, Y.; Gao, Q.; Huang, M.; Liu, Y.; Liu, Z.; Liu, X.; Ma, Z. Characterization of RNA silencing components in the plant pathogenic fungus *Fusarium graminearum*. *Sci. Rep.* **2015**, *5*, 12500. [CrossRef] [PubMed]
46. Torres-Martínez, S.; Ruiz-Vázquez, R.M. The RNAi universe in fungi: A varied landscape of small RNAs and biological functions. *Annu. Rev. Microbiol.* **2017**, *71*, 371–391. [CrossRef] [PubMed]
47. Wilson, R.C.; Doudna, J.A. Molecular mechanisms of RNA interference. *Annu. Rev. Biophys.* **2013**, *42*, 217–239. [CrossRef] [PubMed]
48. Laurie, J.D.; Ali, S.; Linning, R.; Mannhaupt, G.; Wong, P.; Güldener, U.; Münsterkötter, M.; Moore, R.; Kahmann, R.; Bakkeren, G.; et al. Genome comparison of barley and maize smut fungi reveals targeted loss of RNA silencing components and species-specific presence of transposable elements. *Plant Cell* **2012**, *24*, 1733–1745. [CrossRef] [PubMed]
49. Laurie, J.D.; Linning, R.; Bakkeren, G. Hallmarks of RNA silencing are found in the smut fungus *Ustilago hordei* but not in its close relative *Ustilago maydis*. *Curr. Genet.* **2008**, *53*, 49–58. [CrossRef]
50. Nicolás, F.E.; Torres-Martínez, S.; Ruiz-Vázquez, R.M. Loss and retention of RNA interference in fungi and parasites. *PLoS Pathog.* **2013**, *9*, e1003089. [CrossRef] [PubMed]
51. Yoshimoto, R.; Ishida, F.; Yamaguchi, M.; Tanaka, S. The production and secretion of tRNA-derived RNA fragments in the corn smut fungus *Ustilago maydis*. *Front. Fungal. Biol.* **2022**, *3*, 958798. [CrossRef]
52. Lee, H.-C.; Li, L.; Gu, W.; Xue, Z.; Crosthwaite, S.K.; Pertsemlidis, A.; Lewis, Z.A.; Freitag, M.; Selker, E.U.; Mello, C.C.; et al. Diverse pathways generate microRNA-like RNAs and Dicer-independent small interfering RNAs in fungi. *Mol. Cell* **2010**, *38*, 803–814. [CrossRef] [PubMed]
53. Zeng, W.; Wang, J.; Wang, Y.; Lin, J.; Fu, Y.; Xie, J.; Jiang, D.; Chen, T.; Liu, H.; Cheng, J. Dicer-like proteins regulate sexual development via the biogenesis of perithecium-specific microRNA in a plant pathogenic fungus *Fusarium graminearum*. *Front. Microbiol.* **2018**, *9*, 818. [CrossRef] [PubMed]
54. Gaffar, F.Y.; Imani, J.; Karlovsky, P.; Koch, A.; Kogel, K.-H. Different components of the RNA interference machinery are required for conidiation, ascosporogenesis, virulence, deoxynivalenol production, and fungal inhibition by exogenous double-stranded RNA in the head blight pathogen *Fusarium graminearum*. *Front. Microbiol.* **2019**, *10*, 1662. [CrossRef] [PubMed]
55. Song, X.-S.; Gu, K.-X.; Duan, X.-X.; Xiao, X.-M.; Hou, Y.-P.; Duan, Y.-B.; Wang, J.-X.; Yu, N.; Zhou, M.-G. Secondary amplification of siRNA machinery limits the application of spray-induced gene silencing. *Mol. Plant Pathol.* **2018**, *19*, 2543–2560. [CrossRef]
56. Weiberg, A.; Bellinger, M.; Jin, H. Conversations between kingdoms: Small RNAs. *Curr. Opin. Biotechnol.* **2015**, *32*, 207–215. [CrossRef] [PubMed]
57. Weiberg, A.; Jin, H. Small RNAs—The secret agents in the plant-pathogen interactions. *Curr. Opin. Plant Biol.* **2015**, *26*, 87–94. [CrossRef] [PubMed]
58. Wang, M.; Weiberg, A.; Lin, F.-M.; Thomma, B.P.H.J.; Huang, H.-D.; Jin, H. Bidirectional cross-kingdom RNAi and fungal uptake of external RNAs confer plant protection. *Nat. Plants* **2016**, *2*, 16151. [CrossRef] [PubMed]
59. Weiberg, A.; Wang, M.; Lin, F.M.; Zhao, H.; Zhang, Z.; Kaloshian, I.; Huang, H.D.; Jin, H. Fungal small RNAs suppress plant immunity by hijacking host RNA interference pathways. *Science* **2013**, *342*, 118–123. [CrossRef]
60. Wang, M.; Weiberg, A.; Dellota, E., Jr.; Yamane, D.; Jin, H. *Botrytis* small RNA Bc-siR37 suppresses plant defense genes by cross-kingdom RNAi. *RNA Biol.* **2017**, *14*, 421–428. [CrossRef] [PubMed]
61. Shahid, S.; Kim, G.; Johnson, N.R.; Wafula, E.; Wang, F.; Coruh, C.; Bernal-Galeano, V.; Phifer, T.; dePamphilis, C.W.; Westwood, J.H.; et al. MicroRNAs from the parasitic plant *Cuscuta campestris* target host messenger RNAs. *Nature* **2018**, *553*, 82–85. [CrossRef] [PubMed]

62. Zhang, T.; Zhao, Y.-L.; Zhao, J.-H.; Wang, S.; Jin, Y.; Chen, Z.-Q.; Fang, Y.-Y.; Hua, C.L.; Ding, S.-W.; Guo, H.-S. Cotton plants export microRNAs to inhibit virulence gene expression in a fungal pathogen. *Nat. Plants* **2016**, *2*, 16153. [CrossRef]
63. Derbyshire, M.; Mbengue, M.; Barascud, M.; Navaud, O.; Raffaele, S. Small RNAs from the plant pathogenic fungus *Sclerotinia sclerotiorum* highlight host candidate genes associated with quantitative disease resistance. *Mol. Plant Pathol.* **2019**, *20*, 1279–1297. [CrossRef] [PubMed]
64. Ji, H.-M.; Mao, H.-Y.; Li, S.-J.; Feng, T.; Zhang, Z.-Y.; Cheng, L.; Luo, S.-J.; Borkovich, K.A.; Ouyang, S.-Q. *Fol*-milR1, a pathogenicity factor of *Fusarium oxysporum*, confers tomato wilt disease resistance by impairing host immune responses. *New Phytol.* **2021**, *232*, 705–718. [CrossRef] [PubMed]
65. Jian, J.; Liang, X. One small RNA of *Fusarium graminearum* targets and silences *CEBiP* gene in common wheat. *Microorganisms* **2019**, *7*, 425. [CrossRef] [PubMed]
66. Pyott, D.E.; Molnar, A. Going mobile: Non-cell-autonomous small RNAs shape the genetic landscape of plants. *Plant Biotechnol. J.* **2015**, *13*, 306–318. [CrossRef]
67. Tamiru, M.; Hardcastle, T.J.; Lewsey, M.G. Regulation of genome-wide DNA methylation by mobile small RNAs. *New Phytol.* **2018**, *217*, 540–546. [CrossRef]
68. Cai, Q.; Qiao, L.; Wang, M.; He, B.; Lin, F.M.; Palmquist, J.; Huang, S.D.; Jin, H. Plants send small RNAs in extracellular vesicles to fungal pathogen to silence virulence genes. *Science* **2018**, *360*, 1126–1129. [CrossRef]
69. Cai, Q.; He, B.; Weiberg, A.; Buck, A.H.; Jin, H. Small RNAs and extracellular vesicles: New mechanisms of cross-specific communication and innovative tools for disease control. *PLoS Pathog.* **2019**, *15*, e1008090. [CrossRef]
70. Cheng, A.-P.; Kwon, S.; Adeshara, T.; Göhre, V.; Feldbrügge, M.; Weiberg, A. Extracellular RNAs released by plant-associated fungi: From fundamental mechanisms to biotechnological applications. *Appl. Microbiol. Biotechnol.* **2023**, *107*, 5935–5945. [CrossRef]
71. Kwon, S.; Rupp, O.; Brachmann, A.; Blum, C.F.; Kraege, A.; Goesmann, A.; Feldbrügge, M. mRNA inventory of extracellular vesicles from *Ustilago maydis*. *J. Fungi* **2021**, *7*, 562. [CrossRef] [PubMed]
72. Schlemmer, T.; Barth, P.; Weipert, L.; Preußer, C.; Hardt, M.; Möbus, A.; Busche, T.; Koch, A. Isolation and characterization of barley (*Hordeum vulgare*) extracellular vesicles to assess their role in RNA spray-based crop protection. *Int. J. Mol. Sci.* **2021**, *22*, 7212. [CrossRef]
73. Schlemmer, T.; Lischka, R.; Wegner, L.; Ehlers, K.; Biedenkopf, D.; Koch, A. Extracellular vesicles isolated from dsRNA-sprayed barley plants exhibit no growth inhibition or gene silencing in *Fusarium graminearum*. *Fungal Biol. Biotechnol.* **2022**, *9*, 14. [CrossRef]
74. Karimi, H.Z.; Baldrich, P.; Rutter, B.D.; Borniego, L.; Zajt, K.K.; Meyers, B.C.; Innes, R.W. Arabidopsis apoplastic fluid contains sRNA- and circular RNA-protein complexes that are located outside extracellular vesicles. *Plant Cell* **2022**, *34*, 1863–1881. [CrossRef] [PubMed]
75. Qiao, L.; Lan, C.; Capriotti, L.; Ah-Fong, A.; Nino-Sánchez, J.; Hamby, R.; Heller, J.; Zhao, H.; Glass, N.L.; Judelson, H.S.; et al. Spray-induced gene silencing for disease control is dependent on the efficiency of pathogen RNA uptake. *Plant Biotechnol. J.* **2021**, *19*, 1756–1768. [CrossRef]
76. Koch, A.; Biedenkopf, D.; Furch, A.; Weber, L.; Rossbach, O.; Abdellatef, E.; Linicus, L.; Johannsmeier, J.; Jelonek, L.; Goesmann, A.; et al. An RNAi-based control of *Fusarium graminearum* infections through spraying of long dsRNAs involves a plant passage and is controlled by the fungal silencing machinery. *PLoS Pathog.* **2016**, *12*, e1005901. [CrossRef] [PubMed]
77. Song, X.-S.; Gu, K.-X.; Duan, X.-X.; Xiao, X.-M.; Hou, Y.-P.; Duan, Y.-B.; Wang, J.-X.; Zhou, M.-G. A *myosin5* dsRNA that reduces the fungicide resistance and pathogenicity of *Fusarium asiaticum*. *Pestic. Biochem. Physiol.* **2018**, *150*, 1–9. [CrossRef]
78. Kettles, G.J.; Hofinger, B.J.; Hu, P.; Bayon, C.; Rudd, J.J.; Balmer, D.; Courbot, M.; Hammond-Kosack, K.E.; Scalliet, G.; Kanyuka, K. sRNA profiling combined with gene function analysis reveals a lack of evidence for cross-kingdom RNAi in the wheat—*Zymoseptoria tritici* pathosystem. *Front. Plant Sci.* **2019**, *10*, 892. [CrossRef] [PubMed]
79. Wytinck, N.; Sullivan, D.S.; Biggar, K.T.; Crisostomo, L.; Pelka, P.; Belmonte, M.F.; Whyard, S. Clathrin mediated endocytosis is involved in the uptake of exogenous double-stranded RNA in the white mold phytopathogen *Sclerotinia sclerotiorum*. *Sci. Rep.* **2020**, *10*, 12773. [CrossRef] [PubMed]
80. Abdel-Hadi, A.M.; Caley, D.P.; Carter, D.R.F.; Magan, N. Control of aflatoxin production of *Aspergillus flavus* and *Aspergillus parasiticus* using RNA silencing technology by targeting aflD (nor-1) gene. *Toxins* **2011**, *3*, 647–659. [CrossRef]
81. McDonald, T.; Brown, D.; Keller, N.P.; Hammond, T.M. RNA silencing of mycotoxin production in *Aspergillus* and *Fusarium* species. *Mol. Plant-Microbe Interact.* **2005**, *18*, 539–545. [CrossRef]
82. Nowara, D.; Gay, A.; Lacomme, C.; Shaw, J.; Ridout, C.; Douchkov, D.; Hensel, G.; Kumlehn, J.; Schweizer, P. HIGS: Host-induced gene silencing in the obligate biotrophic fungal pathogen *Blumeria graminis*. *Plant Cell* **2010**, *22*, 3130–3141. [CrossRef] [PubMed]
83. Tinoco, M.L.P.; Dias, B.B.A.; Dall'Astta, R.C.; Pamphile, J.A.; Aragão, F.J.L. In vivo trans-specific gene silencing in fungal cells by in planta expression of a double-stranded RNA. *BMC Biol.* **2010**, *8*, 27. [CrossRef]
84. Cheng, W.; Song, X.-S.; Li, H.-P.; Cao, L.-H.; Sun, K.; Qiu, X.-L.; Xu, Y.-B.; Yang, P.; Huang, T.; Zhang, J.-B.; et al. Host-induced gene silencing of an essential chitin synthase gene confers durable resistance to Fusarium head blight and seedling blight in wheat. *Plant Biotechnol. J.* **2015**, *13*, 1335–1345. [CrossRef]
85. Chen, W.; Kastner, C.; Nowara, D.; Oliveira-Garcia, E.; Rutten, T.; Zhao, Y.; Deising, H.B.; Kumlehn, J.; Schweizer, P. Host-induced silencing of *Fusarium culmorum* genes protects wheat from infection. *J. Exp. Bot.* **2016**, *67*, 4979–4991. [CrossRef] [PubMed]

86. Gilbert, M.K.; Majumdar, R.; Rajasekaran, K.; Chen, Z.-Y.; Wei, Q.; Sickler, C.M.; Lebar, M.D.; Cary, J.W.; Frame, B.R.; Wang, K. RNA interference-based silencing of the alpha-amylase (*amy1*) gene in *Aspergillus flavus* decreases fungal growth and aflatoxin production in maize kernels. *Planta* **2018**, *247*, 1465–1473. [CrossRef]
87. Masanga, J.O.; Matheka, J.M.; Omer, R.A.; Ommeh, S.C.; Monda, E.O.; Alakonya, A.E. Downregulation of transcription factor *aflR* in *Aspergillus flavus* confers reduction of aflatoxin accumulation in transgenic maize with alteration of host plant architecture. *Plant Cell Rep.* **2015**, *34*, 1379–1387. [CrossRef] [PubMed]
88. Omolehin, O.; Raruang, Y.; Hu, D.; Han, Z.-Q.; Wei, Q.; Wang, K.; Rajasekaran, K.; Cary, J.W.; Chen, Z.-Y. Resistance to aflatoxin accumulation in maize mediated by host-induced silencing of the *Aspergillus flavus* alkaline protease (*alk*) gene. *J. Fungi* **2021**, *7*, 904. [CrossRef] [PubMed]
89. Raruang, Y.; Omolehin, O.; Hu, D.; Wei, Q.; Han, Z.-Q.; Rajasekaran, K.; Cary, J.W.; Wang, K.; Chen, Z.-Y. Host induced gene silencing targeting *Aspergillus flavus aflM* reduced aflatoxin contamination in transgenic maize under field condition. *Front. Microbiol.* **2020**, *11*, 754. [CrossRef]
90. Raruang, Y.; Omolehin, O.; Hu, D.; Wei, Q.; Promyou, S.; Parekattil, L.J.; Rajasekaran, K.; Cary, J.W.; Wang, K.; Chen, Z.-Y. Targeting the *Aspergillus flavus p2c* gene through host-induced gene silencing reduces *A. flavus* infection and aflatoxin contamination in transgenic maize. *Front. Plant Sci.* **2023**, *14*, 1150086. [CrossRef]
91. Thakare, D.; Zhang, J.; Wing, R.A.; Cotty, P.J.; Schmidt, M.A. Aflatoxin-free transgenic maize using host-induced gene silencing. *Sci. Adv.* **2017**, *3*, e1602382. [CrossRef]
92. Koch, A.; Kumar, N.; Weber, L.; Keller, H.; Imani, J.; Kogel, K.-H. Host-induced gene silencing of cytochrome P450 lanosterol C14α-demethylase-encoding genes confers strong resistance to *Fusarium* species. *Proc. Natl. Acad. Sci. USA* **2013**, *110*, 19324–19329. [CrossRef] [PubMed]
93. Koch, A.; Höfle, L.; Werner, B.T.; Imani, J.; Schmidt, A.; Jelonek, L.; Kogel, K.-H. SIGS vs HIGS: A study on the efficacy of two dsRNA delivery strategies to silence *Fusarium FgCYP51* genes in infected host and non-host plants. *Mol. Plant Pathol.* **2019**, *20*, 1636–1644. [CrossRef] [PubMed]
94. Arias, R.; Dang, P.M.; Sobolev, V.S. RNAi-mediated control of aflotoxins in peanut: Method to analyze mycotoxin production and transgene expression in the peanut/*Aspergillus* pathosystem. *J. Vis. Exp.* **2015**, *106*, e53398. [CrossRef]
95. Prasad, K.; Yogendra, K.; Sanivarapu, H.; Rajasekaran, K.; Cary, J.W.; Sharma, K.K.; Bhatnagar-Mathur, P. Multiplexed host-induced gene silencing of *Aspergillus flavus* genes confers aflatoxin resistance in groundnut. *Toxins* **2023**, *15*, 319. [CrossRef] [PubMed]
96. Panwar, V.; McCallum, B.; Jordan, M.; Loewen, M.; Fobert, P.; McCartney, C.; Bakkeren, G. RNA silencing approaches for identifying pathogenicity and virulence elements towards engineering crop resistance to plant pathogenic fungi. *CAB Rev.* **2016**, *11*, 027. [CrossRef]
97. Mishra, R.; Zhao, K. Genome editing technologies and their applications in crop improvement. *Plant Biotechnol. Rep.* **2018**, *12*, 57–68. [CrossRef]
98. Jang, G.; Joung, Y.H. CRISPR/Cas-mediated genome editing for crop improvement: Current applications and future prospects. *Plant Biotechnol. Rep.* **2019**, *13*, 1–10. [CrossRef]
99. Herman, R.A.; Storer, N.P.; Anderson, J.A.; Amijee, F.; Cnudde, F.; Raybould, A. Transparency in risk-disproportionate regulation of regulation of modern crop-breeding techniques. *GM Crops Food* **2021**, *12*, 376–381. [CrossRef]
100. Scofield, S.R.; Huang, L.; Brandt, A.S.; Gill, B.S. Development of a virus-induced gene-silencing system for hexaploid wheat and its use in functional analysis of the *Lr21*-mediated leaf rust resistance pathway. *Plant Physiol.* **2005**, *138*, 2165–2173. [CrossRef]
101. Yuan, C.; Li, C.; Yan, L.; Jackson, A.O.; Liu, Z.; Han, C.; Yu, J.; Li, D. A high throughput barley stripe mosaic virus vector for virus induced gene silencing in monocots and dicots. *PLoS ONE* **2011**, *6*, e26468. [CrossRef]
102. Tretiakova, P.; Voegele, R.T.; Soloviev, A.; Link, T.I. Successful silencing of the mycotoxin synthesis gene *TRI5* in *Fusarium culmorum* and observation of reduced virulence in VIGS and SIGS experiments. *Genes* **2022**, *13*, 395. [CrossRef] [PubMed]
103. Yin, C.; Jurgenson, J.E.; Hulbert, S.H. Development of a host-induced RNAi system in the wheat stripe rust fungus *Puccinia striiformis* f. sp. *tritici*. *Mol. Plant-Microbe Interact.* **2011**, *24*, 554–561. [CrossRef] [PubMed]
104. Zhang, L.; Wang, S.; Ruan, S.; Nzabanita, C.; Wang, Y.; Guo, L. A mycovirus VIGS vector confers hypovirulence to a plant pathogenic fungus to control wheat FHB. *Adv. Sci.* **2023**, *10*, 2302606. [CrossRef] [PubMed]
105. Biedenkopf, D.; Will, T.; Knauer, T.; Jelonek, L.; Furch, A.C.U.; Busche, T.; Koch, A. Systemic spreading of exogenous applied RNA biopesticides in the crop plant *Hordeum vulgare*. *ExRNA* **2020**, *2*, 12. [CrossRef]
106. Gu, K.-X.; Song, X.-S.; Xiao, X.-M.; Duan, X.-X.; Wang, J.-X.; Duan, Y.-B.; Hou, Y.-P.; Zhou, M.-G. A *β2*-tubulin dsRNA derived from *Fusarium asiaticum* confers plant resistance to multiple phytopathogens and reduces fungicide resistance. *Pestic. Biochem. Physiol.* **2019**, *153*, 36–46. [CrossRef]
107. Hao, G.; McCormick, S.; Vaughan, M.M. Effects of double-stranded RNAs targeting *Fusarium graminearum TRI6* on Fusarium head blight and mycotoxins. *Phytopathology* **2021**, *111*, 2080–2087. [CrossRef] [PubMed]
108. Yang, P.; Yi, S.-Y.; Nian, J.-N.; Yuan, Q.-S.; He, W.-J.; Zhang, J.-B.; Liao, Y.-C. Application of double-strand RNAs targeting chitin synthase, glucan synthase, and protein kinase reduces *Fusarium graminearum* spreading in wheat. *Front. Microbiol.* **2021**, *12*, 660976. [CrossRef] [PubMed]
109. Werner, B.T.; Gaffar, F.Y.; Schuemann, J.; Biedenkopf, D.; Koch, A.M. RNA-spray-mediated silencing of *Fusarium graminearum AGO* and *DCL* genes improve barley disease resistance. *Front. Plant Sci.* **2020**, *11*, 476. [CrossRef] [PubMed]

110. Ui-Tei, K.; Naito, Y.; Takahashi, F.; Haraguchi, T.; Ohki-Hamazaki, H.; Juni, A.; Ueda, R.; Saigo, K. Guidelines for the selection of highly effective siRNA sequences for mammalian and chick RNA interference. *Nucleic Acids Res.* **2004**, *32*, 936–948. [CrossRef]
111. Jackson, A.L.; Bartz, S.R.; Schelter, J.; Kobayashi, S.V.; Burchard, J.; Mao, M.; Li, B.; Cavet, G.; Linsley, P.S. Expression profiling reveals off-target gene regulation by RNAi. *Nat. Biotechnol.* **2003**, *21*, 635–637. [CrossRef]
112. Stakheev, A.A.; Samokhvalova, L.V.; Mikityuk, O.D.; Zavriev, S.K. Phylogenetic analysis and molecular typing of trichothecene-producing *Fusarium* fungi from Russian collections. *Acta Naturae* **2018**, *10*, 79–92. [CrossRef] [PubMed]
113. Ghag, S.B.; Shekhawat, U.K.S.; Ganapathi, T.R. Host-induced post-transcriptional hairpin RNA-mediated gene silencing of vital fungal genes confers efficient resistance against *Fusarium* wilt in banana. *Plant Biotechnol. J.* **2014**, *12*, 541–553. [CrossRef]
114. Altschul, S.F.; Gish, W.; Miller, W.; Myers, E.W.; Lipman, D.J. Basic local alignment tool. *J. Mol. Biol.* **1990**, *215*, 403–410. [CrossRef]
115. Majumdar, R.; Rajasekaran, K.; Cary, J.W. RNA interference (RNAi) as a potential tool for control of mycotoxin contamination in crop plants: Concepts and considerations. *Front. Plant Sci.* **2017**, *8*, 200. [CrossRef] [PubMed]
116. Naito, Y.; Yamada, T.; Matsumiya, T.; Ui-Tei, K.; Saigo, K.; Morishita, S. dsCheck: Highly sensitive off-target search software for double-stranded RNA-mediated RNA interference. *Nucleic Acids Res.* **2005**, *33*, W589–W591. [CrossRef] [PubMed]
117. Horn, T.; Boutros, M. E-RNAi: A web application for the multi-species design of RNAi reagents—2010 update. *Nucleic Acids Res.* **2010**, *38*, W332–W339. [CrossRef] [PubMed]
118. Yilmazel, B.; Hu, Y.; Sigoillot, F.; Smith, J.A.; Shamu, C.E.; Perrimon, N.; Mohr, S.E. Online GESS: Prediction of miRNA-like off-target effects in large-scale RNAi screen data by seed region analysis. *BMC Bioinform.* **2014**, *15*, 192. [CrossRef] [PubMed]
119. Lück, S.; Kreszies, T.; Strickert, M.; Schweizer, P.; Kuhlmann, M.; Douchkov, D. siRNA-finder (si-Fi) software for RNAi-target design and off-target prediction. *Front. Plant Sci.* **2019**, *10*, 1023. [CrossRef] [PubMed]
120. Ahmed, F.; Senthil-Kumar, M.; Dai, X.; Ramu, V.S.; Lee, S.; Mysore, K.S.; Zhao, P.X. pssRNAit: A web server for designing effective and specific plant siRNAs with genome-wide off-target assessment. *Plant Physiol.* **2020**, *184*, 65–81. [CrossRef]
121. Cazares, T.; Higgs, R.E.; Wang, J.; Ozer, H.G. SeedMatchR: Identify off-target effects mediated by siRNA seed regions in RNA-seq experiments. *Bioinformatics* **2024**, *40*, btae011. [CrossRef] [PubMed]
122. Chen, J.; Peng, Y.; Zhang, H.; Wang, K.; Zhao, C.; Zhu, G.; Reddy Palli, S.; Han, Z. Off-target effects of RNAi correlate with the mismatch rate between dsRNA and non-target mRNA. *RNA Biol.* **2021**, *18*, 1747–1759. [CrossRef]
123. Kamola, P.; Nakano, Y.; Takahashi, T.; Wilson, P.A.; Ui-Tei, K. The siRNA non-seed region and its target sequences are auxiliary determinants of off-target effects. *PLoS Comput. Biol.* **2015**, *11*, e1004656. [CrossRef]
124. Chen, J.; Sheng, C.-W.; Peng, Y.; Wang, K.; Jiao, Y.; Reddy Palli, S.; Cao, H. Transcript level and sequence matching are key determinants of off-target effects in RNAi. *J. Agric. Food Chem.* **2024**, *72*, 577–589. [CrossRef] [PubMed]
125. Johnson, E.T.; Proctor, R.H.; Dunlap, C.A.; Busman, M. Reducing production of fumonisin mycotoxins in *Fusarium verticillioides* by RNA interference. *Mycotoxin Res.* **2018**, *34*, 29–37. [CrossRef] [PubMed]
126. Baldwin, T.; Islamovic, E.; Klos, K.; Schwartz, P.; Gillespie, J.; Hunter, S.; Bregitzer, P. Silencing efficiency of dsRNA fragments targeting *Fusarium graminearum TRI6* and patterns of small interfering RNA associated with reduced virulence and mycotoxin production. *PLoS ONE* **2018**, *13*, e0202798. [CrossRef]
127. Power, I.L.; Faustinelli, P.C.; Orner, V.A.; Sobolev, V.S.; Arias, R.S. Analysis of small RNA populations generated in peanut leaves after exogenous application of dsRNA and dsDNA targeting aflatoxin synthesis genes. *Sci. Rep.* **2020**, *10*, 13820. [CrossRef]
128. Dahlmann, T.A.; Kück, U. Dicer-dependent biogenesis of small RNAs and evidence for microRNA-like RNAs in the penicillin producing fungus *Penicillium chrysogenum*. *PLoS ONE* **2015**, *10*, e0125989. [CrossRef]
129. Mi, S.; Cai, T.; Hu, Y.; Chen, Y.; Hodges, E.; Ni, F.; Wu, L.; Li, S.; Zhou, H.; Long, C.; et al. Sorting of small RNAs into Arabidopsis argonaute complexes is directed by the 5′ terminal nucleotide. *Cell* **2008**, *133*, 116–127. [CrossRef]
130. Qi, X.; Bao, F.S.; Xie, Z. Small RNA deep sequencing reveals role for *Arabidopsis thaliana* RNA-dependent RNA polymerases in viral siRNA biogenesis. *PLoS ONE* **2009**, *4*, e4971. [CrossRef]
131. Höfle, L.; Biedenkopf, D.; Werner, B.T.; Shrestha, A.; Jelonek, L.; Koch, A. Study on the efficiency of dsRNAs with increasing length in RNA-based silencing of the *Fusarium CYP51* genes. *RNA Biol.* **2020**, *17*, 463–473. [CrossRef]
132. Beaucage, S.L.; Reese, C.B. Recent advances in the chemical synthesis of RNA. *Curr. Protoc. Nucleic Acid Chem.* **2009**, *38*, 2–16. [CrossRef] [PubMed]
133. Papić, L.; Rivas, J.; Toledo, S.; Romero, J. Double-stranded RNA production and kinetics of recombinant *Escherichia coli* HT115 in fed-batch culture. *Biotechnol. Rep.* **2018**, *20*, e00292. [CrossRef]
134. Guan, R.; Chu, D.; Han, X.; Miao, X.; Li, H. Advances in the development of microbial double-stranded RNA production systems for application of RNA interference in agricultural pest control. *Front. Bioeng. Biotechnol.* **2021**, *9*, 753790. [CrossRef] [PubMed]
135. Ahn, S.-J.; Donahue, K.; Koh, Y.; Martin, R.R.; Choi, M.-Y. Microbial-based double-stranded RNA production to develop cost-effective RNA interference application for insect pest management. *Int. J. Insect Sci.* **2019**, *11*, 1179543319840323. [CrossRef]
136. Prates, L.H.F.; Merlau, M.; Rühl-Teichner, J.; Schetelig, M.F.; Häcker, I. An optimized/scale up-ready protocol for extraction of bacterially produced dsRNA at good yield and low costs. *Int. J. Mol. Sci.* **2023**, *24*, 9266. [CrossRef] [PubMed]
137. Verdonckt, T.-W.; Broeck, J.V. Methods for the cost-effective production of bacteria-derived double-stranded RNA for in vitro knockdown studies. *Front. Physiol.* **2022**, *13*, 836106. [CrossRef]
138. Niehl, A.; Soininen, M.; Poranen, M.M.; Heinlein, M. Synthetic biology approach for plant protection using dsRNA. *Plant Biotechnol. J.* **2018**, *16*, 1679–1687. [CrossRef] [PubMed]

139. Álvarez-Sánchez, A.R.; Romo-Quinones, C.; Rosas-Quijano, R.; Reyes, A.G.; Barraza, A.; Magallón-Barajas, F.; Angulo, C.; Mejía-Ruíz, C.H. Production of specific dsRNA against white spot syndrome virus in the yeast *Yarrowia lipolytica*. *Aquac. Res.* **2018**, *49*, 480–491. [CrossRef]
140. Duman-Scheel, M. *Saccharomyces cerevisiae* (Baker's yeast) as an interfering RNA expression and delivery system. *Curr. Drug Targets* **2019**, *20*, 942–952. [CrossRef] [PubMed]
141. Holeva, M.C.; Sklavounos, A.; Rajeswaran, R.; Pooggin, M.M.; Voloudakis, A.E. Topical application of double-stranded RNA targeting 2b and CP genes of Cucumber mosaic virus protects plants against local and systemic viral infection. *Plants* **2021**, *10*, 963. [CrossRef] [PubMed]
142. Nerva, L.; Sandrini, M.; Gambino, G.; Chitarra, W. Double-stranded RNAs (dsRNAs) as a sustainable tool against gray mold (*Botrytis cinerea*) in grapevine: Effectiveness of different application methods in an open-air environment. *Biomolecules* **2020**, *10*, 200. [CrossRef] [PubMed]
143. Zhu, J.; Dong, Y.C.; Li, P.; Niu, C.Y. The effect of silencing 20E biosynthesis relative genes by feeding bacterially expressed dsRNA on the larval development of *Chilo suppressalis*. *Sci. Rep.* **2016**, *6*, 28697. [CrossRef]
144. Hough, J.; Howard, J.D.; Brown, S.; Portwood, D.E.; Kilby, P.M.; Dickman, M.J. Strategies for the production of dsRNA biocontrols as alternatives to chemical pesticides. *Front. Bioeng. Biotechnol.* **2022**, *10*, 980592. [CrossRef]
145. Bennett, M.; Deikman, J.; Hendrix, B.; Iandolino, A. Barriers to efficient foliar uptake of dsRNA and molecular barriers to dsRNA activity in plant cells. *Front. Plant Sci.* **2020**, *11*, 816. [CrossRef]
146. Dalakouras, A.; Wassenegger, M.; McMillan, J.; Cardoza, V.; Maegele, I.; Dadami, E.; Runne, M.; Krczal, G.; Wassenegger, M. Induction of silencing in plants by high-pressure spraying of *in vitro*-synthesized small RNAs. *Front. Plant Sci.* **2016**, *7*, 1327. [CrossRef]
147. Jibrin, M.O.; Liu, Q.; Jones, J.B.; Zhang, S. Surfactants in plant disease management: A brief review and case studies. *Plant Pathol.* **2021**, *70*, 495–510. [CrossRef]
148. Mitter, N.; Worrall, E.A.; Robinson, K.E.; Li, P.; Jain, R.G.; Taochy, C.; Fletcher, S.J.; Carroll, B.J.; Lu, G.Q.; Xu, Z.P. Clay nanosheets for topical delivery of RNAi for sustained protection against plant viruses. *Nat. Plants* **2017**, *3*, 16207. [CrossRef]
149. Mosa, M.A.; Yuossef, K. Topical delivery of host induced RNAi silencing by layered hydroxide nanosheets: An efficient tool to decipher pathogenicity gene function of Fusarium crown and root rot in tomato. *Physiol. Mol. Plant Pathol.* **2021**, *115*, 101684. [CrossRef]
150. Chen, X.; Shi, T.; Tang, T.; Chen, C.; Liang, Y.; Zuo, S. Nanosheet-facilitated spray delivery of dsRNAs represents a potential tool to control *Rhizoctonia solani* infection. *Int. J. Mol. Sci.* **2022**, *23*, 12922. [CrossRef]
151. Lichtenberg, S.S.; Nuti, K.; DeRouchey, J.E.; Tsyuko, O.V.; Unrine, J.M. Efficacy of chitosan/double-stranded RNA polyplex nanoparticles for gene silencing under variable environmental conditions. *Environ. Sci. Nano* **2020**, *7*, 1582–1592. [CrossRef]
152. Negm, N.A.; Hefni, H.H.H.; Abd-Elaal, A.A.A.; Badr, E.A.; Abou Kana, M.T.H. Advancement on modification of chitosan biopolymer and its potential applications. *Int. J. Biol. Macromol.* **2020**, *152*, 681–702. [CrossRef] [PubMed]
153. Kolge, H.; Kadam, K.; Galande, S.; Lanjekar, V.; Ghormade, V. New frontiers in pest control: Chitosan nanoparticles-shielded dsRNA as an effective topical RNAi spray for gram podborer biocontrol. *ACS Appl. Bio Mater.* **2021**, *4*, 5145–5157. [CrossRef] [PubMed]
154. Wang, K.; Peng, Y.; Chen, J.J.; Peng, Y.; Wang, X.; Shen, Z.; Han, Z. Comparison of efficacy of RNAi mediated by various nanoparticles in the rice striped stem borer (*Chilo supressalis*). *Pestic. Biochem. Physiol.* **2020**, *165*, 104467. [CrossRef] [PubMed]
155. Wang, Y.; Yan, Q.; Lan, C.; Tang, T.; Wang, K.; Shen, J.; Niu, D. Nanoparticle carriers enhance RNA stability and uptake efficiency and prolong the protection against *Rhizoctonia solani*. *Phytopathol. Res.* **2023**, *5*, 2. [CrossRef]
156. Qiao, L.; Niño-Sánchez, J.; Hamby, R.; Capriotti, L.; Chen, A.; Mezzetti, B.; Jin, H. Artificial nanovesicles for dsRNA delivery in spray-induced gene silencing for crop protection. *Plant Biotechnol. J.* **2023**, *21*, 854–865. [CrossRef]
157. Islam, M.T.; Davis, Z.; Chen, L.; Englaender, J.; Zomorodi, S.; Frank, J.; Bartlett, K.; Somers, E.; Carballo, S.M.; Kester, M.; et al. Minicell-based fungal RNAi delivery for sustainable crop protection. *Microb. Biotechnol.* **2021**, *14*, 1847–1856. [CrossRef]
158. Niño-Sánchez, J.; Chen, L.-H.; De Souza, J.T.; Mosquera, S.; Stergiopoulos, I. Targeted delivery of gene silencing in fungi using genetically engineered bacteria. *J. Fungi* **2021**, *7*, 125. [CrossRef]
159. Mujtaba, M.; Wang, D.; Carvalho, L.B.; Oliveira, J.L.; do Espirito Santo Pereira, A.; Sharif, R.; Jogaiah, S.; Paidi, M.K.; Wang, L.; Ali, Q.; et al. Nanocarrier-mediated delivery of miRNA, RNAi, and CRISPR-Cas for plant protection: Current trends and future directions. *ACS Agric. Sci. Technol.* **2021**, *1*, 417–435. [CrossRef]
160. Ray, P.; Sahu, D.; Aminedi, R.; Chandran, D. Concepts and considerations for enhancing RNAi efficiency in phytopathogenic fungi for RNAi-based crop protection using nanocarrier-mediated dsRNA delivery systems. *Front. Fungal Biol.* **2022**, *3*, 977502. [CrossRef]
161. Ghosh, S.; Patra, S.; Ray, S. A combinational nanobased spray-induced gene silencing technique for crop protection and improvement. *ACS Omega* **2023**, *8*, 22345–22351. [CrossRef] [PubMed]
162. Bonner, J.C. Nanoparticles as a potential cause of pleural and interstitial lung disease. *Proc. Am. Thorac. Soc.* **2010**, *7*, 138–141. [CrossRef] [PubMed]
163. Olivier, J.C.; Fenart, L.; Chauvet, R.; Pariat, C.; Cecchelli, R.; Couet, W. Indirect evidence that drug brain targeting using polysorbate 80-coated polybutylcyanoacrylate nanoparticles is related to toxicity. *Pharm. Res.* **1999**, *16*, 1836–1842. [CrossRef]

164. Singewar, K.; Fladung, M. Double-stranded RNA (dsRNA) technology to control forest insect pests and fungal pathogens: Challenges and opportunities. *Funct. Integr. Genom.* **2023**, *23*, 185. [CrossRef] [PubMed]
165. Shen, X.; Corey, D.R. Chemistry, mechanism and clinical status of antisense oligonucleotides and duplex RNAs. *Nucleic Acids Res.* **2018**, *46*, 1584–1600. [CrossRef] [PubMed]
166. Peacock, H.; Kannan, A.; Beal, P.A.; Burrows, C.J. Chemical modification of siRNA bases to probe and enhance RNA interference. *J. Org. Chem.* **2011**, *76*, 7295–7300. [CrossRef]
167. Evers, M.M.; Toonen, L.J.A.; van Roon-Mom, W.M.C. Antisense oligonucleotides in therapy for neurodegenerative disorders. *Adv. Drug Deliv. Rev.* **2015**, *87*, 90–103. [CrossRef]
168. Howard, J.D.; Beghyn, M.; Dewulf, N.; De Vos, Y.; Phillips, A.; Portwood, D.; Kilby, P.M.; Oliver, D.; Maddelein, W.; Brown, S.; et al. Chemically modified dsRNA induces RNAi effects in insects in vitro and *in vivo*: A potential new tool for improving RNA-based plant protection. *J. Biol. Chem.* **2022**, *298*, 102311. [CrossRef]

Journal of *Fungi*

MDPI

Communication

Molecular Characterization and Expression Analysis of a Gene Encoding 3-Hydroxy-3-Methylglutaryl-CoA Reductase (HMGR) from *Bipolaris eleusines*, an Ophiobolin A-Producing Fungus

Jianping Zhang [1,*], Ke Yang [2], Wei Tang [1], Yongjie Yang [1], Xiaoyue Yu [1], Yongliang Lu [1] and Liuqing Yu [1]

1 State Key Laboratory of Rice Biology and Breeding, China National Rice Research Institute, Hangzhou 310006, China
2 Department of Industrial Engineering, University of Arkansas, Fayetteville, AR 72701, USA; ky013@uark.edu
* Correspondence: nkzhang_jp@163.com

Abstract: Ophibolin A, a fungal sesterterpene, exerts a pivotal influence in a diverse array of biological processes, encompassing herbicidal, bactericidal, fungicidal, and cytotoxic activities. Sixty genes associated with sesterterpene compound biosynthesis were obtained from *Bipolaris eleusines* via transcriptome sequencing, and those closely linked to ophiobolin A biosynthesis were subsequently filtered. A gene encoding 3-hydroxy-3-methylglutaryl-CoA reductase (HMGR) that catalyzes the first committed step of ophiobolin biosynthesis in the mevalonic acid (MVA) pathway was isolated and characterized using RACE (Rapid Amplification of cDNA Ends) technology from ophiobolin A-producing fungus, *B. eleusines*. The full-length cDNA of the *B. eleusines HMGR* gene (*BeHMGR*) was 3906 bp and contained a 3474 bp open reading frame (ORF) encoding 1157 amino acids. Sequence analysis revealed that deduced BeHMGR had high homology to the known HMGRs from *Pyrenophora tritici-repentis* and *Leptosphaeria maculans*. It had a calculated molecular mass of about 124.65 kDa and an isoelectric point (pI) of 6.90. It contained two putative HMG-CoA-binding motifs and two NADP(H)-binding motifs. Induced expression analysis of the *BeHMGR* gene by methyl jasmonate treatment using quantitative fluorescence PCR showed that it significantly elevated after 3 h of methyl jasmonate treatment, peaked at 6 h, and then gradually decreased. This demonstrates that *BeHMGR* gene expression is induced by methyl jasmonate.

Keywords: cloning; RACE technology; transcriptome sequencing; methyl jasmonate (MJ); phytotoxin

Citation: Zhang, J.; Yang, K.; Tang, W.; Yang, Y.; Yu, X.; Lu, Y.; Yu, L. Molecular Characterization and Expression Analysis of a Gene Encoding 3-Hydroxy-3-Methylglutaryl-CoA Reductase (HMGR) from *Bipolaris eleusines*, an Ophiobolin A-Producing Fungus. *J. Fungi* **2024**, *10*, 445. https://doi.org/10.3390/jof10070445

Academic Editor: Ofir Degani

Received: 24 May 2024
Revised: 24 June 2024
Accepted: 25 June 2024
Published: 26 June 2024

1. Introduction

Bipolaris eleusines, obtained from naturally infected barnyard grass (*Echinochloa crus-galli*), was evaluated as a potential biological control agent for barnyard grass [1]. It produces ophiobolin A, a sesquiterpenoid with a tricyclic structure known for its herbicidal [2,3], antimicrobial [4,5], and anti-tumor cell activities [6–9], holding promising applications in agriculture and medicine. Previous studies have focused on isolating ophiobolin toxins and investigating their functions [2–5,10–12].

Ophiobolin A, a secondary metabolite produced by plant pathogenic fungi, is synthesized in small quantities, limiting its commercial production. Chemical synthesis could address the challenge of low microbial secondary metabolite production, but its complex compound structure and high synthesis costs pose limitations to its industrial-scale production [13–15]. Alternatively, toxin production can be enhanced through methods such as mutagenesis. Our laboratory has employed various techniques, including chemical mutagenesis, ultraviolet mutagenesis, protoplast fusion technology, and Restriction enzyme-mediated integration (REMI) technology, to increase toxin production [16,17]. Despite screening some mutant and fusion strains with improved toxin production, they have not yet met the requirements for industrial-scale production.

Gene regulation might enhance toxin production in *B. eleusines*. As a significant terpenoid, the biosynthesis pathway of ophiobolin A closely resembles that of other terpenes in fungi, operating via the mevalonate pathway [18,19]. This pathway initiates from acetyl-CoA and proceeds through a series of enzymatic reactions to produce 3-hydroxy-3-methylglutaryl coenzyme A (HMG-CoA), farnesyl pyrophosphate (FPP), geranylgeranyl pyrophosphate (GGPP), and geranyl farnesyl pyrophosphate (GFPP). These intermediates then undergo cyclization and a sequence of redox reactions to form mature terpenoid structures.

Among these, 3-hydroxy-3-methylglutaryl-CoA reductase (HMGR) catalyzes the NADPH-dependent conversion of HMG-CoA to mevalonate (MVA) [20]. As the synthesis of mevalonate is an irreversible process, HMGR is recognized as the primary rate-limiting enzyme in the mevalonate biosynthetic pathway across animals, plants, and fungi [21–23]. It also serves as a crucial target enzyme in intracellular terpenoid metabolism pathways. Due to HMGR's pivotal role in the biosynthesis of significant terpenoids, such as paclitaxel, artemisinin, tanshinone, and ganoderic acid, this gene has been extensively cloned and characterized from many organisms [24–27]. However, no reports have concerned an ophiobolin A-producing fungus, *B. eleusines*. Hence, this study utilized Rapid Amplification of cDNA Ends (RACE) technology to clone the *HMGR* gene (referred to as *BeHMGR*) from the ophiobolin A-producing plant pathogenic fungus, *B. eleusines*. Additionally, gene expression was analyzed using fluorescence quantitative PCR after methyl jasmonate treatment. The objective of this work is to provide a pivotal enzyme target for further investigation into the molecular mechanisms and metabolic regulation of the ophiobolin A biosynthetic pathway, thereby laying the groundwork for future studies on large-scale toxin synthesis through metabolic regulation.

2. Materials and Methods

2.1. Fungal Inoculum

Bipolaris eleusines (Alcorn & Shivas) was isolated from severely diseased barnyard grass and stored in PDA (potato 20%, glucose 2%, and 1.5% agar powder) at 4 °C in the Weed Laboratory at the China National Rice Research Institute, Hangzhou, China [1]. *Escherichia coli*, JM109 was purchased from Sangon Biotech (Shanghai) Co., Ltd., Shanghai, China.

A fungal-activated inoculum was transferred from the vigorous edge growth of *B. eleusines* into the center of a fresh soybean medium (glucose 3%, soybean flour 4%, $MgSO_4$ 0.1%, $Na_3PO_4 \cdot 12H_2O$ 0.2%, KNO_3 0.5%, and agar powder 1.8%) plate (φ9 cm) using a fungal block (φ5.5 mm). The inoculum was incubated in the dark at 28 °C for 4 days to extract total RNA for subsequent steps.

2.2. Transcriptome Sequencing

RNA preparation, transcriptome sequencing, and De Novo analysis were conducted at the *BGI Genomics* Co., Ltd., Shenzhen, in China using Illumina HiSeq™ 2000.

2.3. Cloning of the Full-Length cDNA of BeHMGR by RACE

2.3.1. Confirmation of EST Amplification from *B. eleusines*

Total RNA was extracted from *B. eluesines* using the PureLink™ RNA Mini Kit following the manufacturer's instructions (Invitrogen, Carlsbad, CA, USA). Single-stranded cDNAs were synthesized from 5 µg of total RNA with an oligo (dT)17 primer according to the manufacturer's protocol (SuperScript® II Reverse Transcriptase, Invitrogen, Carlsbad, CA, USA). Following S.N.A.P.™ column treatment, the single-stranded cDNA mixture served as the template for PCR to confirm the amplification of the EST from *B. eleusines*. This verification employed the primers *BeHMGR*-EST-F and *BeHMGR*-EST-R (Table 1) designed based on the EST sequences obtained from transcriptome sequencing.

Table 1. All primers used in this study.

Primer Name	Primer Sequences (5′-3′)
BeHMGR-ECT-F	CGTACCCCCGGCCCAGATGA
BeHMGR-ECT-R	CGCGCGAAGTTGAAGCGACG
*BeHMGR*3-1	TCTACCTCTCGCTTCGCCAGGCTACAA
*BeHMGR*3-2	TACCGACAAGAAGTCTGCCGCCATCAA
*BeHMGR*5-GSP1	TGGCGTTGCTGCTCTT
*BeHMGR*5-GSP2	AATCTCTGGCTGGGGTCTTGGC
*BeHMGR*5-GSP3	TACGGGGCGGGGTAGGCATGTG
BeHMGR-FL-F	ACAATGCTAGGATCACTCGCCA
BeHMGR-FL-R	ACTTTCTCTATCGCTTGGGCAC
*Actin*177F	CATCAACCCCAAGTCCAACC
*Actin*177R	CCCTCGTAGATGGGGACAAC
*HMGR*158F	TGTCCCCGGAACCCCTCGCA
*HMGR*158R	GGCGTTGCTGCTCTTCCGTTG

2.3.2. 5′ End Amplification of *BeHMGR* Gene

The 5′ RACE method was utilized to amplify the 5′ end of the *BeHMGR* gene. Initially, the SuperScript® II Reverse Transcriptase (Invitrogen™) enzyme and primer *BeHMGR*5GSP-1 (Table 1) were employed to synthesize the first-strand cDNA of the target gene from total RNA. Following treatment with RNase Mix and subsequent purification using the DNA Purification System (GLASSMAX DNA isolation spin cartridges), PolyC residues were added to the ends of the purified single-stranded cDNA using the TdT enzyme and dCTP. In conjunction with the specific primer *BeHMGR*5GSP-2 (Table 1), the bridging primer AAP provided in the kit facilitated the first round of PCR amplification of the cDNA extended with a dC tail. For the second round of nested PCR amplification, the bridging amplification primer AUAP included in the kit and specific primer *BeHMGR*5GSP-3 (Table 1) were utilized. The PCR products from the second round were subjected to electrophoresis, and the target band was isolated and purified from the gel. Subsequently, the purified PCR product was sub-cloned into the pMD18-T vector, and positive clones were selected for sequencing.

2.3.3. 3′ End Amplification of *BeHMGR* Gene

The reverse transcriptase SMARTScribe™ Reverse Transcriptase and primer 3′ CDS primer A were employed to reverse-transcribe total RNA, synthesizing cDNA. For the first round of PCR amplification, the specific primer *BeHMGR*3-1 (Table 1) and universal UPM were utilized, with the synthesized cDNA as the template. Subsequently, the products from the first round of PCR amplification were diluted 50 times, and the second round of PCR amplification was carried out using the primer *BeHMGR*3-2 (Table 1) and universal UPM. The purified PCR product from the second round was sub-cloned into the pMD18-T vector for sequencing.

2.3.4. Obtaining Full-Length cDNA of *BeHMGR* Gene

The amplified 3′ and 5′ RACE cDNA sequences, along with the EST sequences, were assembled using Vector NTI Advance 10 software to generate the full-length cDNA sequence of the target gene. Subsequently, based on this assembled sequence, a pair of gene-specific primers, *BeHMGR*-F and *BeHMGR*-R (Table 1), were designed to amplify the full-length cDNA. The PCR product was confirmed by sequencing. The sequence obtained through sequencing was submitted to the National Center for Biotechnology Information (NCBI), and blast analysis was conducted on its open reading frame (ORF), start codon, stop codon, and other information. Once the analysis was completed, the sequence was then submitted to the NCBI (GenBank accession No. JQ780844).

2.4. Bioinformatics Analysis

Bioinformatic analysis of *BeHMGR* was conducted online using the websites http://www.ncbi.nlm.nih.gov (accessed on 1 February 2024) and http://cn.expasy.org (accessed on 3 February 2024). Sequence alignment and assembly were performed using Vector NTI Advance 10 software, while multiple sequence alignment was carried out using DNAMAN 8.0 software. ORF, GenBank BLAST, and protein sequence searches were executed on the NCBI website [http://www.ncbi.nlm.nih.gov (accessed on 1 February 2024)]. ORF translation was completed using DNAStar Lasergene 14.1 software.

2.5. Expression Analysis of BeHMGR Gene under Methyl Jasmonate Treatment

2.5.1. Methyl Jasmonate Treatment Procedure for *B. eleusines*

A fungal block (φ5.5 mm) was used to transfer an inoculum from the vigorous edge growth of *B. eleusines* into a 250 mL Erlenmeyer flask containing 100 mL of PDB medium. The inoculum was cultured on a 120 rpm shaker at 28 °C for 7 days. Then, a 5% inoculation volume of *B. eleusines* was taken from the above culture and inoculated into 50 mL of PDB liquid medium, where it was cultured for 60 h. Methyl jasmonate (a final concentration of 2 mM) was added to the fungal culture fluid as induction treatment, and culturing was continued. Mycelia were collected at intervals of 0.25 days, 1 day, 2 days, 3 days, 4 days, and 5 days for the analysis of the *BeHMGR* gene expression level.

2.5.2. Fluorescent Quantitative PCR Detection of *BeHMGR* Gene Expression Level

Total RNA was extracted from *B. eleusines* strains treated with methyl jasmonate at different time intervals and reverse-transcribed into first-strand cDNA. The expression levels of the *BeHMGR* gene were detected using fluorescent quantitative PCR. The primer sequences used can be found in Table 1. The real-time PCR amplification protocol was as follows: Stage 1: Pre-denaturation at 95 °C for 2 min, followed by 95 °C for 10 min. Stage 2: PCR reaction, consisting of 40 cycles of 95 °C for 15 s, 60 °C for 40 s, and 72 °C for 40 s. Stage 3: Dissociation curve analysis, consisting of 95 °C for 15 s, 60 °C for 1 min, 95 °C for 15 s, and 60 °C for 1 min. Three biological replicates were set for each treatment, and each sample was analyzed thrice.

2.5.3. Data Analysis and Statistics

The real-time quantitative PCR data were analyzed using the $2^{-\Delta\Delta ct}$ method. Statistical analysis was performed using an SPSS 13.0 statistical package. The statistical data, presented as mean ± standard deviation (SD), were analyzed for the significance of difference ($p < 0.05$) using a standard variance analysis with a completely randomized experimental design, followed by Tukey's multiple-range tests at a 5% significance level.

3. Results and Analysis

3.1. Results of Transcriptome Sequence Analysis of B. eleusines

Transcriptome sequencing yielded a total of 26,555,560 high-quality ESTs (i.e., total reads) with a total nucleotide count of 2,390,000,400 and a Q20 score of 92.91%, indicating a very high sequencing quality (Table 2). After clustering and assembly, 32,100 high-quality consensus sequences (Unigenes) were obtained, and the total nucleotide count of the consensus sequences reached 17,959,906 (Table 3). Among these, there were 20,075 consensus sequences in the range of 100-500 nt, representing 62.54% of the total, and 466 consensus sequences greater than 2000 nt, accounting for 1.45% of the total. The N50 of the consensus sequences is 743 nt, and the average length is 559 nt.

Table 2. Sequencing yield statistics.

Total Reads	Total Nucleotides * (nt)	Q20 Percent	N Percent	GC Percent
26,555,560	2,390,000,400	92.91%	0.00%	50.77%

* Total nucleotides = Total Reads 1 × Reads 1 size + Total Reads 2 × Reads 2 size.

Table 3. Consistency sequence (Unigene) statistics obtained by Blast annotation.

Nucleotide Length	100–500 nt	500–1000 nt	1000–1500 nt	1500–2000 nt	≥2000 nt
Number	20,075	7460	3021	1078	466
Percent (%)	62.54	23.24	9.41	3.36	1.45

The consensus sequences obtained from clustering and assembly with the COG database were compared to predict the possible functions of the consensus sequences and perform functional classification statistics on them. A total of 31,432 sequences had homologous sequences, with an annotation rate of 97.9%. The remaining 668 sequences were of an unknown function, from which new genes may be discovered. The annotated genes were functionally classified into 26 categories (Figure 1). The largest number of sequences were mainly concentrated in protein sequences with only general function predictions (2341), transcription-related sequences (1221), amino acid transport and metabolic sequences (1169), and sequences involved in translation, ribosome structure, and biogenesis (1165). Additionally, there were 579 sequences consistent with the categories we focused on for secondary metabolite biosynthesis, transport, and catalysis, as well as some sequences related to protein modification, folding, and molecular chaperones (865), coenzyme transport and metabolism (435 sequences), and sequences related to signal transduction mechanisms (650 sequences), which also attracted our attention.

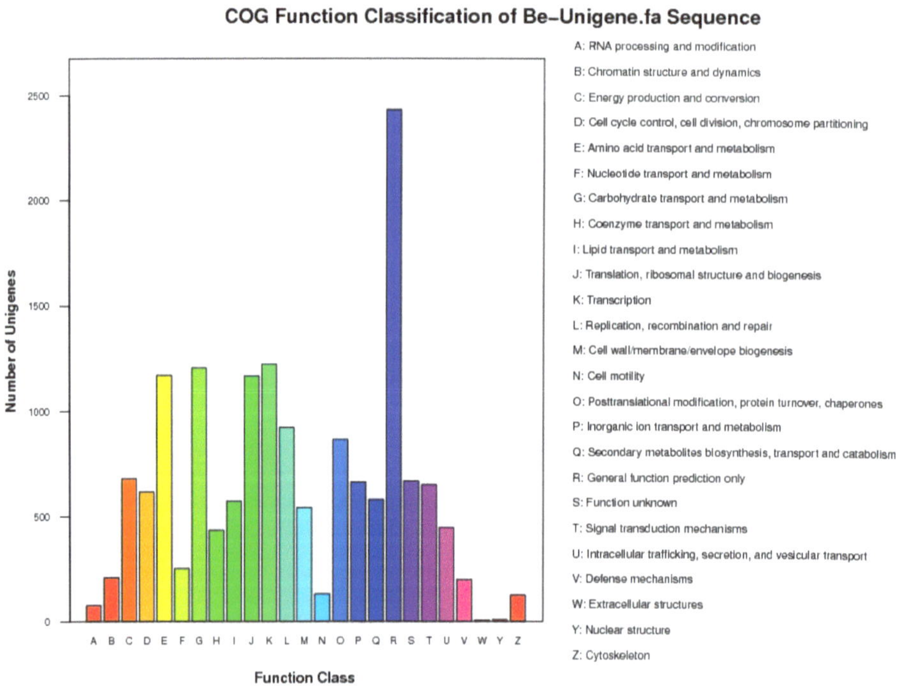

Figure 1. Data of COG classification.

KEGG Pathway analysis identified 60 genes in the terpenoid backbone biosynthesis pathway within the transcriptome sequence, which are the gene sequences essential for our research. The genes in this pathway include Farnesyl pyrophosphate synthase (*FPPS*), Acetyl-CoA acetyltransferase (*AAT*), Hexaprenyl pyrophosphate synthase (*HPS*), Isopentenyl-diphosphate delta-isomerase (*IDDI*), Geranylgeranyl pyrophosphate synthase

(*GGPPS*), 3-hydroxy-3-methylglutaryl CoA synthase (*HMGS*), 3-hydroxy-3-methylglutaryl CoA reductase (*HMGR*), Diphosphomevalonate decarboxylase (*DMD*), Mevalonate kinase (*MK*), Phosphomevalonate kinase (*PMK*), 3-ketoacyl-CoA thiolase B (*KT*), and 17 other genes.

Additionally, 161 other genes related to terpenoid metabolism pathways were found, including 90 genes involved in the sterol biosynthesis pathway and genes related to quinone and another terpenoid–quinone biosynthesis. Sixty-four genes were found, including seven genes in the carotenoid biosynthesis pathway. All these compounds share isoprene as their basic unit. Moreover, 76 transcription factor-related sequences were discovered.

3.2. Molecular Characterization of HMGR Gene of *B. eleusines*

3′RACE and 5′RACE were employed to clone the 3′ end and 5′ end of the *HMGR* gene. Primers were designed based on the EST sequence of the *HMGR* gene in the *B. eleusines* transcriptome, and 5′RACE and 3′RACE kits were used to amplify the 5′ and 3′ ends of *HMGR*, respectively. The PCR product bands obtained from the second round of PCR amplification of 5′RACE and 3′RACE are greater than 2000 bp and about 500 bp, respectively. The target bands from electrophoresis were excised from the gel, recovered, purified, and sequenced. The BLAST results of the obtained sequences in the GenBank database show high homology with the fungal *HMGR* gene, confirming that they are the correct 3′ and 5′ ends of the *BeHMGR* gene.

The EST sequence, the 3′ end, and the 5′ end of the *BeHMGR* gene amplified by RACE were assembled to obtain the deduced full-length cDNA sequence, which was then confirmed by sequencing, along with its open reading frame, start codon, and stop codon. The analysis reveals that the full-length *BeHMGR* cDNA sequence (GenBank accession No. JQ780844) is 3906 bp long and contains an ORF of 3474 bp, which translates into a polypeptide chain of 1157 amino acids (Supplementary File S1), similar in length to polypeptide chains in other fungi. The full-length *BeHMGR* cDNA includes a 189 bp 5′-untranslated region (UTR, from 1 bp to 189 bp) and a 223 bp 3′-untranslated region downstream of the ORF (3664 bp to 3886 bp), followed by a 20 bp poly(dA) tail.

3.3. Bioinformatics Analysis of BeHMGR Protein

3.3.1. Analysis of Physical and Chemical Properties of BeHMGR Protein

The Computer pI/Mw Tool [https://web.expasy.org/compute_pi/ (accessed on 12 February 2024)] was used to analyze and estimate that the molecular weight (MW) and theoretical isoelectric point (pI) of the BeHMGR protein are 124.65 kDa and 6.90, respectively. The half-life of the protein is 10–30 h, and the instability index is 43.12, indicating that it is an unstable protein. The aliphatic index is 92.59, and the Grand Average of Hydropathy (GRAVY) is 0.066. The number of negatively charged amino acid residues (Asp + Glu) is 105, and the number of positively charged amino acid residues (Arg + Lys) is 103.

3.3.2. Conservation Analysis of BeHMGR Protein Sequence

The isolated and amplified full-length cDNA sequence of *BeHMGR* and its deduced amino acid sequence were deposited in the GenBank database. The protein–protein BLAST analysis results show that the deduced amino acid sequence of BeHMGR shares extensive and varying degrees of similarity with HMGRs from other fungi. It has 93% similarity and 86% identity with *Pyrenophora tritici-repentis* XP001941036, 90% similarity and 82% identity with *Leptosphaeria maculans* CBX91449, and 86% similarity and 78% identity with *Phaeosphaeria nodorum* XP001800116. The multiple sequence alignment analysis of BeHMGR and HMGR amino acid sequences from ten other fungi is shown in Supplementary File S2.

By analyzing multiple alignments of the amino acid sequences of BeHMGR and ten other fungal HMGRs, it was found that the N-terminal composition and length of the BeHMGR protein vary significantly, while the functions of some amino acids at the C-terminal are highly conserved, affecting the conformation and catalytic properties of the protein.

The highly conserved regions of the C-terminal amino acids include two NADPH-binding motifs, DAMGMNM and GTIGGGT, and two HMG-CoA-binding motifs, ENV(V/I) GY(L/M)PLP and TTEGVLVA (see Supplementary File S2). Among them, glutamic acid in TTEGVLVA plays a crucial role in HMGR catalysis. The NADPH-binding domain and HMG-CoA-binding domain of BeHMGR share almost identical amino acid composition with the substrate-binding domains of other fungi, indicating high conservation. Through comparative analysis, it was also found that four conserved amino acid residues exist in the corresponding positions of BeHMGR: Glu811, Asp1021, His1117, and Ser1123. The presence of these conserved regions and conserved amino acids indicates that the BeHMGR enzyme is likely to be a functional protein.

3.3.3. Phylogenetic Analysis of BeHMGR Proteins

Phylogenetic trees are useful for analyzing the evolutionary relationships between different organisms. Although the structure and function of HMGR proteins are highly conserved during evolution, amino acid counts may vary, allowing for the analysis of evolutionary relationships among different organisms.

To analyze the molecular evolutionary relationship between BeHMGR and HMGRs from other species, we constructed a molecular phylogenetic tree using the full-length amino acid sequences of HMGRs from bacteria, fungi, animals, and plants. As shown in Figure 2, HMGR proteins share common ancestors, which then diversified into different branches: bacteria, fungi, animals, and plants. Within the fungal branch, yeast evolved earlier than other filamentous fungi. BeHMGR forms a well-defined group with the HMGR of filamentous fungi, and its closest genetic relationship is with *Pyrenophora tritici-repentis*.

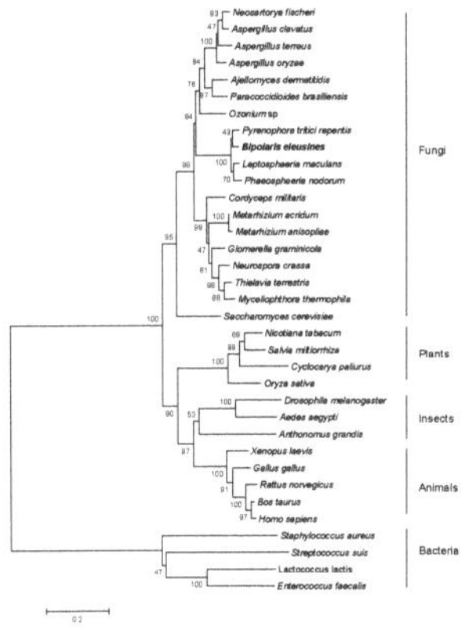

Figure 2. Phylogenetic tree of HMGRs from various species using the Mega 3.1 program by the neighbor-joining method. The bars represent evolutionary distance, and the numbers at each node represent the bootstrap values (with 1000 replicates). The scale bar represents 0.2 nucleotide substitutions per site, indicating the genetic distance between species. Branch lengths are proportional to genetic divergence, allowing for a comparison of the evolutionary genetic distances among the species. The sequences used are listed below with their GenBank accession numbers: *Ozonium* sp., ABU95054; *Leptosphaeria maculans*, CBX91449; *Glomerella graminicola*, EFQ33622; *Metarhizium acridum*, EFY91120; *Metarhizium anisopliae*, EFY99198; *Ajellomyces dermatitidis*, EGE84234; *Cordyceps militaris*, EGX93619; *Neurospora crassa*, XP964546; *Aspergillus terreus*, XP001218142; *Neosartorya fischeri*, XP001265930; *Aspergillus clavatus*, XP001272815; *Phaeosphaeria nodorum*, XP001800116; *Aspergillus oryzae*, XP001823959; *Pyrenophora tritici-repentis*, XP001941036; *Paracoccidioides brasiliensis*, XP002792508; *Thielavia terrestris*, XP003656898; *Myceliophthora thermophila*, XP003658928; *Saccharomyces cerevisiae*, EDV08752; *Anthonomus grandis*, AF162705; *Drosophila melanogaster*, AAA28608; *Aedes aegypti*, XP001659923; *Rattus norvegicus*, AAA40608; *Bos taurus*, DAA25922; *Homo sapiens*, NP000850; *Cyclocarya paliurus*, EU296534; *Nicotiana tabacum*, U60452; *Salvia miltiorrhiza*, FJ747636; *Gallus gallus*, NP989816; *Oryza sativa*, NM001070076; *Xenopus laevis*, AAH74197; *Lactococcus lactis*, NP267726; *Enterococcus faecalis*, ZP05563025; *Staphylococcus aureus*, NP373069; and *Streptococcus suis*, AER19871.

3.4. Effect of Methyl Jasmonate on BeHMGR Gene Expression

To investigate the expression of the *BeHMGR* gene in the terpenoid synthesis pathway of *B. eleusines* in response to methyl jasmonate treatment, this study employed quantitative fluorescence PCR to measure the gene's induced expression. The results demonstrate that *BeHMGR* gene expression was significantly affected by methyl jasmonate at $p < 0.05$. Specifically, the transcript levels of *BeHMGR* increased markedly, showing a significant rise at 3 h post-treatment, peaking at 6 h, and achieving a maximum expression level of 9.2-fold compared to the baseline (0 h) (see Figure 3). Following this peak, the expression levels gradually declined.

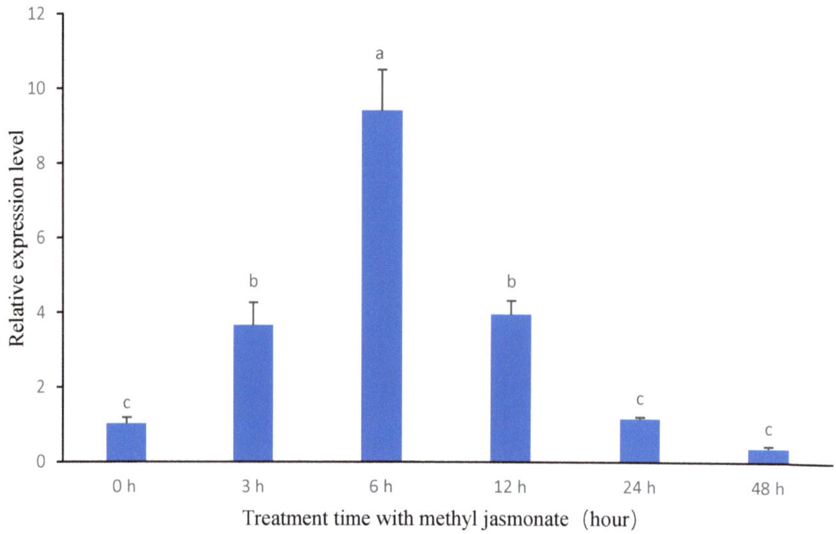

Figure 3. Quantitative real-time PCR analysis of *BeHMGR* gene expression associated with the ophiobolin A biosynthesis pathway. Total RNA was extracted from *Bipolaris eleusnies* mycelia at various time points (0 h, 3 h, 6 h, 12 h, 24 h, and 48 h) following treatment with methyl jasmonate. The actin gene served as the control to demonstrate the normalization of RNA loading in the PCR reaction. Data are expressed using the $2^{-\Delta\Delta Ct}$ method. A standard variance analysis with a completely randomized experimental design, followed by Tukey's multiple-range tests, was conducted to evaluate statistical differences between treatment times. Data are presented as mean ± SD (n = 3). Means with different letters represent Tukey's honest significant difference at $p < 0.05$.

4. Discussion

Although whole genome sequence data for several *Cochliobolus* (anamorph: *Bipolaris*) species exist in international databases, Condon et al. (2013) reported that the genes encoding polyketide synthase (PKS), which catalyze the synthesis of a class of secondary metabolites closely related to virulence, are remarkably diverse among *Cochliobolus* species but conserved among isolates of the same species [28]. Presently, the complete genome sequence of *B. eleusines* has not been determined, and the genetic biology of this fungus remains largely unknown. To obtain specific information on genes associated with ophiobolin biosynthesis, we conducted transcriptome sequencing of *B. eleusines*. This yielded a total of 26,555,560 high-quality ESTs, and KEGG Pathway analysis identified 60 genes in the terpenoid backbone biosynthesis pathway. A comparison with the database reveals that these pathway gene sequences offer abundant genetic insights for gene cloning and research on the ophiobolin biosynthetic pathway. We identified several key enzyme genes, including *HMGR, IDDI, FPPS*, and *GGPPS*, in the ophiobolin biosynthetic pathway and first selected the *HMGR* gene for cloning analysis.

The *HMGR* gene plays a critical role in the mevalonate pathway, synthesizing isoprenoids and sterols, including ophibolin A. It has been cloned in numerous species of fungi [23,29–34]. In this study, we report the isolation and characterization of the *HMGR* gene from *B. eleusines* using RACE technology. The full-length cDNA of this gene was 3906 base-pairs long, with a 3474 base-pair open reading frame (ORF) encoding 1157 amino acids. Sequence analysis reveals that the deduced BeHMGR shares high similarity with

HMGRs from other fungi, including *Pyrenophora tritici-repentis* (93%), *Leptosphaeria maculans* (90%), and *Phaeosphaeria nodorum* (86%).

In eukaryotes, HMGR is an integral membrane glycoprotein located in the endoplasmic reticulum. The N-terminal composition and length of the HMGR protein vary significantly, whereas the functions of certain amino acids at the C-terminal are highly conserved and play a crucial role in the protein's conformation and catalytic properties [34,35]. These conserved amino acid residues of HMGRs include His924, Glu618, Asp828, and Ser930, which regulate HMGR activity [36–38]. Nevertheless, variations in the conserved amino acids among HMGR proteins across various fungal species exist. The C-terminal amino acid sequence of the cassava pathogen *Sphaceloma manihoticola* HMGR contains all conserved amino acids [31], while no serine residue was found at the corresponding position in yeast HMGR [23]. Through comparative analysis, we found that four conserved amino acid residues exist in the corresponding positions of BeHMGR: Glu811, Asp1021, His1117, and Ser1123 (see Supplementary File S2). The number of intervening amino acids between them is entirely consistent with that reported in the literature [31].

Ophiobolin A, like most fungal secondary metabolites, is synthesized in very small amounts under normal physiological conditions. Inducing and regulating the biosynthetic pathway is an effective method for significantly increasing its production. Methyl jasmonate is an important signaling molecule that activates the transcription and translation of specific defense genes, ultimately promoting the synthesis of the secondary metabolite ophiobolin. Numerous studies have reported that the expression of terpenoid synthesis genes can be induced by methyl jasmonate [39–41]. This study finds that the expression levels of *BeHMGR* genes were significantly increased after treatment with methyl jasmonate, indicating that they are induced by methyl jasmonate and can be effectively regulated, at least at the transcriptional level.

5. Conclusions

The transcriptome sequence of *B. eleusines* was determined, and the first rate-limiting enzyme gene in the ophiobolin A synthesis pathway of *B. eleusines*, *BeHMGR*, was isolated and cloned. The full-length cDNA sequence of the *BeHMGR* gene is 3906 bp, contains an open reading frame of 3474 bp, and encodes a protein with 1157 amino acid residues. Fluorescence quantitative PCR analysis reveals that the expression of the *BeHMGR* gene was induced after methyl jasmonate treatment.

Supplementary Materials: The following supporting information can be downloaded at: https://www.mdpi.com/article/10.3390/jof10070445/s1, Supplementary File S1: *BeHMGR* gene full-length cDNA sequence and its deduced amino acid sequence; Supplementary File S2: Multiple alignment of the deduced amino acid sequences of BeHMGR and other fungal HMGR proteins.

Author Contributions: Conceptualization, J.Z.; methodology, J.Z.; software, K.Y.; validation, X.Y.; formal analysis, J.Z.; resources, Y.Y.; data curation, J.Z.; writing—original draft preparation, J.Z.; writing—review and editing, J.Z. and W.T.; supervision, L.Y.; project administration, J.Z. and Y.L. All authors have read and agreed to the published version of the manuscript.

Funding: This research was funded by National Key Research and Development Program (2023YFD1401100); National Natural Science Foundation of China (31401453); China Academy of Agricultural Science (CARS-01−02A) and Agricultural Science and Technology Innovation Program of China Academy of Agricultural Science (Rice Pest Management Research Group).

Institutional Review Board Statement: Not applicable.

Informed Consent Statement: Not applicable.

Data Availability Statement: Data are contained within the article.

Conflicts of Interest: The authors declare no conflict of interest.

References

1. Zhang, J.; Peng, G.; Duan, G.; Zhou, Y.; Yang, S.; Yu, L. *Bipolaris eleusines*, a potential mycoherbicide candidate for control of barnyardgrass (*Echinochloa crus-galli*). *Biocontrol. Sci. Technol.* **2014**, *24*, 839–846. [CrossRef]
2. Duan, G.; Zhou, Y.; Yuan, Q.; Yu, L. Phytotoxic ophiobolins produced by *Helminthosporium gramineum* Rabenh, a potential bioherbicide for control of barnyard grass (*Echinochloa crus-galli*). *Nat. Prod. Indian. J.* **2007**, *3*, 11–17.
3. Evidente, A.; Andolfi, A.; Cimmino, A.; Vurro, M.; Fracchiolla, M.; Charudattan, R. Herbicidal potential of ophiobolins produced by *Drechslera gigantean*. *J. Agric. Food Chem.* **2006**, *54*, 1779–1783. [CrossRef] [PubMed]
4. Duan, G.; Zhang, Z.; Zhang, J.; Zhou, Y.; Yu, L.; Yuan, Q. Evaluation of crude toxin and metabolite produced by *Helminthosporium gramineum* Rabenh for the control of rice sheath blight in paddy field. *Crop Prot.* **2007**, *26*, 1036–1041. [CrossRef]
5. Li, E.; Clark, A.M.; Rotella, D.P.; Hufford, C.D. Microbial metabolites of ophiobolin A and antimicrobial evaluation of ophiobolins. *J. Nat. Prod.* **1995**, *58*, 74–81. [CrossRef] [PubMed]
6. Morrison, R.; Gardiner, C.; Evidente, A.; Kiss, R.; Townley, H. Incorporation of ophiobolin A into novel chemoembolization particles for cancer cell treatment. *Pharm. Res.* **2014**, *31*, 2904–2917. [CrossRef] [PubMed]
7. Chidley, C.; Trauger, S.A.; Birsoy, K.; O'Shea, E.K. The anticancer natural product ophiobolin A induces cytotoxicity by covalent modification of phosphatidylethanolamine. *eLife* **2016**, *5*, e14601. [CrossRef] [PubMed]
8. Bury, M.; Novo-Uzal, E.; Andolfi, A.; Cimini, S.; Wauthoz, N.; Heffeter, P.; Lallemand, B.; Avolio, F.; Delporte, C.; Cimmino, A.; et al. Ophiobolin A, a sesterterpenoid fungal phytotoxin, displays higher in vitro growth-inhibitory effects in mammalian than in plant cells and displays in vivo antitumor activity. *Int. J. Oncol.* **2013**, *43*, 575–585. [CrossRef]
9. Au, T.K.; Chick, W.S.; Leung, P.C. The biology of ophiobolins. *Life Sci.* **2000**, *67*, 733–742. [CrossRef]
10. Sugawara, F.; Strobel, G.; Strange, R.N.; Siedow, J.N.; Vanduyne, G.D.; Clardyvi, J. Phytotoxins from the pathogenic fungi *Drechslera maydis* and *Drechslera sorghicola*. *Proc. Natl. Acad. Sci. USA* **1987**, *84*, 3081–3085. [CrossRef]
11. Singh, S.B.; Smith, J.L.; Sabnis, G.S.; Dombrowski, A.W.; Schaeffer, J.M.; Goetz, M.A.; Bills, G.F. Structure and conformation of ophiobolinK and 6- epiophiobolinK from *Aspergillus ustus* as anematocidal agent. *Tetrahedron* **1991**, *32*, 6931–6938. [CrossRef]
12. Sugawara, F.; Takahashi, N.; Strobel, G.; Yun, C.H.; Gray, G.; Fu, Y.; Clardy, J. Some new phytotoxic ophiobolins produced by *Drechslera oryzae*. *J. Org. Chem.* **1988**, *53*, 2170–2172. [CrossRef]
13. Brill, Z.G.; Grover, H.K.; Maimone, T.J. Enantioselective synthesis of an ophiobolin sesterterpene via a programmed radical cascade. *Science* **2016**, *352*, 1078–1082. [CrossRef] [PubMed]
14. Tsuna, K.; Noguchi, N.; Nakada, M. Convergent total synthesis of (+)-ophiobolin A. *Angew Chem Int Ed.* **2011**, *123*, 9624–9627. [CrossRef]
15. Thach, D.Q.; Brill, Z.G.; Grover, H.K.; Esguerra, K.V.; Thompson, J.K.; Maimone, T.J. Total synthesis of (+)-6-epi-ophiobolin A. *Angew Chem. Int. Ed.* **2020**, *59*, 1532–1536. [CrossRef] [PubMed]
16. Zhang, J.; Duan, G.; Zhu, K.; Zhou, Y.; Lu, Y.; Yu, L. Screening and identification of insertion mutants from *Bipolaris eleusines* by mutagenesis based on restriction enzyme-mediated integration. *FEMS Microbiol. Lett.* **2012**, *330*, 90–97.
17. Zhang, Z.; Burgos, N.; Zhang, J.; Yu, L. Biological control agent for rice weeds from protoplast fusion between *Curvularia lunata* and *Helminthosporium gramineum*. *Weed Sci.* **2007**, *55*, 603–609. [CrossRef]
18. Shigeo, N.; Masuo, M.; Shigenobu, O.; Kyosuke, T. Biosynthesis of ophiobolins from the doubly labeled mevalonate. *Tetrahedron Lett.* **1968**, *9*, 2347–2349.
19. Shigeo, N.; Masuo, M. Enzymic formation of a tricyclic sesterterpene alcohol from mevalonic acid and all-trans-geranylfarnesyl pyrophosphate. *J. Chem. Soc. D.* **1969**, *22*, 1319–1320.
20. Istvan, E.S.; Deisenhofer, J. The structure of the catalytic portion of human HMG-CoA reductase. *BBA-Mol. Cell. Biol. L* **2000**, *1529*, 9–18. [CrossRef]
21. Chen, X.; Wang, X.; Li, Z.; Kong, L.; Liu, G.; Fu, J.; Wang, A. Molecular cloning, tissue expression and protein structure prediction of the porcine 3-hydroxy-3-methylglutaryl-coenzyme A reductase (HMGR) gene. *Gene* **2012**, *495*, 170–177. [CrossRef] [PubMed]
22. Wang, S.; Feng, Y.; Lou, Y.; Niu, J.; Yin, C.; Zhao, J.; Du, W.; Yue, A. 3-Hydroxy-3-methylglutaryl coenzyme A reductase genes from *Glycine max* regulate plant growth and isoprenoid biosynthesis. *Sci. Rep.* **2023**, *13*, 1–14.
23. Basson, M.E.; Thorsness, M.; Rine, J. *Saccharomyces cerevisiae* contains two functional genes encoding 3-hydroxy-3-methylglutaryl-coenzyme A reductase. *Proc. Natl. Acad. Sci. USA* **1986**, *83*, 5563–5567. [CrossRef] [PubMed]
24. Xu, W.; Xu, N.; Zhong, J. Enhancement of Ganoderic Acid Accumulation by Overexpression of an N-Terminally Truncated 3-Hydroxy-3-Methylglutaryl Coenzyme A Reductase Gene in the Basidiomycete *Ganoderma lucidum*. *Appl. Environ. Microbiol.* **2012**, *78*, 7968–7976. [CrossRef] [PubMed]
25. Dai, Z.; Cui, G.; Zhou, S.; Zhang, X.; Huang, L. Cloning and characterization of a novel 3-hydroxy-3-methylglutaryl coenzyme A reductase gene from *Salvia miltiorrhiza* involved in diterpenoid tanshinone accumulation. *J. Plant Physiol.* **2011**, *168*, 148–157. [CrossRef] [PubMed]
26. Liao, Z.; Tan, Q.; Chai, Y.; Zuo, K.; Chen, M.; Gong, Y.; Wang, P.; Pi, Y.; Tan, F.; Sun, X.; et al. Cloning and characterisation of the gene encoding HMG-CoA reductase from *Taxus media* and its functional identification in yeast. *Funct. Plant Biol.* **2004**, *31*, 73–81. [CrossRef] [PubMed]
27. Rao, S.; Meng, X.; Liao, Y.; Yu, T.; Cao, J.; Tan, J.; Xu, F.; Cheng, S. Characterization and functional analysis of two novel 3-hydroxy-3-methylglutaryl-coenzyme A reductase genes (*GbHMGR2* and *GbHMGR3*) from *Ginkgo biloba*. *Sci. Rep.* **2019**, *9*, 1–13.

28. Condon, B.J.; Leng, Y.; Wu, D.; Bushley, K.E.; Ohm, R.A.; Otillar, R.; Martin, J.; Schackwitz, W.; Grimwood, J.; MohdZainudin, N.; et al. Comparative genome structure, secondary metabolite, and effector coding capacity across *Cochliobolus* pathogens. *PLoS Genet.* **2013**, *9*, e1003233. [CrossRef] [PubMed]
29. Lum, P.Y.; Edwards, S.; Wright, R. Molecular, functional and evolutionary characterization of the gene encoding HMG-CoA reductase in the fission yeast, *Schizosaccharomyces pombe*. *Yeast* **1996**, *12*, 1107–1124. [CrossRef]
30. Burmester, A.; Czempinski, K. Sequence comparison of a segment of the gene for 3-hydroxy-3-methylglutaryl-coenzyme A reductase in zygomycetes. *Eur. J. Biochem.* **1994**, *220*, 403–408. [CrossRef]
31. Woitek, S.; Unkles, S.E.; Kinghorn, J.R.; Tudzynski, B. 3-Hydroxy-3-methylglutaryl-CoA reductase gene of *Gibberella fujikuroi*: Isolation and characterization. *Curr. Genet.* **1997**, *31*, 38–47. [CrossRef] [PubMed]
32. Croxen, R.; Goosey, M.W.; Keon, J.P.; Hargreaves, J.A. Isolation of a *Ustilago maydis* gene encoding 3-hydroxy-3-methylglutaryl coenzyme A reductase and expression of a C-terminal-truncated form in *Escherichia coli*. *Microbiology* **1994**, *140*, 2363–2370. [CrossRef] [PubMed]
33. Seong, K.; Li, L.; Hou, Z.; Tracy, M.; Kistler, H.C.; Xu, J. Cryptic promoter activity in the coding region of the HMG-CoA reductase gene in *Fusarium graminearum*. *Fungal Genet. Biol.* **2006**, *43*, 34–41. [CrossRef] [PubMed]
34. Shang, C.H.; Zhu, F.; Li, N.; Ou-Yang, X.; Shi, L.; Zhao, M.W.; Li, Y.X. Cloning and characterization of a gene encoding HMG-CoA reductase from *Ganoderma lucidum* and its functional identification in yeast. *Biosci. Biotechnol. Biochem.* **2008**, *72*, 1333–1339. [CrossRef] [PubMed]
35. Dugan, R.E.; Katiyar, S.S. Evidence for catalytic site cysteine and histidine by chemical modification of 3-hydroxy-3-methylglutaryl-coenzyme A reductase. *Biochem. Biophys. Res. Commun.* **1986**, *41*, 278–284. [CrossRef] [PubMed]
36. Darnay, B.G.; Rodwell, V.W. His865 is the catalytically important histidyl residue of Syrian hamster 3-hydroxy-3-methylglutaryl-coenzyme A reductase. *J. Biol. Chem.* **1993**, *268*, 8429–8435. [CrossRef] [PubMed]
37. Darnay, B.G.; Wang, Y.; Rodwell, V.W. Identification of the catalytically important histidine of 3-hydroxy-3-methylglutaryl-coenzyme A reductase. *J. Biol. Chem.* **1992**, *267*, 15064–15070. [CrossRef] [PubMed]
38. Frimpong, K.; Rodwell, V.W. Catalysis by Syrian hamster 3-hydroxy-3-methylglutaryl-coenzyme A reductase: Proposed roles of histidine 865, glutamate 558 and aspartate 766. *J. Biol. Chem.* **1994**, *259*, 11478–11483. [CrossRef]
39. Ren, A.; Qin, L.; Shi, L.; Dong, X.; Mu, D.S.; Li, Y.X.; Zhao, M.W. Methyl jasmonate induces ganoderic acid biosynthesis in the basidiomycetous fungus *Ganoderma lucidum*. *Bioresour. Technol.* **2010**, *101*, 6785–6790. [CrossRef]
40. Zhao, M.W.; Zhong, J.Y.; Liang, W.Q.; Wang, N.; Chen, M.J.; Zhang, D.B.; Pan, Y.J.; Jong, S.C. Analysis of squalene synthase expression during the development of *Ganoderma lucidum*. *J. Microbiol. Biotech.* **2004**, *14*, 116–120.
41. Nimsa, E.; Duboisb, C.P.; Robertsa, S.C.; Walker, E.L. Expression profiling of genes involved in paclitaxel biosynthesis for targeted metabolic engineering. *Metab. Eng.* **2006**, *8*, 385–394. [CrossRef]

Journal of *Fungi*

MDPI

Article

The Cysteine Protease CfAtg4 Interacts with CfAtg8 to Govern the Growth, Autophagy and Pathogenicity of *Colletotrichum fructicola*

Shufeng Guo [1,2,3] and Shengpei Zhang [1,2,3,*]

1 College of Forestry, Central South University of Forestry and Technology, Changsha 410004, China; gsf120312361@163.com
2 Key Laboratory of Forest Bio-Resources and Integrated Pest Management for Higher Education in Hunan Province, Changsha 410004, China
3 Hunan Provincial Key Laboratory for Control of Forest Diseases and Pests, Changsha 410004, China
* Correspondence: csufzsp@163.com; Tel.: +86-731-85623450

Abstract: *Camellia oleifera* is a native woody oil plant in southern China and is infected with anthracnose wherever it is grown. We previously identified *Colletotrichum fructicola* as the major causal agent of anthracnose on *C. oleifera* and found that CfAtg8 regulates the pathogenicity and development of *C. fructicola*. Here, we revealed that CfAtg4 interacts with CfAtg8, contributing to the formation of autophagosomes. The CfAtg8$^{1-160}$ allele, which only contains 1–160 amino acids of the CfAtg8, partially recovered the autophagosome numbers and autophagy defects of the Δ*Cfatg4* mutant. Consequently, these recoveries resulted in the restoration of the defects of the Δ*Cfatg4* mutant in growth and responses to different external stresses, albeit to an extent. Importantly, we illustrated the critical roles of CfAtg8$^{1-160}$ in appressoria formation, and pathogenicity. Collectively, our findings provide new insights into the importance of the interaction between CfAtg8 and CfAtg4 in the growth, autophagy and pathogenicity of the phytopathogenic fungi.

Keywords: autophagy; CfAtg4; cleavage; pathogenicity; *C. fructicola*

Citation: Guo, S.; Zhang, S. The Cysteine Protease CfAtg4 Interacts with CfAtg8 to Govern the Growth, Autophagy and Pathogenicity of *Colletotrichum fructicola*. *J. Fungi* **2024**, *10*, 431. https://doi.org/10.3390/jof10060431

Academic Editor: Ofir Degani

Received: 28 April 2024
Revised: 7 June 2024
Accepted: 8 June 2024
Published: 18 June 2024

1. Introduction

Camellia oleifera is a Chinese native woody plant that is widely planted in southern China [1]. The tea oil extracted from its seed is highly appreciated for its rich unsaturated fatty acids and for use as raw materials in the cosmetics industry, contributing to the health and economic benefits [2,3]. However, anthracnose occurs wherever *C. oleifera* is grown and greatly impacts the tea oil yield and quality. We previously found that *Colletotrichum fructicola* is the dominant pathogen of anthracnose [4], but its molecular mechanisms underlying pathogenicity remain poorly understood.

Over the past 20 years, accumulated evidence supported the critical roles in autophagy to pathogenicity in phytopathogenic fungi [5–9]. Recently, we also revealed that the autophagy-related proteins CfAtg8, CfAtg9, and CfAtg5 are crucial for the pathogenicity of *C. fructicola* [10,11], initially illustrating the importance of autophagy in the pathogenicity of *Colletotrichum* spp. However, the precise regulatory mechanism is unknown.

Autophagy is a conserved pathway for the degradation and recycling of proteins and organelles in eukaryotes [12,13]. Atg8, which serves as the core protein of autophagy, mediates the expansion, closure, and fusion of the autophagosome membrane during autophagy and is a key protein in the formation of autophagosomes [14]. The reversible Atg8-phosphatidylethanolamine (PE) conjugation on the autophagosome membrane is a hallmark event in the autophagic process [15]. The cysteine protease Atg4 mainly acts on the coupling system of Atg8-PE. The newly synthesized Atg8 was modified by Atg4, exposing the glycine residues at the carbon terminal, which is specifically bound to the amine group of phosphatidylethanolamines (PEs) to form the Atg8-PE conjugation system [16–18]. The conjugated autophagosomes were integrated into lysosomes or vacuoles to complete the degradation steps in autophagy [19]. In addition, Atg4 also cleaves the amide bond of Atg8-PE in conjugation with autophagosomes and recovers Atg8 for the next round of the conjugation reaction [20,21]. Consequently, the protease activity of Atg4 is essential for autophagy and is highly specific, and many studies have listed Atg4 as a prospective target for the development of autophagy-specific inhibitors [22].

In this paper, we demonstrated that CfAtg4 interacts with CfAtg8 to control autophagosome formation and the subsequent autophagy flux. And we also provided evidence showing that CfAtg4-mediated autophagy governs the growth, development, and pathogenicity of *C. fructicola*.

2. Results

2.1. CfAtg4 Interacts with CfAtg8 in C. fructicola

To demonstrate whether CfAtg4 interacts with CfAtg8 in *C. fructicola*, we conducted the yeast two-hybrid (Y2H) assay. The yeast cells carrying pGADT7-CfAtg4 and pGBKT7-CfAtg8 were grown on SD-His/-Leu/-Trp medium. In contrast, yeast carrying the empty vector and the negative control could not be grown on the same medium (Figure 1A,B). The results demonstrate that CfAtg4 interacts with CfAtg8 in *C. fructicola*.

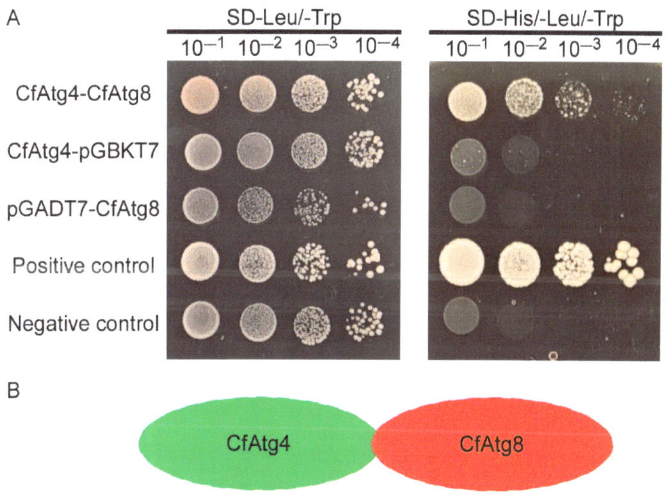

Figure 1. CfAtg4 interacts with CfAtg8. (**A**) The Y2H assays between CfAtg4 and CfAtg8. The pGADT7 and pGBKT7 vectors fused the related genes and were co-transformed into the yeast strain AH109. Leu: leucine; Trp: tryptophan; and His: histidine. (**B**) The schematic of the interaction between CfAtg4 and CfAtg8.

2.2. The Acquisition of ΔCfatg4/CfAtg8$^{1-160}$ Strains

The cleavage of Atg8 in the carboxy-terminal glycine residue by Atg4 is a prerequisite for Atg8-PE formation and the latter autophagosome formation [20,23]. Accordingly, when the cleaved Atg8 instead of full-length Atg8 was expressed in the *ATG4* gene deletion mutant, the requirement for Atg8 cleavage could be bypassed [19]. Therefore, we identified the 160th glycine of CfAtg8 as the conserved cleavage site by the alignment of Atg8 proteins among different species using CLUSTAL_W programs (Figure 2). Then, we transferred a plasmid expressing the GFP-CfAtg8$^{1-160}$ protein exposing glycine at site 160 to the Δ*Cfatg4* and obtained the Δ*Cfatg4*/CfAtg8$^{1-160}$ strain.

Figure 2. The alignment of Atg8 proteins among different species. The Atg8 proteins were aligned by the BioEdit v7.0 software and an asterisk indicates the conserved glycine of Atg8 proteins among fungi. The related species names are as follows: *C. fructicola*, *C. gloeosporioides*, *C. siamense*, *S. cerevisiae*, *M. oryzae*, *A. oryzae*, *A. nidulans*, *U. maydis*, *F. graminearum*, *N. crassa*, *R. norvegicus* and *A. thaliana*.

2.3. *CfAtg8$^{1-160}$ Partially Restores Growth and Conidia Production of ΔCfatg4*

To investigate whether Δ*Cfatg4*/CfAtg8$^{1-160}$ could compensate for the growth defect of Δ*Cfatg4*, we cultured WT (wide-type), Δ*Cfatg4*, Δ*Cfatg4*/CfAtg8$^{1-160}$, and complemented strain Δ*Cfatg4*/*CfATG4* on complete medium (CM) and minimal medium (MM) for 3 days. The results showed that the colony diameters of Δ*Cfatg4*/CfAtg8$^{1-160}$ were significantly larger than those of the Δ*Cfatg4* strain but still significantly smaller than those of the WT strain and Δ*Cfatg4*/*CfATG4* strain on CM and MM media (Figure 3A,B). These results suggest that Δ*Cfatg4*/CfAtg8$^{1-160}$ can partially compensate for the growth defect of Δ*Cfatg4*.

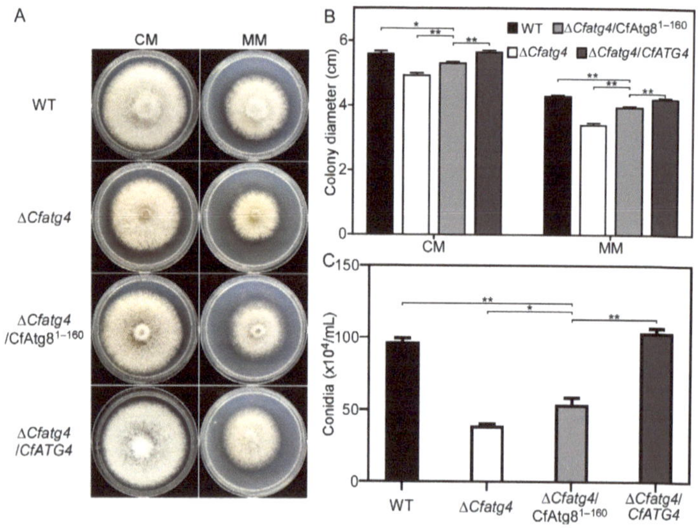

Figure 3. CfAtg8$^{1-160}$ partially restores the growth and asexual reproduction of Δ*Cfatg4*. (**A**) The growth of WT, Δ*Cfatg4*, Δ*Cfatg4*/CfAtg8$^{1-160}$, and Δ*Cfatg4*/*CfATG4* strains in CM and MM medium for 3 days. (**B**) The colony diameters were measured and statistically analyzed. (**C**) Statistical analysis of conidia. Asterisks mean the difference is significant (**, $p < 0.01$; *, $0.01 \leq p < 0.05$). The experiments were performed with 3 biological replications. The data were analyzed by Student's *t*-test, and the error bars indicate standard deviation (SD).

We next explored the roles of CfAtg8$^{1-160}$ in conidia. By incubating WT, Δ*Cfatg4*, Δ*Cfatg4*/CfAtg8$^{1-160}$, and Δ*Cfatg4*/*CfATG4* strains in liquid CM medium for 2 days, we found that Δ*Cfatg4*/CfAtg8$^{1-160}$ showed a significant increase in conidia numbers compared to Δ*Cfatg4* but could not be fully restored to the level of WT, exhibiting only half of that to WT (Figure 3C). The results show that CfAtg8$^{1-160}$ partially compensates for the conidia defects of Δ*Cfatg4*.

2.4. CfAtg8$^{1-160}$ Partially Restores the Resistance of ΔCfatg4 to Rapamycin

Rapamycin is an agent that regulates autophagy by inhibiting the TOR kinase [24]. We inoculated WT, Δ*Cfatg4*, Δ*Cfatg4*/CfAtg8$^{1-160}$, and Δ*Cfatg4*/*CfATG4* strains in CM medium supplemented with 50 nM rapamycin, which was used to initially explore the roles of CfAtg8$^{1-160}$ in autophagy. The results showed that the Δ*Cfatg4*/CfAtg8$^{1-160}$ strain was more tolerant compared to Δ*Cfatg4* (Figure 4A,B). However, the tolerance of the Δ*Cfatg4*/CfAtg8$^{1-160}$ strain could not be restored to the level of the WT strain (Figure 4A,B). The above results indicate that CfAtg8$^{1-160}$ makes Δ*Cfatg4* more resistant to rapamycin.

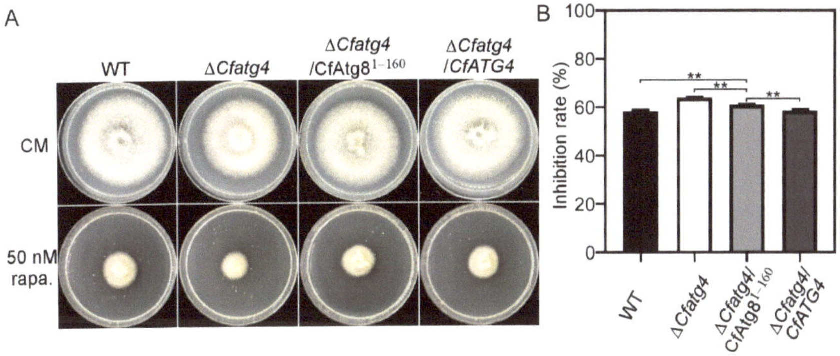

Figure 4. CfAtg8$^{1-160}$ partially restores the resistance of Δ*Cfatg4* to rapamycin. (**A**) The WT, Δ*Cfatg4*, Δ*Cfatg4*/CfAtg8$^{1-160}$, and Δ*Cfatg4*/*CfATG4* strains were incubated on CM plates with 50 nM of rapamycin at 28 °C for 3 days. (**B**) Statistical analysis of growth inhibition rates of the strains to rapamycin stress. The data were analyzed by Student's *t*-test, and the error bars indicate SD. Asterisks indicate the difference is significant (**, $p < 0.01$).

2.5. CfAtg8$^{1-160}$ Is Important for Autophagy

We acquired the Δ*Cfatg4* strain expressing GFP-CfAtg8$^{1-160}$, and we further acquired the strains expressing GFP-CfAtg8 in the WT and Δ*Cfatg4* mutants, respectively. MM-N (minimal medium without $NaNO_3$) treatment was another way to induce autophagy by inactivating TOR [25]. Similar to Δ*Cfatg4*, the GFP fluorescence of Δ*Cfatg4*/CfAtg8$^{1-160}$ still could not be transferred into the vacuoles during MM-N-induced autophagy conditions compared with the GFP-contained vacuoles in WT (Figure 5A). However, the number of autophagosomes in Δ*Cfatg4*/CfAtg8$^{1-160}$ was more than those in Δ*Cfatg4* before and after MM-N induction (Figure 5B).

The process of autophagic flux was further tested by the subsequent breakdown of GFP-Atg8. After the autophagic body was transferred into the vacuoles, GFP-Atg8 was exposed to vacuolar hydrolases for degradation, whereas the intact GFP moiety cleaved from GFP-Atg8 was resistant to vacuolar proteolysis [26]. Thus, we can evaluate the autophagic flux by calculating the ratios of free GFPs relative to the total amount of intact GFP-CfAtg8 plus free GFP. We found that ratios were elevated as the induction time extended in WT expressing GFP-CfAtg8 and Δ*Cfatg4* expressing GFP-CfAtg8 strains showed no GFP for the whole time, whereas Δ*Cfatg4* expressing GFP-CfAtg8$^{1-160}$ strains could produce free GFP despite a little increase (Figure 5C). The above results suggest that the cleavage of CfAtg8 by CfAtg4 is important for autophagy.

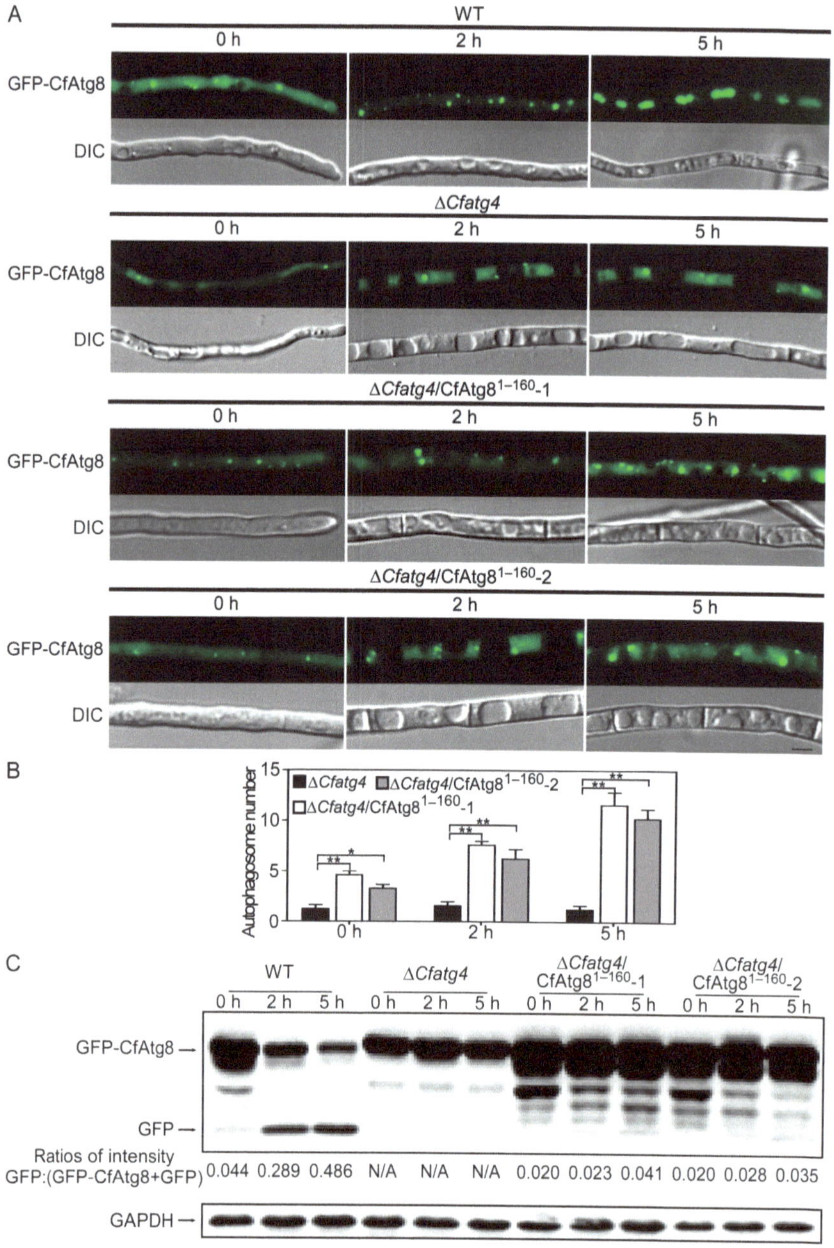

Figure 5. CfAtg8$^{1–160}$ is important for autophagy. (**A**) Micrographs of GFP-CfAtg8-labeled autophagosomes in the WT, Δ*Cfatg4*, and Δ*Cfatg4*/CfAtg8$^{1–160}$. (**B**) Statistical analysis of autophagosome number in Δ*Cfatg4* and Δ*Cfatg4*/CfAtg8$^{1–160}$ after being induced for 0 h, 2 h, and 5 h in MM-N liquid medium. The data were analyzed by Student's *t*-test, and the error bars indicate SD. Asterisks indicate the difference is significant (**, $p < 0.01$; *, $0.01 \leq p < 0.05$) (**C**) Immunoblot analysis of GFP-CfAtg8 proteolysis. The upper and lower lanes point to the intact GFP-Atg8 (46 kDa) and free GFP (26 kDa), respectively.

2.6. CfAtg8$^{1-160}$ Partially Restores the Pathogenicity Defects of ΔCfatg4

As CfAtg8$^{1-160}$ partially restores autophagy, we speculate that it may also recover autophagy-related pathogenicity. The assay on the pathogenicity of wounded tea oil leaves showed that the Δ*Cfatg4*/CfAtg8$^{1-160}$ strain restores the pathogenicity of the Δ*Cfatg4* mutant but could not conduct this to the same level as WT and Δ*Cfatg4*/*CfATG4* strains (Figure 6A,B). The pathogenicity assays were further tested on apples; the Δ*Cfatg4*/CfAtg8$^{1-160}$ strain was inoculated on apples and showed significantly increased lesion areas compared to the Δ*Cfatg4* strain, and it was also unable to recover to the level of the WT strain (Figure 6C,D). The above results suggest that CfAtg8$^{1-160}$ can partially restore the pathogenicity of Δ*Cfatg4*.

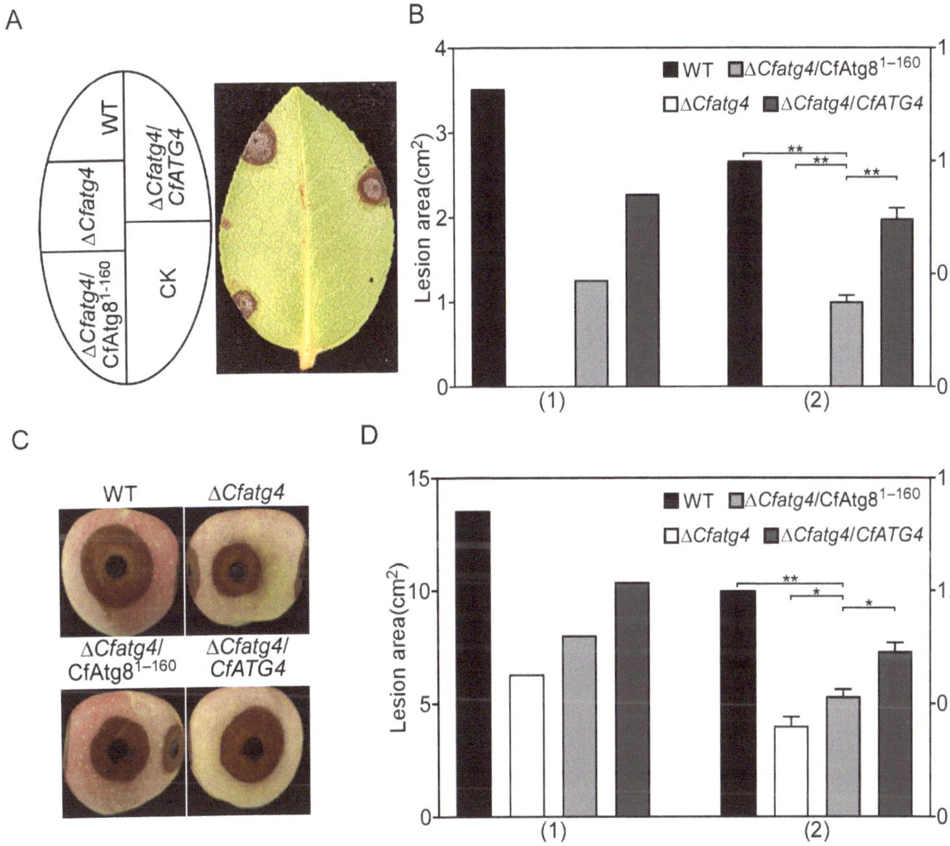

Figure 6. CfAtg8$^{1-160}$ partially restores the pathogenicity of Δ*Cfatg4*. (**A**) The WT, Δ*Cfatg4*, Δ*Cfatg4*/CfAtg8$^{1-160}$ and Δ*Cfatg4*/*CfATG4* strains were inoculated on wounded *C. oleifera* leaves. (**B**) The disease spot areas of strains on wounded *C. oleifera* leaves were measured by ImageJ, and the lesion ratios of the related strains to WT were statistically analyzed. (**C**) The strains were inoculated on wounded apples and photographed. (**D**) Statistical analysis of the related strains to WT on apples. CK: compared control. The data were analyzed by Student's *t*-test, and the error bars indicate SD. Asterisks indicate the difference is significant (**, $p < 0.01$; *, $0.01 \leq p < 0.05$).

2.7. CfAtg8$^{1-160}$ Partially Restores Conidia Germination and Appressoria Formation in ΔCfatg4

The appressoria of *Colletotrichum* spp. play critical roles in infesting hosts [27]. To explore the underlying mechanism of the partially recovered pathogenicity, the functions of CfAtg8$^{1-160}$ in appressoria formation were examined. Therefore, we examined conidial germination and the appressoria formation of WT, Δ*Cfatg4*, Δ*Cfatg4*/CfAtg8$^{1-160}$, and

Δ*Cfatg4*/*CfATG4* after 12 and 24 h of incubation on hydrophobic glass slides. The results showed a significant increase in the germination rates of Δ*Cfatg4*/CfAtg8$^{1-160}$ compared to Δ*Cfatg4*, but which was still significantly lower than those in WT and Δ*Cfatg4*/*CfATG4* (Figure 7A,B). A similar partial recovery was also observed in the appressoria formation among the strains above (Figure 7A,B). These results suggest that CfAtg8$^{1-160}$ partially restores conidia germination and appressoria formation.

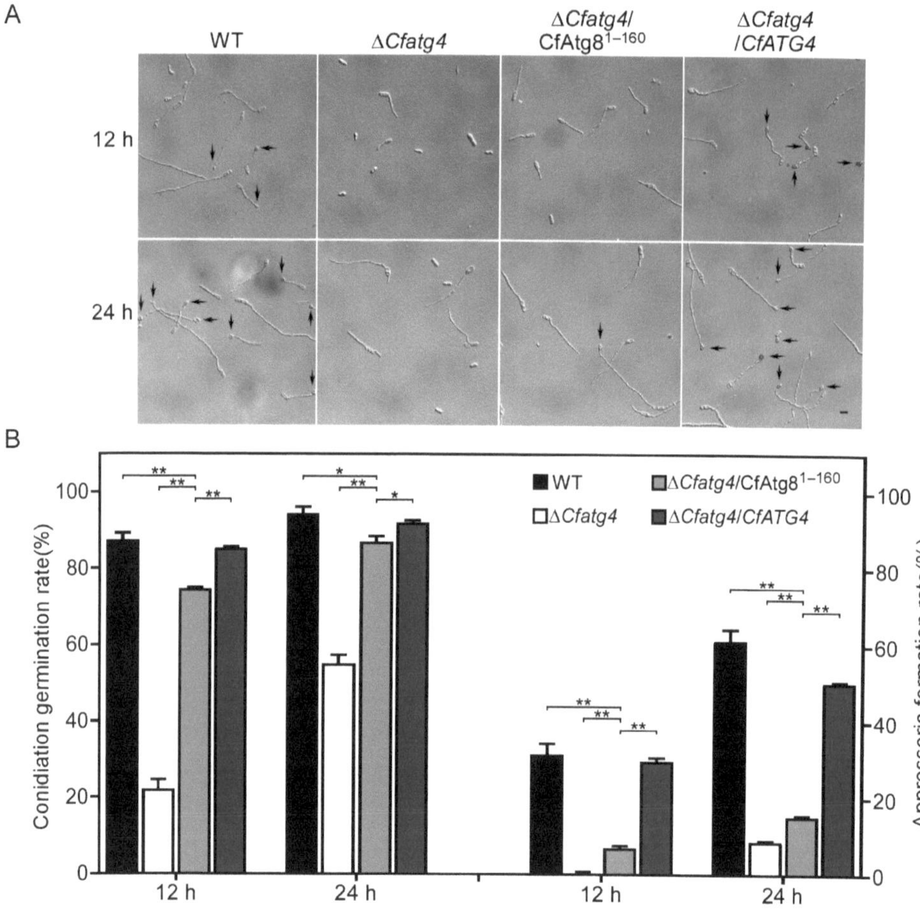

Figure 7. CfAtg8$^{1-160}$ partially restores conidia germination and appressoria formation in Δ*Cfatg4*. (**A**) The conidia of the WT, Δ*Cfatg4*, Δ*Cfatg4*/CfAtg8$^{1-160}$, and Δ*Cfatg4*/*CfATG4* strains were cultured on hydrophobic glass for 12 h and 24 h, and the appressoria formation was observed. (**B**) Conidia germination and appressoria formation were statistically analyzed. The data were analyzed by Student's *t*-test and the error bars indicate SD. Asterisks indicate the difference is significant (**, $p < 0.01$; *, $0.01 \leq p < 0.05$). Arrows represent appressoria. Bar = 5 μm.

2.8. *CfAtg8$^{1-160}$ Helps ΔCfatg4 Respond to External Environmental Stresses*

During the interaction between pathogen and micro-environments, the cell wall, as the first barrier to contact with the outside world, is subject to external environmental stresses [28]. Furthermore, pathogens also face host-derived ER stress during infection [29,30]. Thus, we used several reagents to mimic the cell wall stress and ER stress. We inoculated WT, Δ*Cfatg4*, Δ*Cfatg4*/CfAtg8$^{1-160}$, and Δ*Cfatg4*/*CfATG4* on the CM medium with 1 M NaCl, 1 M KCl, 0.1% SDS, and 2.5 mM DTT to explore the importance of

Δ*Cfatg4*/CfAtg8$^{1–160}$ in response to different environmental stresses. The results showed that CfAtg8$^{1–160}$ partially recovers the osmotic stress (1 M NaCl and 1 M KCl) and cell wall stress (0.1% SDS) defects of the Δ*Cfatg4* mutant, while the inhibition rates showed no significant difference between Δ*Cfatg4* and Δ*Cfatg4*/CfAtg8$^{1–160}$ on the ER stress (2.5 mM DTT) (Figure 8A,B). These results suggest that CfAtg8$^{1–160}$ helps Δ*Cfatg4* respond to external environmental stresses.

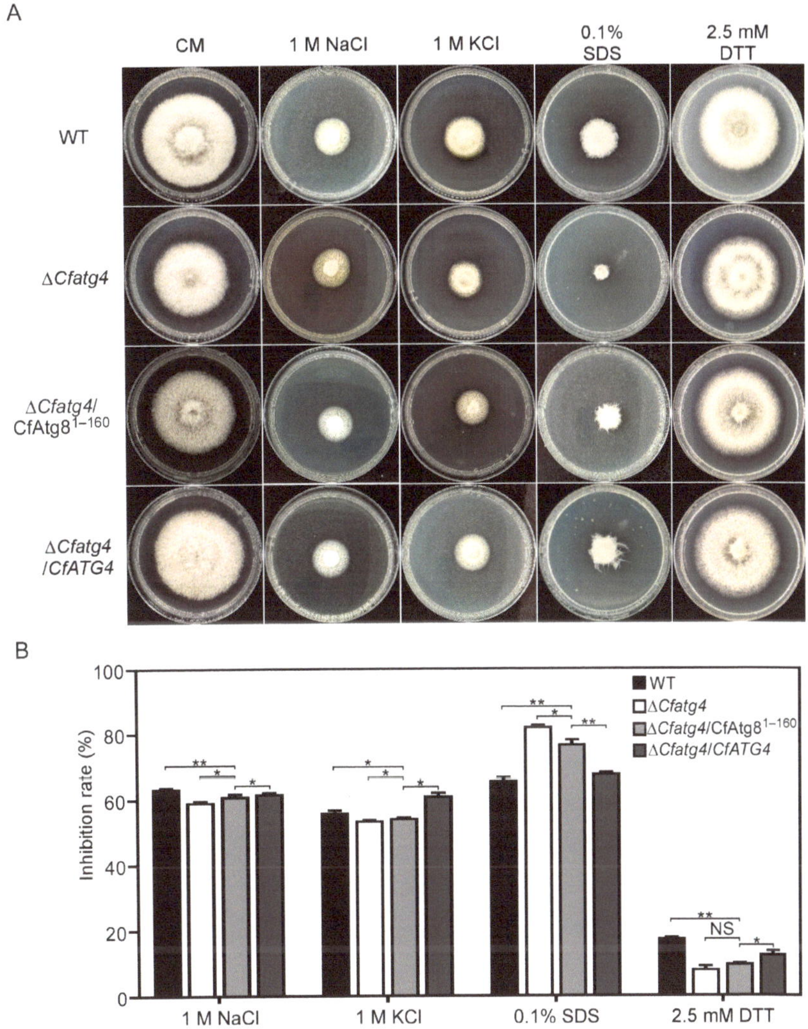

Figure 8. CfAtg8$^{1–160}$ helps Δ*Cfatg4* respond to external environmental stresses. (**A**) The strains of the WT, Δ*Cfatg4*, Δ*Cfatg4*/CfAtg8$^{1–160}$, and Δ*Cfatg4*/*CfATG4* were cultured on CM and CM plus 1 M of NaCl, 1 M of KCl, 0.1% SDS, and 2.5 mM of DTT for 3 days. (**B**) The inhibition rate of strains to stresses was statistically analyzed against an untreated control. The data were analyzed by Student's *t*-test, and the error bars indicate SD. Asterisks indicate the difference is significant (**, $p < 0.01$; *, $0.01 \leq p < 0.05$).

3. Discussion

In our previous study, we initially found that CfAtg8-mediated autophagy regulates the growth, conidia production, appressoria formation, and pathogenicity of *C. fructicola* [11]. However, the molecular mechanisms underlying the regulatory roles of CfAtg8-mediated autophagy on pathogenesis remain largely unknown. Here, we revealed that CfAtg4 interacts with CfAtg8 to regulate the formation of autophagosomes and the autophagy-mediated pathogenicity in *C. fructicola*.

Through the yeast two-hybrid assay, we demonstrated that CfAtg4 interacts with CfAtg8 directly, which is consistent with studies performed using Atg4 and Atg8 orthologs in *S. cerevisiae*, *A. thaliana*, and *Sordaria macrospora* [20,31,32]. The study of *M. oryzae* further revealed the interaction between MoAtg4 and MoAtg8 under autophagy-induced conditions in vivo [33]. Thus, we hypothesized that Atg4 proteins share conserved functions in autophagy among different organisms.

In *S. cerevisiae* and *S. macrospora*, Atg8 proteins were C-terminally cleaved by Atg4 to generate Atg8-PE, mediating the autophagosome formation [20,23,32]. If Atg4 proteins share conserved roles in autophagy, CfAtg4 functions when autophagy occurs in *C. fructicola*. The targeted gene deletion of *CfATG4* caused a defect in autophagosome formation and autophagy flux, which is in line with the studies in *S. cerevisiae*, *M. oryzae*, and *Fusarium verticillioides* [33–35]. Moreover, we identified the 160th glycine of CfAtg8 as the conserved cleavage site and constructed an Atg8 variant GFP-CfAtg8$^{1-160}$. When GFP-CfAtg8$^{1-160}$ was introduced into the Δ*Cfatg4* mutant, it partially restored the response defects to rapamycin stress, which forecasted the roles of CfAtg8$^{1-160}$ in autophagy. As expected, GFP-CfAtg8$^{1-160}$ partially restores the formation of autophagosomes. Furthermore, the immunoblot analysis also revealed a slightly higher autophagy level in Δ*Cfatg4*/CfAtg8$^{1-160}$ than that in the Δ*Cfatg4* mutant, demonstrating its partial recovery in the autophagy flux. It can be considered that except for the roles in the lipidation of Atg8, Atg4 also mediates the deconjugation of Atg8-PE to maintain a proper supply of Atg8 and facilitates the maturation of autophagosomes [19,35,36]. Consequently, though CfAtg8$^{1-160}$ bypasses the cleavage steps, it is still defective in the deconjugation of CfAtg8-PE, resulting in defects in autophagy.

Nevertheless, CfAtg8$^{1-160}$ partially restores the autophagy flux, which might foretell its roles in pathogenicity. Indeed, the lesions caused by CfAtg8$^{1-160}$ are significantly larger than those caused by the Δ*Cfatg4* mutant, but still lesser than the typical lesions caused by WT and Δ*Cfatg4*/*CfATG4* strains. We reason that the partial recovery in the pathogenicity defects of CfAtg8$^{1-160}$ is directly caused by the increased conidia germination and appressoria formation, which are important infectious processes for *Colletotrichum* to infect host plants [27,37,38]. Furthermore, based on the fact that the cell wall is the first barrier for fungi facing the host [28,30,39,40] and that the Δ*Cfatg4*/CfAtg8$^{1-160}$ strain shows a slight recovery for the cell wall stresses, we concluded that this might also be a reason for its partial recovery in the pathogenicity. It might be very interesting to investigate how CfAtg4-mediated autophagy governs pathogenicity through conidia germination, appressoria formation, and cell wall stress responses. Thus, further studies are highly warranted.

In conclusion, our study not only illustrates the importance of the interaction between CfAtg4 and CfAtg8 in autophagy but also sheds light on the pleiotropic functions of the CfAtg4-mediated autophagy in conidiation germination, appressoria formation, cell wall stress responses, and the pathogenicity of *C. fructicola*. Therefore, we reveal that CfAtg4 is a prospective target in the development of new fungicides for anthracnose control.

4. Materials and Methods

4.1. Strains and Culture Conditions

In this study, CFLH16 was used as the wild type (WT). The *CfATG4* gene deletion mutant Δ*Cfatg4* and the complemented strains were obtained in our unpublished Chinese article. Briefly, two 1.0 kb of sequences flanking the *CfATG4* gene were PCR-amplified and overlapped to the flanks of the hygromycin resistance cassette (1.4 kb), respectively. After sequencing validation, the ~3.4 kb fragment was transformed into the protoplasts of WT to obtain the Δ*Cfatg4* mutant. The complement fragment, which contains the entire *CfATG4* gene and its native promoter region, was amplified by a PCR and inserted into pYF11 (bleomycin resistance) to complement the Δ*Cfatg4* mutant. Δ*Cfatg4*/CfAtg8$^{1–160}$ was the strain expressing GFP-CfAtg8$^{1–160}$ in the Δ*Cfatg4* mutant. For generating GFP-CfAtg8$^{1–160}$, the GFP and the truncated CfAtg8 were amplified using the GFP-CfAtg8 plasmid as the template. Then, the fragment of GFP-CfAtg8$^{1–160}$ was inserted into a pYF11 vector, as previously described [41].

All the strains were cultured in CM medium (10 g D-glucose, 2 g peptone, 1 g yeast extract, 1 g casamino acid, 1 mL vitamin solution [0.01 g biotin, 0.01 g pyridoxine, 0.01 g thiamine, 0.01 g riboflavin, 0.01 g p-aminobenzoic acid, and 0.01 g nicotinic acid in 1 L ddH_2O), 1 mL trace elements [1.1 g H_3BO_3, 2.2 g $ZnSO_4{\bullet}7H_2O$, 0.5 g $FeSO_4{\bullet}7H_2O$, 0.5 g $MnCl_2{\bullet}4H_2O$, 0.17 g $CoCl_2{\bullet}6H_2O$, 0.15 g $Na_2MnO_4{\bullet}2H_2O$, 0.16 g $CuSO_4{\bullet}5H_2O$, and 5 g Na_4EDTA in 100 mL ddH_2O, pH5.8], and 50 mL 20× nitrate salts [120 g $NaNO_3$, 10.4 g $MgSO_4{\bullet}7H_2O$, 10.4 g KCl, and 30.4 g KH_2PO_4 in 1 L ddH_2O]) at 28 °C in dark conditions. The liquid CM was used to prepare mycelia for DNA and protein extraction. The MM-N medium (1.52 g KH2PO4, 0.52 g KCl, 0.152 g $MgSO_4{\cdot}7H_2O$, 0.01 g vitamin B1, 1 mL trace elements, and 10 g D-glucose in 1 L of distilled water) was used to induce autophagy.

4.2. The Alignment of Atg8 Proteins

The Atg8 proteins in *C. fructicola*, *C. gloeosporioides*, *C. siamense*, *S. cerevisiae*, *M. oryzae*, *A. oryzae*, *A. nidulans*, *U. maydis*, *F. graminearum*, *N. crassa*, *R. norvegicus* and *A. thaliana* were obtained from the NCBI database. Then, the proteins were aligned using CLUSTAL_W programs by BioEdit v7.0 software.

4.3. Yeast Two-Hybrid Assays

The yeast two-hybrid assay was performed according to the instructions of the Alkali-Cation™ Yeast Transformation Kit (Mpbio, Santa Ana, CA, USA). The coding sequences of each candidate gene were amplified from the cDNA library of WT and then inserted into the bait vector pGBKT7 digested by BamHI and prey vector pGADT7 digested by BamHI, respectively, using the one-step ligase of Novozymes. After sequencing, pairs of plasmids were co-transformed into the yeast strain AH109 and then cultured on an SD-Leu-Trp medium. After growing colonies for 4 days, it was transferred on an SD-His/-Leu/-Trp medium. The positive control is ADRECT + BD (+), and the negative control is ADRECT + BD (−).

4.4. Growth, Conidiation and Appressoria Formation Assays

Small mycelial blocks of WT, Δ*Cfatg4*, Δ*Cfatg4*/*CfATG4*, and Δ*Cfatg4*/CfAtg8$^{1–160}$ were cut from the edge of 3-day-old cultures and were cultured on CM and MM media for 3 days in the dark at 28 °C. Then, the diameters of the strains were measured and photographed. In the conidia production experiments, the strains above were cultured on a liquid CM in a shaker for 48 h, and the conidia were observed after filtering with three layers of filter paper. For appressoria formation assays, the collected conidia were dropped onto hydrophobic glass slides and incubated at 28 °C; then, they were observed and photographed under the microscope (ZEISS, Axio Observer. A1, Jena, Germany) at 12 h and 24 h. The germ tubes represent conidial germination, and the dome-shaped cells represent appressoria formation.

4.5. Stress Response Assays

The strains of WT, Δ*Cfatg4*, Δ*Cfatg4*/*CfATG4* and Δ*Cfatg4*/CfAtg8$^{1–160}$ were cultured on a CM and CM medium with 50 nM of rapamycin, osmotic stress (1 M NaCl, 1 M KCl), endoplasmic reticulum stress (2.5 mM DTT) and cell wall integrity stress (0.1% SDS) for 3 days; the colony diameters were measured, and the inhibition rates were statistically analyzed.

4.6. Pathogenicity Assays

The WT, Δ*Cfatg4*, Δ*Cfatg4*/*CfATG4*, and Δ*Cfatg4*/CfAtg8$^{1-160}$ strains were inoculated on the edge of injured tea oil leaves in a moisturizing culture at 28 °C for 2 days; then its lesions were observed and measured. The above strains were also inoculated on wounded apples, as in our previous description [42]. First, several 4 mm diameter holes were punched evenly on the apples; then, strains of the same diameter were inoculated onto them. Finally, the apples were cultured in an incubator for 4~5 days, and the lesions were measured and statistically analyzed.

4.7. Autophagy Induction and Western Blotting Assays

The WT strains that transformed the GFP-CfAtg8 and the Δ*Cfatg4* strains and which transformed the GFP-CfAtg8 or GFP-CfAtg8$^{1-160}$ strains were cultured on a liquid CM medium for 36 h; then the mycelia were washed with ddH$_2$O and collected, followed by treatment with MM-N for 2 h or 5 h. The mean autophagosome numbers were calculated from at least 25 hyphae under a fluorescence microscope (ZEISS, Axio Observer. A1) and statistically analyzed.

The Western blotting analyses were carried out according to our previous description [10]. First, the mycelia of the above strains, treated with CM and MM-N, were frozen in liquid nitrogen before protein extraction. Then, the frozen mycelia were ground into a fine powder in liquid nitrogen and re-suspended in a 1 mL lysis buffer (EpiZyme, PC101, Shanghai, China) with a 10 μL proteinase inhibitor cocktail (EpiZyme, GRF101). For protein lysing, the lysates were placed in the ice for 30 min before re-suspending with a vortex-genie once every 10 min, followed by centrifugation at 12,000 rpm for 20 min at 4 °C, and the supernatant proteins were harvested. The proteins were analyzed by 10% SDS-PAGE and Western blotting. The primary antibodies of protein analysis by Western blotting were anti-GFP (rabbit, 1:5000, Abways, AB0045) and anti-GAPDH (mouse, 1:20,000, jiahebio, M070101), and the second antibodies were HRP-labeled goat anti-rabbit lgG (H + L) (1:10,000, Abways, AB0101) or HRP-labeled goat anti-mouse lgG (H + L) (1:20,000, HUABIO, HA1006). After that, the proteins were detected by the Omni-ECL Femto Light Chemiluminescence kit (EpiZyme, SQ201) and analyzed by ImageJ v1.48.

4.8. Accession Number

The sequence data of this research can be acquired in the GenBank database (https://www.ncbi.nlm.nih.gov/genbank/, accessed on 20 February 2023) with the following accession numbers: KAF4481970.1 (CfAtg8), XP_045259742.1 (CgAtg8), XP_036500530.1 (CsAtg8), NP_009475.1 (ScAtg8), KAI6350462.1 (MoAtg8), XP_001727929.1 (AoAtg8), XP_662735.1 (AnAtg8), XP_011391873.1 (UmAtg8), XP_011325668.1 (FgAtg8), XP_956248.1 (NcAtg8), 1KJT_A (RnAtg8) and AAM64870.1 (AtAtg8).

4.9. Statistical Analysis

All data are presented as the mean ± standard deviation and analyzed by Student's *t*-test, $p < 0.01$ or $0.01 \leq p < 0.05$.

Author Contributions: Conceptualization, S.Z.; Validation, S.Z.; Investigation, S.G.; Writing—original draft, S.G.; Writing—review & editing, S.Z.; Funding acquisition, S.Z. All authors have read and agreed to the published version of the manuscript.

Funding: This research was funded by the Natural Science Foundation of Hunan Province (2023JJ20100, S.Z.) and National Natural Science Foundation of China (32001317, S.Z.).

Institutional Review Board Statement: Not applicable.

Informed Consent Statement: Not applicable.

Data Availability Statement: The original contributions presented in the study are included in the article, further inquiries can be directed to the corresponding author.

Conflicts of Interest: The authors declare no conflict of interest.

References

1. Chen, Y.; Wang, B.; Chen, J.; Wang, X.; Wang, R.; Peng, S.; Chen, L.; Ma, L.; Luo, J. Identification of Rubisco rbcL and rbcS in *Camellia oleifera* and their potential as molecular markers for selection of high tea oil cultivars. *Front. Plant Sci.* **2015**, *6*, 131993. [CrossRef] [PubMed]
2. Feás Sánchez, X.; Estevinho, L.M.; Salinero, C.; Vela, P.; Sainz Oses, M.J.; Vázquez Tato, M.d.P.; Seijas Vázquez, J.A. Triacylglyceride, antioxidant and antimicrobial features of virgin *Camellia oleifera*, *C. reticulata* and *C. sasanqua* oils. *Molecules* **2013**, *18*, 4573–4587. [CrossRef] [PubMed]
3. Wang, X.; Zeng, Q.; del Mar Contreras, M.; Wang, L. Profiling and quantification of phenolic compounds in *Camellia* seed oils: Natural tea polyphenols in vegetable oil. *Food Res. Int.* **2017**, *102*, 184–194. [CrossRef] [PubMed]
4. Li, H.; Zhou, G.-Y.; Liu, J.-A.; Xu, J. Population genetic analyses of the fungal pathogen *Colletotrichum fructicola* on tea-oil trees in China. *PLoS ONE* **2016**, *11*, e0156841. [CrossRef] [PubMed]
5. Kershaw, M.J.; Talbot, N.J. Genome-wide functional analysis reveals that infection-associated fungal autophagy is necessary for rice blast disease. *Proc. Natl. Acad. Sci. USA* **2009**, *106*, 15967–15972. [CrossRef] [PubMed]
6. Talbot, N.J.; Kershaw, M.J. The emerging role of autophagy in plant pathogen attack and host defence. *Curr. Opin. Plant Biol.* **2009**, *12*, 444–450. [CrossRef] [PubMed]
7. Zhu, X.-M.; Li, L.; Wu, M.; Liang, S.; Shi, H.-B.; Liu, X.-H.; Lin, F.-C. Current opinions on autophagy in pathogenicity of fungi. *Virulence* **2019**, *10*, 481–489. [CrossRef] [PubMed]
8. Veneault-Fourrey, C.; Barooah, M.; Egan, M.; Wakley, G.; Talbot, N.J. Autophagic fungal cell death is necessary for infection by the rice blast fungus. *Science* **2006**, *312*, 580–583. [CrossRef] [PubMed]
9. Zheng, H.W.; Miao, P.F.; Lin, X.L.; Li, L.P.; Wu, C.X.; Chen, X.M.; Abubakar, Y.S.; Norvienyeku, J.; Li, G.P.; Zhou, J.; et al. Small GTPase Rab7-mediated FgAtg9 trafficking is essential for autophagy-dependent development and pathogenicity in *Fusarium graminearum*. *PLoS Genet.* **2018**, *14*, e1007546. [CrossRef]
10. Chen, Y.; Jin, J.; Li, Y.; Jiao, H.; Luo, L.; Chen, Q.; Li, H.; Zhang, S. The CfAtg5 regulates the autophagy and pathogenicity of *Colletotrichum fructicola* on *Camellia oleifera*. *Agronomy* **2023**, *13*, 1237. [CrossRef]
11. Zhang, S.; Guo, Y.; Li, S.; Li, H. Histone acetyltransferase CfGcn5-mediated autophagy governs the pathogenicity of *Colletotrichum fructicola*. *mBio* **2022**, *13*, e0195622. [CrossRef] [PubMed]
12. Mizushima, N.; Komatsu, M. Autophagy: Renovation of cells and tissues. *Cell* **2011**, *147*, 728–741. [CrossRef] [PubMed]
13. Ohsumi, Y. Historical landmarks of autophagy research. *Cell Res.* **2014**, *24*, 9–23. [CrossRef] [PubMed]
14. Nakatogawa, H.; Ichimura, Y.; Ohsumi, Y. Atg8, a ubiquitin-like protein required for autophagosome formation, mediates membrane tethering and hemifusion. *Cell* **2007**, *130*, 165–178. [CrossRef]
15. Ichimura, Y.; Kirisako, T.; Takao, T.; Satomi, Y.; Shimonishi, Y.; Ishihara, N.; Mizushima, N.; Tanida, I.; Kominami, E.; Ohsumi, M.; et al. A ubiquitin-like system mediates protein lipidation. *Nature* **2000**, *408*, 488–492. [CrossRef] [PubMed]
16. Mizushima, N.; Yoshimori, T.; Ohsumi, Y. The role of Atg proteins in autophagosome formation. *Annu. Rev. Cell Dev. Biol.* **2011**, *27*, 107–132. [CrossRef] [PubMed]
17. Ohsumi, Y. Molecular dissection of autophagy: Two ubiquitin-like systems. *Nat. Rev. Mol. Cell Biol.* **2001**, *2*, 211–216. [CrossRef]
18. Mizushima, N.; Yoshimori, T.; Ohsumi, Y. Role of the Apg12 conjugation system in mammalian autophagy. *Int. J. Biochem. Cell Biol.* **2003**, *35*, 553–561. [CrossRef]
19. Hirata, E.; Ohya, Y.; Suzuki, K. Atg4 plays an important role in efficient expansion of autophagic isolation membranes by cleaving lipidated Atg8 in *Saccharomyces cerevisiae*. *PLoS ONE* **2017**, *12*, e0181047. [CrossRef]
20. Kirisako, T.; Ichimura, Y.; Okada, H.; Kabeya, Y.; Mizushima, N.; Yoshimori, T.; Ohsumi, M.; Takao, T.; Noda, T.; Ohsumi, Y. The reversible modification regulates the membrane-binding state of Apg8/Aut7 essential for autophagy and the cytoplasm to vacuole targeting pathway. *J. Cell Biol.* **2000**, *151*, 263–276. [CrossRef]
21. Kim, J.; Huang, W.P.; Klionsky, D.J. Membrane recruitment of Aut7p in the autophagy and cytoplasm to vacuole targeting pathways requires Aut1p, Aut2p, and the autophagy conjugation complex. *J. Cell Biol.* **2001**, *152*, 51–64. [CrossRef] [PubMed]
22. Maruyama, T.; Noda, N.N. Autophagy-regulating protease Atg4: Structure, function, regulation and inhibition. *J. Antibiot.* **2018**, *71*, 72–78. [CrossRef]
23. Suzuki, K.; Kubota, Y.; Sekito, T.; Ohsumi, Y. Hierarchy of Atg proteins in pre-autophagosomal structure organization. *Genes Cells* **2007**, *12*, 209–218. [CrossRef] [PubMed]
24. Zhang, W.; Han, L.; Huang, Y.; He, J.; He, F.; Dong, Y. Unveiling the significance of Target of Rapamycin (TOR) signalling in grafting. *Veg. Res.* **2024**, *4*, e004. [CrossRef]
25. Dobrenel, T.; Caldana, C.; Hanson, J.; Robaglia, C.; Vincentz, M.; Veit, B.; Meyer, C. TOR signaling and nutrient sensing. *Annu. Rev. Plant Biol.* **2016**, *67*, 261–285. [CrossRef] [PubMed]
26. Cheong, H.; Klionsky, D.J. Biochemical methods to monitor autophagy-related processes in yeast. *Methods Enzymol.* **2008**, *451*, 1–26.
27. Fukada, F.; Kodama, S.; Nishiuchi, T.; Kajikawa, N.; Kubo, Y. Plant pathogenic fungi *Colletotrichum* and *Magnaporthe* share a common G1 phase monitoring strategy for proper appressorium development. *New Phytol.* **2019**, *222*, 1909–1923. [CrossRef]

28. Malavazi, I.; Goldman, G.H.; Brown, N.A. The importance of connections between the cell wall integrity pathway and the unfolded protein response in filamentous fungi. *Brief. Funct. Genom.* **2014**, *13*, 456–470. [CrossRef] [PubMed]
29. Tang, W.; Jiang, H.; Aron, O.; Wang, M.; Wang, X.; Chen, J.; Lin, B.; Chen, X.; Zheng, Q.; Gao, X.; et al. Endoplasmic reticulum-associated degradation mediated by MoHrd1 and MoDer1 is pivotal for appressorium development and pathogenicity of *Magnaporthe oryzae*. *Environ. Microbiol.* **2020**, *22*, 4953–4973. [CrossRef]
30. Yin, Z.; Feng, W.; Chen, C.; Xu, J.; Li, Y.; Yang, L.; Wang, J.; Liu, X.; Wang, W.; Gao, C.; et al. Shedding light on autophagy coordinating with cell wall integrity signaling to govern pathogenicity of *Magnaporthe oryzae*. *Autophagy* **2020**, *16*, 900–916. [CrossRef]
31. Ketelaar, T.; Voss, C.; Dimmock, S.A.; Thumm, M.; Hussey, P.J. Arabidopsis homologues of the autophagy protein Atg8 are a novel family of microtubule binding proteins. *FEBS Lett.* **2004**, *567*, 302–306. [CrossRef] [PubMed]
32. Voigt, O.; Poggeler, S. Autophagy genes Smatg8 and Smatg4 are required for fruiting-body development, vegetative growth and ascospore germination in the filamentous ascomycete *Sordaria macrospora*. *Autophagy* **2013**, *9*, 33–49. [CrossRef] [PubMed]
33. Liu, T.B.; Liu, X.H.; Lu, J.P.; Zhang, L.; Min, H.; Lin, F.C. The cysteine protease MoAtg4 interacts with MoAtg8 and is required for differentiation and pathogenesis in *Magnaporthe oryzae*. *Autophagy* **2010**, *6*, 74–85. [CrossRef] [PubMed]
34. Wang, Y.; Liu, X.; Xu, Y.; Gu, Y.; Zhang, X.; Zhang, M.; Wen, W.; Lee, Y.W.; Shi, J.; Mohamed, S.R.; et al. The autophagy-related proteins FvAtg4 and FvAtg8 are involved in virulence and fumonisin biosynthesis in *Fusarium verticillioides*. *Virulence* **2022**, *13*, 764–780. [CrossRef] [PubMed]
35. Yu, Z.Q.; Ni, T.; Hong, B.; Wang, H.Y.; Jiang, F.J.; Zou, S.; Chen, Y.; Zheng, X.L.; Klionsky, D.J.; Liang, Y.; et al. Dual roles of Atg8-PE deconjugation by Atg4 in autophagy. *Autophagy* **2012**, *8*, 883–892. [CrossRef]
36. Nair, U.; Yen, W.L.; Mari, M.; Cao, Y.; Xie, Z.; Baba, M.; Reggiori, F.; Klionsky, D.J. A role for Atg8-PE deconjugation in autophagosome biogenesis. *Autophagy* **2012**, *8*, 780–793. [CrossRef]
37. Fukada, F.; Kubo, Y. *Colletotrichum orbiculare* regulates cell cycle G1/S progression via a two-component GAP and a GTPase to establish plant infection. *Plant Cell* **2015**, *27*, 2530–2544. [CrossRef]
38. Ryder, L.S.; Cruz-Mireles, N.; Molinari, C.; Eisermann, I.; Eseola, A.B.; Talbot, N.J. The appressorium at a glance. *J. Cell Sci.* **2022**, *135*, jcs259857. [CrossRef]
39. Levin, D.E. Regulation of cell wall biogenesis in *Saccharomyces cerevisiae*: The cell wall integrity signaling pathway. *Genetics* **2011**, *189*, 1145–1175. [CrossRef]
40. Feng, W.Z.; Yin, Z.Y.; Wu, H.W.; Liu, P.; Liu, X.Y.; Liu, M.X.; Yu, R.; Gao, C.Y.; Zhang, H.F.; Zheng, X.B.; et al. Balancing of the mitotic exit network and cell wall integrity signaling governs the development and pathogenicity in *Magnaporthe oryzae*. *PLoS Pathog* **2021**, *17*, e1009080. [CrossRef]
41. Bruno, K.S.; Tenjo, F.; Li, L.; Hamer, J.E.; Xu, J.R. Cellular localization and role of kinase activity of PMK1 in *Magnaporthe grisea*. *Eukaryot. Cell* **2004**, *3*, 1525–1532. [CrossRef] [PubMed]
42. Zhang, S.; Li, H. The histone acetyltransferase CfGcn5 regulates growth, development, and pathogenicity in the anthracnose fungus *Colletotrichum fructicola* on the tea-oil tree. *Front. Microbiol.* **2021**, *12*, 680415. [CrossRef] [PubMed]

Journal of *Fungi*

MDPI

Article

Microbial-Based Biofungicides Mitigate the Damage Caused by *Fusarium oxysporum* f. sp. *cubense* Race 1 and Improve the Physiological Performance in Banana

Luisa Fernanda Izquierdo-García *, Sandra Lorena Carmona-Gutiérrez, Carlos Andrés Moreno-Velandia *, Andrea del Pilar Villarreal-Navarrete, Diana Marcela Burbano-David, Ruth Yesenia Quiroga-Mateus, Magda Rocío Gómez-Marroquín, Gustavo Adolfo Rodríguez-Yzquierdo and Mónica Betancourt-Vásquez

Corporación Colombiana de Investigación Agropecuaria, AGROSAVIA, Centro de Investigación Tibaitatá, Km 14 vía Bogotá a Mosquera, Mosquera, Cundinamarca 250047, Colombia; scarmona@agrosavia.co (S.L.C.-G.); avillarreal@agrosavia.co (A.d.P.V.-N.); dmburbano@agrosavia.co (D.M.B.-D.); ryquiroga@agrosavia.co (R.Y.Q.-M.); mrgomez@agrosavia.co (M.R.G.-M.); grodriguezy@agrosavia.co (G.A.R.-Y.); mbetancourtv@agrosavia.co (M.B.-V.)

* Correspondence: lfizquierdo@agrosavia.co (L.F.I.-G.); cmoreno@agrosavia.co (C.A.M.-V.); Tel.: +57-601-4227373 (ext. 1303) (L.F.I.-G. & C.A.M.-V.)

Abstract: Fusarium wilt of banana (FWB) is the most limiting disease in this crop. The phytosanitary emergency caused by FWB since 2019 in Colombia has required the development of ecofriendly control methods. The aim of this study was to test the effectiveness of microbial-based biofungicides against FWB caused by *Fusarium oxysporum* f. sp. *cubense* race 1 (Foc R1) and correlate such effect with plant physiological parameters. Five *Trichoderma* (T1 to T4 and T9) and four *Bacillus* (T5 to T8)-based biofungicides were evaluated in pot experiments. In vitro, dual confrontation tests were also carried out to test whether the in vitro effects on Foc growth were consistent with the in vivo effects. While *Trichoderma*-based T3, T4, and T9, and *Bacillus*-based T8, significantly reduced the growth of Foc R1 in vitro, *Trichoderma*-based T1, T3, T4, and T9 temporarily reduced the Foc population in the soil. However, the incidence progress of FWB was significantly reduced by Bacterial-based T7 (74% efficacy) and *Trichoderma*-based T2 (50% efficacy). The molecular analysis showed that T7 prevented the inner tissue colonization by Foc R1 in 80% of inoculated plants. The T2, T4, T7, and T9 treatments mitigated the negative effects caused by Foc R1 on plant physiology and growth. Our data allowed us to identify three promising treatments to control FWB, reducing the progress of the disease, delaying the colonization of inner tissue, and mitigating physiological damages. Further studies should be addressed to determine the modes of action of the biocontrol agents against Foc and validate the utilization in the field.

Keywords: biological control; microbial consortia; bioproducts; Panama disease

Citation: Izquierdo-García, L.F.; Carmona-Gutiérrez, S.L.; Moreno-Velandia, C.A.; Villarreal-Navarrete, A.d.P.; Burbano-David, D.M.; Quiroga-Mateus, R.Y.; Gómez-Marroquín, M.R.; Rodríguez-Yzquierdo, G.A.; Betancourt-Vásquez, M. Microbial-Based Biofungicides Mitigate the Damage Caused by *Fusarium oxysporum* f. sp. *cubense* Race 1 and Improve the Physiological Performance in Banana. *J. Fungi* **2024**, *10*, 419. https://doi.org/10.3390/jof10060419

Academic Editor: Ofir Degani

Received: 7 May 2024
Revised: 25 May 2024
Accepted: 30 May 2024
Published: 12 June 2024

1. Introduction

The banana is one of the most consumed and traded foods worldwide [1]. This fruit is part of the family basket in the countries where it is grown, due to its nutritional characteristics, thus playing an essential role in food security [2] (pp. 15–19), [3] (pp. 73–76), [4]. On the other hand, banana exports from developing countries substantially contribute to their economic income [2].

However, worldwide, banana-producing farmers have had to face the threat of fungal diseases to fruit production throughout history. The Fusarium wilt of banana (FWB), also known as Panama disease, caused by *Fusarium oxysporum* f. sp. *cubense* (Foc), is the most destructive and difficult to control [5,6]. Three races of Foc have been identified according to their ability to cause disease in specific banana cultivars. Race 1 (Foc R1) affects banana varieties such as Gros Michel (AAA), Manzano (AAB), and Pisang Awak (ABB). Race 2 (Foc R2) affects cooking bananas, especially those of the Bluggoe (ABB) subgroup. Lastly, race 4

(Foc R4) affects all those varieties susceptible to Foc R1 and Foc R2, especially cultivars of the Cavendish subgroup [7,8].

F. oxysporum f. sp. *cubense* Tropical Race 4 (Foc TR4) has spread inexorably through the banana-producing regions around the world, including South America, where symptomatic plants from the cv. Cavendish, resistant to Foc R1, were detected in 2019 [9]. The same year, Colombia's national plant protection organization (Instituto Colombiano Agropecuario—ICA) declared a phytosanitary emergency, since Foc TR4 was isolated from Cavendish plantations located in the north of the country (i.e., La Guajira Department) showing typical symptoms of wilt disease [10]. This phytosanitary problem has been treated as a quarantine disease, and the official institutions, in cooperation with banana growers, are working together to contain and eradicate the pathogen in the affected area, implementing dispersal prevention measures and developing control methods [11].

The typical external symptoms of FWB include yellowing of the leaf blade, pseudostem breakage, and wilting of the oldest leaves [12]. These symptoms are caused by the obstruction of the vascular bundles exerted by the proliferation of Foc structures and the formation of gums and tyloses secreted by the plant in the vascular tissue. This plugging increases the resistance in the translocation of water and nutrients and decreases the water potential in the leaves. As a consequence, stomatal conductance decreases and, by reducing CO_2 absorption, the photosynthetic activity and quantum efficiency of photosystem II (PSII) are limited, ultimately affecting biomass accumulation and plant growth [13,14].

The dispersal of Foc TR4 is particularly difficult to control since this fungus spreads easily through contaminated soil, water, and agricultural tools [15]. To deal with this pathogen, the research has been focusing on diagnosis and early detection, containment of the pathogen through the destruction of infected plants and declaration of quarantine areas, implementation of biosafety schemes [7,16], development of resistant varieties [17,18], the implementation and application of biological control [19–23], and crop management practices [24–27].

Considering that the indiscriminate use of chemical pesticides to control plant pathogens has caused a negative impact on the environment [28,29], biological control as an alternative tool for the management of Fusarium wilt is projected as a very effective and environmentally friendly strategy [30]. Biological control with antagonistic fungi and bacteria mainly from the genera *Trichoderma* and *Bacillus*, respectively, has been focused on reducing the negative effects caused by soil-borne phytopathogens on crop yields [20,31,32] through the stimulation of plant defense mechanisms or exerting direct mechanisms of action on the target phytopathogens. In all of this, complex interactions between plant-pathogen antagonists and the resident microbiota take place.

Several well-known species of fungi from the genus *Trichoderma* have been recognized for their high biocontrol potential against diseases and insect pests, as was shown in the recent review of Sharma et al. [33]. Accordingly, different *Trichoderma* species have the ability to multiply rapidly and advantageously inhibit the growth of fungi through the production of different enzymes (viz., glucanases, chitobiosidases, and chitinases) used mainly during the parasitizing action on pathogenic fungi; low-molecular-weight volatile or nonvolatile antibiotics or secondary metabolites (viz., viridin, gliotoxin, and peptaibols) that restrict the growth of pathogens; and the induction of local or systemic defense mechanisms in host plants. Some of the most known species of *Trichoderma* with antagonistic properties are *T. harzianum*, *T. asperellum*, *T. koningii*, *T. koningiopsis*, *T. longibrachiatum*, *T. hamatum*, *T. viride*, *T. pseudokoningii*, *T. polysporum*, and *T. virens* [33,34], which are the active ingredient of a vast amount of *Trichoderma*-based biofungicides [34]. These species of fungi have a significant impact on the development of plant diseases caused by *Rhizoctonia solani*, *Sclerotium rolfsii*, *Pythium aphanidermatum*, *Gaeumannomyces graminis* var. *tritici*, *Verticillium dahliae*, *Fusarium culmorum*, and *F. oxysporum* [34].

On the other hand, the potential of biological control and plant growth promotion of many members of rhizosphere-associated bacteria have also been explored for many decades, of which species belonging to the *Pseudomonas* and *Bacillus* genera have attracted the most attention. Particularly, the soil-ubiquitous *Bacillus* genus constitutes one of the main groups of beneficial microorganisms used against diseases caused by soil-borne phytopathogens [35]. Except for the pathogenic species *B. cereus* and *B. anthracis*, the genus *Bacillus* includes species with properties generally recognized as safe or with a classification of supposedly safe (GRAS/QPS) [36]. The species *B. subtilis*, and those from the operational group *B. amyloliquefaciens* (consisting of the soil-borne *B. amyloliquefaciens*, and plant-associated *B. siamensis* and *B. velezensis*) [37], are among the most used rhizobacteria in agriculture and have been extensively studied. These closely related species share properties, such as the production of cyclic lipopeptides with antimicrobial activity and the ability to colonize the rhizosphere and protect plants against phytopathogens. For instance, it has been determined that surfactin production is a determinant of root colonization [38,39], iturins and fengycins are known for their direct effect on plant pathogens [40–42], whereas systemic resistance can be induced by surfactins [43,44] and fengycins [45]. The importance of the relationship between beneficial strains of *Bacillus* spp. with plants and pathogens has merited the recent publication of several reviews viz., [28,46–51], in which it is possible to notice that *F. oxysporum* in one of the most commonly targeted pathogens.

With the arrival of Foc TR4 in Colombia, the need arose to investigate methods to counteract its dispersion, reduce its multiplication in the soil, and control the disease. Since there are no registered biopesticides for controlling FWB in Colombia, and in consonance with the emergency treatment that has been given to this phytosanitary problem, priority was given to determining the potential of biopesticides in the local market as alternatives to control FWB. Different publications have reported promising results by strains of different species from *Trichoderma* and *Bacillus* genera against both Foc R1 [52,53] and Foc TR4 [54,55], but, as far as we know, no studies have been carried out on biocontrol of Foc R1 and Foc TR4 with *Trichoderma* and *Bacillus*-based bioproducts in Colombia.

Currently, 395 bioproducts are registered with the Instituto Colombiano Agropecuario—ICA, of which 66 are manufactured based on microbial agents for the control of phytopathogens, which are mostly based on different *Trichoderma* (59%) and *Bacillus* species (23%). Other antagonists used in the bioproducts registered in the country are mainly *Paecilomyces lilacinus* and *Streptomyces racemochromogenes*. However, other *Trichoderma* and *Bacillus*-based bioproducts are registered as microbial inoculants with biofertilizer effects [56]. *F. oxysporum* is the biological target of 17% of these 66 bioproducts, but none are authorized to control Foc in bananas and not all registered bioproducts are available on the market.

With this context, the aim of this study was to determine the potential of available biofungicides in the Colombian market to control Foc R1. We tested the hypothesis of whether the *Trichoderma*-based bioproducts are more efficient than *Bacillus*-based ones against Foc, and we evaluated the effects of the biocontrol treatments on the physiological parameters of the plants growing in the presence of Foc R1. This work combined in vitro and in vivo evaluations of the antagonistic and biocontrol activity of bioproducts against Foc R1. The results of this study allow us to generate recommendations for further in vivo validations of potential biocontrol tools against Foc TR4.

2. Materials and Methods

2.1. Selection of Biofungicides

Three available *Trichoderma* and two *Bacillus*-based biofungicides registered with ICA to control *F. oxysporum* in various crops, and four more with known efficacy against Fusarium diseases through previous experiments with carnation, cape gooseberry, and tomato (Unpublished data), were included in this study (Table 1). The dosages of the bioproducts were chosen by considering the application of the same concentration of the active ingredient; thus, 1×10^6 conidia mL^{-1} for *Trichoderma*-based bioproducts, and 1×10^7 cells mL^{-1} for *Bacillus*-based bioproducts. The scheme for applying the bioproducts under in vivo experiments was designed according to their possible modes of action. Thus, the bioproducts were applied to the seedlings in the seedbed to facilitate the possible induction of resistance in the host, to stimulate the rhizospheric competence, and to promote endophytic colonization. The application of treatments to the soil before the transplant could favor the direct competition with the pathogen, the soil colonization, and facilitate direct modes of action against Foc.

Table 1. Description of biofungicides selected to be evaluated against Foc R1.

Treatment	Active Ingredient	Target Pathosystem	Formulation	Concentration (CFU g^{-1} or mL^{-1})
T1	*T. harzianum*	*F. oxysporum*—Carnation	WP	1.0×10^8
T2	*T. harzianum*	*F. oxysporum*—Carnation	WP	2.0×10^8
T3	*T. koningiopsis*	*F. oxysporum*—Tomato	WDG	1.0×10^9
T4	*T. harzianum, T. koningii, T. viride*	Various soil-borne phytopathogens	WP	1.0×10^8
T5	*B. subtilis*	*F. oxysporum*—tomato	SC	5.0×10^9
T6	*B. subtilis, B. pumilus, B. thuringensis* var. *kurstaki, B. amyloliquefaciens*	*F. oxysporum*—tomato, cape gooseberry, lulo, tree tomato, chili pepper, pepper, and egg plant	SC	4.0×10^8
T7	*B. amyloliquefaciens, Agrobacterium radiobacter, B. pumilus*	Known performance against *F. oxysporum*—Carnation at experimental level	SC	2.5×10^8
T8	*B. amyloliquefaciens*	*Mycosphaerella fijiensis*—banana	SC	1.0×10^9
T9	Mixture of *Trichoderma harzianum* strains (no formulated)	Applied by commercial banana growers	Unformulated conidia	

WP: Wettable Powder; WDG: Water-Dispersible Granule; SC: Suspension Concentrate.

2.2. Plant Material and Foc R1 Strain

Two-month-old banana seedlings of the susceptible Gros Michel type were used to evaluate the efficacy of bioproducts against the Fusarium wilt disease development in a pot experimental model under greenhouse conditions. Seedlings from Foc R1-resistant Cavendish type cv. Williams were included as a positive control. Plant propagation and supply were performed by the Eastern Catholic University (UCO), Rionegro, Antioquia, Colombia.

F. oxysporum f. sp. *cubense* race 1 IB strain (Foc R1 IB), supplied by the national banana research center (CENIBANANO), was used in the in vitro and in vivo experiments. The IB strain was isolated from Foc R1-susceptible Gros Michel banana plants showing typical symptoms of Fusarium wilt from plantations in Urabá, Colombia. PDA-plugs with mycelium and conidia of the IB strain were preserved in a glycerol sterile solution (20%) at −70 °C. The strain was activated in a potato dextrose agar culture medium (PDA—Oxoid®, Hampshire, UK) at 27 °C, and the second subculture was used as inoculum for the in vitro experiments, and as pre-inoculum to multiply the pathogen for the in vivo experiments.

2.3. Production of Foc R1 Inoculum for In Vivo Experiment

Mass production of Foc R1 IB was made following a solid-state fermentation model with wheat bran (72 g) in autoclavable plastic bags and moistened with 52 mL of sterile distilled water (SDW). The substrate in the bags was sterilized in an autoclave (two continuous cycles at 121 °C and 15 PSI). A suspension of microconidia from the IB strain was prepared by scraping a ten-day-old subculture in PDA, and the harvested biomass was added to sterile Potato Dextrose Broth (PDB—Difco®, Leeuwarden, The Netherlands) in 50 mL sterile Falcon® (Mexico City, Mexico) tubes. The suspension was vortexed at 3000 rpm for 2 min, then it was filtrated by muslin cloth (0.5 mm mesh) and the microconidial suspension was adjusted to 1×10^7 microconidia·mL^{-1}. This suspension was injected into the sterile solid substrate (10 mL per bag) with the aid of a sterile syringe, then the substrate was massaged to homogenize the inoculum. The inoculated bags were incubated for 20 days at 27 °C and massaged every 5 days to promote uniform colonization. To obtain the inoculum of Foc R1 IB for the in vivo experiment, the colonized substrate from a bag was poured into 500 mL of SDW and was blended at low speed in a home-blender (Oster®, Wickliffe, OH, USA), then filtered through sterile muslin cloth to harvest the microconidial suspension free of mycelium. A total of 200 mL of conidial suspension (1×10^6 microconidia·mL^{-1}) was drenched into 2000 g of substrate (soil and rice husk mixture, 3:1) contained in a pot (4 Mil poly bags 35 cm height × 25 cm width) at 14 days before transplant (dbt) in order to obtain a final concentration of 1×10^5 microconidia·g^{-1} of soil.

2.4. Treatments Application Scheme

The treatments (T1 to T9 described in Table 1) were applied in four moments: first, to the seedlings in the seedbed five days before transplant (5 mL seedling^{-1}), second, to the soil contained in the pots five days before transplant (100 mL pot^{-1}), third, to the soil again, immediately after transplanting (100 mL plant^{-1}), and fourth, to the soil at one week after transplant (100 mL plant^{-1}). While the treatments of *Trichoderma*-based biofungicides were applied adjusting the concentration shown in each label to a final concentration of 1×10^6 conidia mL^{-1}, the dosage of the *Bacillus*-based biofungicides was adjusted to 1×10^7 cells mL^{-1}. Plants growing in soil artificially inoculated with Foc R1 IB but not treated against the disease were included in the experiment as negative controls (T10), and plants growing in soil free of both Foc R1 IB and biological treatments were included as absolute controls (T12).

Although the doses of the biofungicides were calculated according to the label, quality control was carried out to estimate the concentration of viable active ingredients in terms of colony-forming units (CFU mL^{-1} or CFU g^{-1}). To this end, 1 g or 1 mL, according to the formulation of the bioproducts, was diluted in 9 mL of Tween 80 sterile solution (0.1% v/v), it was vortexed at maximum speed for 2 min, and serial dilutions (100 µL in 900 µL of Tween 80 sterile solution at 0.1% v/v) were made and plated on solid culture media. Dilutions, from which colony counts of from 20 to 200 colonies for fungi and from 30 to 300 for bacteria were obtained, were taken into account for the estimations. While the suspensions from *Trichoderma*-based bioproducts were plated on PDA supplemented with Triton (0.1% v/v) and chloramphenicol (100 mg L^{-1}), the suspensions from *Bacillus*-based bioproducts were plated on nutrient agar medium. Inoculated media were incubated for 48 h at 28 °C for bacteria counts, and seven days at 25 °C for fungi counts.

2.5. Experimental Design and Measured Disease Variables

The experiment was carried out under an unheated greenhouse with a temperature of 21.4 °C on average, 40.9 °C maximum, and 12.5 °C minimum, relative humidity of 60% on average, and photoperiod of 12 h light: 12 h darkness. The treatments were arranged under a randomized complete block experimental design. The experimental unit consisted of five plants and there were three repetitions per treatment. The entire experiment was replicated two times more over time, from 24 November 2021 to 9 February 2022 (first replicate), from 15 December 2021 to 25 February 2022 (second replicate), and from 11 February to 27 April 2022 (third replicate). The disease incidence, which expresses the proportion of plants showing typical symptoms of Fusarium wilt, and the disease severity, which expresses the intensity of the wilt symptoms, were recorded periodically. The scale from Dita et al. [57] with modifications was used for recording the disease severity (Supplementary Table S1). The area under the disease incidence and severity progress curves (AUDPC) was calculated using the trapezoidal rule.

With the aim to identify the inner symptoms of the disease and to measure its severity in the rhizome, all the alive plants were gently uprooted at the end of the experiment, and a longitudinal cut was made from the pseudostem to the rhizome on each plant, as was described by Dita et al. [57].

2.6. Physiological and Plant Growth Variables

To test the hypothesis that the application of biofungicides mitigates the negative effects caused by Foc R1 on the physiological processes of banana plants, the photosynthesis and gas exchange-related variables were measured in the plants of the biocontrol experiment under greenhouse as follows:

Chlorophyll *a* fluorescence. The maximum quantum efficiency of photosystem II-PSII (Fv/Fm), the PSII photochemical yield (Y(II)), and the electron transfer rate (ETR) were measured every 20 days through a modulated fluorometer miniPAM II (Heinz Walz®, Effeltrich, Germany). The measures were always taken from the third youngest expanded leaf on the central region of the leaf limb [58]. The readings were taken early in the morning, after a 20 min dark adaptation period.

Net assimilation rate (A). The variables associated with gas exchange [photosynthesis A (μmol m^{-2} s^{-1}) and transpiration rate E (mol m^{-2} s^{-1})] were measured every 20 days using LI-COR® LI-6800 portable photosynthesis system equipment (USA). To determine the optimal intensity of photosynthetically active radiation (PAR) for photosynthesis measuring in the banana plants, a light curve was made previously, applying increasing light intensities and measuring the photosynthesis response. The measurement conditions were as follows: 900 μmol·m^2 s^{-1} of CO_2 flux; 1200 mol·m^2 s^{-1} of light and 60% of relative humidity.

Fluorescence and gas exchange variables were measured on three plants per treatment, sampling the representative plant per block (n = 3) at the beginning of the measurement, and always measuring the same plants.

Chlorophyll content was determined weekly based on the SPAD index using Minolta SPAD-502 leaf chlorophyll meter (Konica Minolta®, Osaka, Japan) equipment.

Growth parameters. The plant height from the pseudostem base to the insertion of the first expanded leaf, and the stem diameter measured with an analogous calibrator (Mitutoyo®, Kawasaki, Japan) at 5 cm height from the stem base, were recorded weekly. The measurements of chlorophyll index and growth variables were carried out weekly on three representative plants per block (nine plants per treatment, n = 9).

2.7. Molecular Detection of Foc in Plant Tissues

Fragments of rhizome tissue from both symptomatic and non-symptomatic plants were taken to confirm the presence of the pathogen in the inner tissue through a standardized procedure for the molecular detection of Foc R1 [59]. Fragments of tissue (nearly 5 mm^2) were stored in sterile polypropylene tubes at −20 °C, lyophilized for 48 h, and then macerated in liquid nitrogen and the DNA extraction was performed using the protocol proposed by Daire et al. [60]. The DNA quality was verified by agarose gel electrophoresis and a quantitative estimation of DNA was made using the NanoDrop spectrophotometer (NanoDrop 2000, Thermo Scientific, Waltham, MA, USA). To determine the quality of the samples for PCR, the amplification of a gene that amplifies specific repetitive elements of musaceae (BrepF: 5′ GATTTTGTAGATTTTGGACACCG 3′ BrepR: 5′ GAATAACAAATATGCTCCAATACCC 3′) was performed following the methods described by Mansoor et al. [61]. The DNA extracted from Foc R1 pure culture and banana plant tissue inoculated with Foc R1 was used as a positive control, and DNA extracted from Kikuyu grass was considered a negative control.

The molecular marker SIX6b_210 (SIX6b_210_F: 5′ ACGCTTCCCAATACCGTCTGT 3′, SIX6b_210_R: 5′ AAGTTGGTGAGTATCAATGC 3′) was amplified by conventional PCR [59] on DNA samples extracted from plant tissue for the detection of Foc R1 in the rhizome samples. DNA extracted from the pure culture of Foc R1 and DNA extracted from the rhizome tissue of banana plants inoculated with Foc R1 were used as positive controls, and DNA from the pure culture of Foc R4T was included as a negative control. Amplification was performed under the following conditions: initial denaturation at 95 °C for 3 min, followed by 32 denaturation cycles at 95 °C for 15 s, banding at 56 °C for 30 s, an extension phase at 72 °C for 60 s, and a final extension cycle of 5 min at 72 °C. The PCR products were verified by agarose gel electrophoresis (1.4%), and the band size was determined with 1 Kb Plus DNA Ladder® (Thermo Fisher Scientific, Waltham, MA, USA).

2.8. In Vitro Antagonism Test of Biopesticides against Foc Race 1

Dual confrontation tests were carried out on PDA medium (90 mm Petri dishes) between the antagonists, i.e., *Trichoderma* and *Bacillus*-based biofungicides and Foc R1. A suspension of each biopesticide was prepared in Tween 80 sterile solution (0.1% v/v) using the concentrations mentioned above for the greenhouse experiment and was vortexed at maximum speed for 2 min. A total of 10 µL suspension of Foc R1 IB (1×10^5 microconidia·mL^{-1}) harvested from a 7-day old solid culture, was dropped in the center of the PDA surface. Once the Foc drop dried, 10 µL suspensions of biopesticides were inoculated at two opposite ends 1 cm from the edge of the Petri dish, perpendicular to the Foc R1 inoculum. After the droplets dried, the dishes were incubated for 7 days under darkness and 25 °C.

Five Petri dishes were inoculated per treatment, and the dishes inoculated only with Foc R1 were the control group. A completely randomized design was used to arrange the plates in the incubator. The diameter of the Foc R1 colony was measured after a week of incubation, and the inhibition of Foc growth was calculated in comparison with the control.

2.9. Population of Microorganisms in the Soil

Rhizospheric-soil samples were taken one week after each application of treatments. to determine the viable population of the Foc R1 IB and the *Trichoderma* and *Bacillus* applied as bioproducts. A composite soil sample was collected from all pots for each treatment in 50 mL Falcon® tubes and was stored at 4 °C until the time of sample processing. The sample was homogenized, and 5 g subsamples were placed in tubes with 45 mL of Tween 80 sterile solution (0.1% v/v), these were vortexed at maximum speed for 2 min, then serial dilutions in sterile Tween 80 were plated on Komada-agar medium for counting *Fusarium* colonies and on PDA supplemented with Triton X-100 (0.1% v/v) and chloramphenicol (100 mg L^{-1}) for counting total fungi and *Trichoderma* spp. colonies. Lastly, to count similar colonies of *Bacillus* spp. on nutrient agar medium, the soil suspension sample was thermally shocked in a water bath at 90 °C for 15 min before being plated.

2.10. Data Analysis

Incidence data were submitted to a t-test for equality of variances between pairs of experiments, and, after the verification, the data from the three replicates of the experiment were pooled. Then, normality (Shapiro–Wilk, $\alpha = 0.05$) and homoscedasticity tests (Bartlett, $\alpha = 0.05$) were performed to determine the final incidence, AUDPC, and efficacy in reduction of AUDPC data before being submitted to ANOVA; here, each replicate of the experiment was considered a blocking factor. The media of the treatments were compared through Tukey's test ($\alpha = 0.05$). The exponential, logistic, and linear growth models were tested to describe the progress of the FW incidence curves using the PROC NLIN procedure of SAS Enterprise Guide v. 8.3, and the growth rate was calculated at the exponential phase of the incidence progress using the PROC REG procedure. The AUDPC of the *DII* from each replicate of the experiment was also submitted to the ANOVA and Tukey's test for mean comparisons.

Data from the physiological and plant growth variables were also submitted to normality and homoscedasticity tests before the ANOVA, as described above. Then, a comparison of the media between treatments was carried out using Fisher's least significant difference (LSD). In some cases of non-normality, the Kruskal–Wallis no parametrical test ($\alpha = 0.05$) was performed using the Statgraphics Centurion XV software.

Data from the antagonism test between bioproducts and Foc R1 were square root-transformed to accomplish the assumptions of normality and homoscedasticity. These transformed data were submitted to an ANOVA test and the media of growth between treatments were compared through a Tukey's test ($\alpha = 0.05$).

Log_{10} data of the microorganism population in the soil were tested for normality and homoscedasticity. ANOVA was performed, and media of the treatments were compared through the Dunnet's test ($\alpha = 0.05$) using the population of Foc R1 as the control.

3. Results

3.1. *Biocontrol of FWB under Greenhouse*

3.1.1. Concentration of the Active Ingredients in Bioproducts

The quality control to verify the concentration of the viable conidia of *Trichoderma* and viable endospores of *Bacillus* in the respective bioproducts showed a higher concentration than indicated on the label of the *Bacillus*-based biofungicides designated as T5, T6, T7, and T8. On the other hand, while a lower concentration of viable conidia was found in the *Trichoderma*-based biofungicides T1 and T2, the concentration of viable conidia in T3 was slightly higher, and that of T4 was similar to what was reported on the label, as well as in T9, in which the concentration corresponded to the concentration reported by the supplier company (Supplementary Figure S1).

3.1.2. Biocontrol Activity of Biofungicides against FWB

As shown in Table 2 and Figure 1, a high variation of the FWB incidence was observed in most of the treatments among the replicates of the entire experiment. However, the verification of equality of variances among the replicates allowed the pooling of the incidence data from the three replications of the experiment. Since the time after transplant in which the disease was recorded was not the same in the three replications of the experiment, the analysis of the incidence was performed in terms of progress per week. The logistic growth model $I(t) = \frac{K}{1+e^{-r(t-t0)}}$, in which *I*(*t*) is the incidence of the disease at time t, *K* represents the maximum incidence that the population can reach, *r* is the growth rate of the incidence, and t_0 represents the time of inflection, fitted well to describe the progress of the incidence for all the treatments (Table 2). Thus, the average incidence from the nine experimental units and the predicted data from the model showed that T2, T3, and T7, highlighted by a less pronounced incidence curve S-shape (Figure 1), had a lower incidence progress rate (17, 21, and 10%, respectively), calculated at the exponential phase of the curves, and a high reduction in the $AUDPC_{INC}$ (50, 37, and 74%, respectively) compared with the negative control (Table 2).

With respect to the analysis of the disease intensity index (*DII*), which considers the incidence as well as the severity of the disease for the calculation, the ANOVA of the first replicate of the experiment showed significantly lower $AUDPC_{DII}$ in banana plants treated with T2 (*T. harzianum*), T4 (*T. harzianum*, *T. koningii*, and *T. viride* consortium), and T7 (*B. amyloliquefaciens*, *Agrobacterium radiobacter*, and *B. pumilus*.consortium), with efficacies of 71%, 48%, and 91% respectively, which represent the percentage of reduction of the FW intensity. In the second replicate, the progress of the *DII* was significantly reduced, only by T7 treatment, by 81%. Additionally, treatments T3 (*T. koningiopsis* Th003), T5 (*B. subtilis*), and T7 significantly reduced the progress of the *DII* in the third replicate by 69%, 59%, and 60%, respectively (Table 3). Although the high variation in the efficacy to reduce the FW intensity index by most biofungicides was also observed among the replications of the experiment, only the T7 treatment showed consistency and effectiveness among the replications (Figure 2, Supplementary Figure S2).

Table 2. Effect of soil treatment with biofungicides on Fusarium wilt incidence in banana plants.

Treatment	Parameters [a]			Statistics [b]			Growth Rate (%) [c]	Final INC (%) [d]	CV [e] INC	AUDPC-INC [f] (% x Weeks)	CV [g] AUDPC	AUDPC Reduction (%) [h]
	K	r	t_0	F	Pr > F	R^2						
T1	89.3	1.28	6.6	264.38	<0.0001	0.999	22.7	88.89	16.35	322.70	24.88	29.90 (4.3) bc
T2	73.7	1.2	7.0	90.13	<0.0001	1.000	17.3	71.11	28.51	243.65	51.32	49.72 (7.0) b
T3	68.2	1.4	6.1	95.88	<0.0001	1.000	21.1	68.89	38.71	292.06	41.31	37.12 (7.0) bc
T4	88.54	1.19	6.82	127.38	<0.0001	1.000	20.8	86.67	19.99	298.89	47.42	37.38 (7.6) bc
T5	85.11	1.23	6.36	190.47	<0.0001	0.997	24.8	84.44	25.88	337.94	35.61	28.22 (8.3) c
T6	93.60	1.23	6.20	300.01	<0.0001	0.999	22.2	95.56	9.23	379.37	27.95	20.77 (6.4) c
T7	64.48	0.77	8.22	37.53	<0.0001	0.991	10.1	53.33	62.19	128.10	77.33	73.90 (6.5) a
T8	93.32	1.30	6.17	301.30	<0.0001	1.000	23.1	91.11	15.95	394.44	21.52	19.47 (5.2) c
T9	89.69	1.09	5.81	216.27	<0.0001	0.999	23.5	88.89	19.84	419.05	34.05	14.57 (5.1) c
T10	98.74	1.30	5.83	556.26	<0.0001	0.999	29.1	97.78	6.82	462.06	20.37	-

[a] Parameters describing the logistic growth model for FW incidence. K is the maximum incidence that the population can reach, r is the growth rate of the incidence per week, and t0 represents the week in which the curve reaches the inflection. [b] Statistics describing the fitting of the logistic growth model of the incidence. F value, significance of the model, and determination coefficient. [c] The growth rate of the incidence curve at the exponential phase calculated based on the lineal model $Y = b \times X + a$. [d] The average incidence of banana plants showing typical symptoms of Fusarium wilt at 10 weeks after transplant (n = 9). [e] Coefficient of variation of the final incidence of Fusarium wilt. [f] The average of the area under the Fusarium wilt incidence progress curve calculated based on the percentage of the incidence per week (n = 9). [g] Coefficient of variation of the area under the FW incidence progress curve. [h] The average reduction in the progress of the FW incidence compared with the negative control (T10) calculated based on the AUDPC value. Value in parenthesis represents the standard error (n = 9). Treatments sharing the same letter are not significantly different according to Tukey's test (α = 0.05).

Table 3. Effect of soil treatment with biofungicides on Fusarium wilt incidence on banana plants.

Treatment	Reduction of AUDPC $_{DII}$ (%) per Replicate *		
	1st	2nd	3rd
T1	18.6 ± 36.2	8.1 ± 23.3	47.3 ± 10.7
T2	70.9 ± 18.8	28.8 ± 22.3	46.3 ± 5.8
T3	30.7 ± 34.2	2.1 ± 11.3	69.4 ± 12.6
T4	47.8 ± 31.4	−5.4 ± 28.5	50.7 ± 17.6
T5	6.5 ± 68.1	21.2 ± 17.0	59.2 ± 11.7
T6	−0.9 ± 14.1	19.6 ± 41.5	−5.2 ± 77.9
T7	90.9 ± 14.6	81.4 ± 14.9	60.0 ± 10.3
T8	−46.1 ± 54.5	15.9 ± 14.2	41.1 ± 25.4
T9	26.9 ± 25.0	−5.2 ± 24.5	1.8 ± 20.1

* 1st, 2nd, and 3rd replication of the entire experiment. Each value represents the average efficacy compared to the negative control ± the standard deviation (n = 3).

3.1.3. Molecular Detection of the Pathogen in Plant Tissue

A total of 316 plant samples from treatments with biofungicides (T1 to T9), the negative control (T10), and the absolute controls (T11, T12, and T13) from the first (172 samples) and second (144 samples) replicates of the entire experiment were analyzed. The extracted DNA from plant material showed good quality for amplification by conventional PCR, furthermore, when the Musaceae constituent gene was amplified, it was observed that 99% of the samples showed a band of 415 bp, corresponding to the expected size with the BrepI marker (Supplementary Figure S3).

A total of 71.7% of samples from Foc R1-inoculated plants (T1 to T10) in replicate 1 showed a band of approximately 210 bp, corresponding to the amplification of the SIX6b marker (specific for the detection of Foc R1). The remaining 28.3% of samples did not amplify the SIX6b gene, despite having been inoculated with Foc R1, so, the pathogen was not detected in these plant material samples. It should be noted that only 20% of samples from T7 amplified the SIX6b gene (Figure 3).

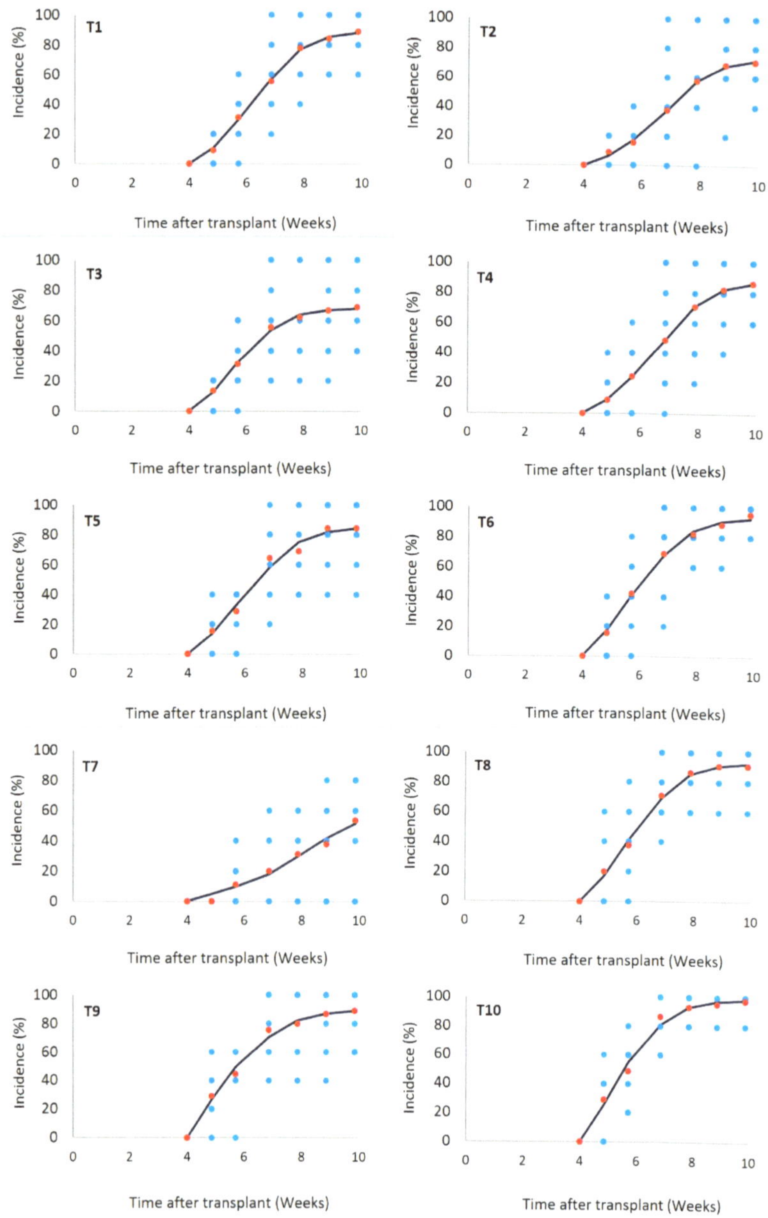

Figure 1. Effect of microbial biofungicides (T1–T9) on Fusarium wilt incidence progress in banana plants. Blue marks represent the original data. Red marks represent the average incidence (n = 9). The black line shows the predicted incidence by the logistic growth model as fitted for each treatment. The parameters of the model are shown in Table 2. T10: negative control.

Figure 2. Effect of microbial biofungicides on Fusarium wilt severity in banana plants. The graphics in the top panel show the effects of biofungicides on the area under the disease intensity index progress curve (AUDPC DII) at the end of each replicate (R1–R3) of the experiment (77, 72, and 68 days after transplant for R1, R2, and R3, respectively). Bars on the columns represent the standard deviation of the mean ($n = 3$). Columns with the same letter are not significantly different according to Tukey's test ($\alpha = 0.05$). Biofungicides: T1–T9. Negative control: T10. The images show the representative state of the plants at the end of the experiment in replicates 2 (R2) and 3 (R3) of the entire experiment.

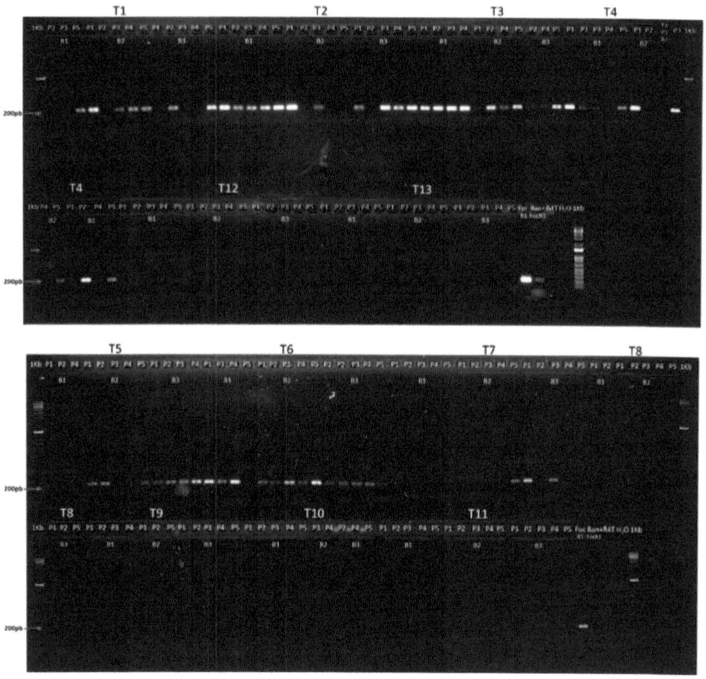

Figure 3. Electrophoresis in agarose gel (1.4%) used for verification of PCR amplification of the marker SIX6b (specific for Foc R1) on DNA extracted from banana plant samples in the first replicate of the experiment.

On the other hand, 75.3% of samples from replicate 2 amplified the specific marker for Foc R1 SIX6b. In this case, as in replicate 1, the treatment T7 presented the highest number of negative samples (46.6%) for the SIX6b marker (Supplementary Figure S4). In the case of plant 4 in treatment 6, two samples were evaluated, one corresponding to a symptomatic mother plant and another to an asymptomatic daughter plant, finding that the latter was negative for the specific marker for Foc R1.

At the end of the first replicate of the experiment, no healthy plants were found in treatments T4, T8, and T10 (negative control) (Supplementary Table S2). On the other hand, 80% of plants in the T7 treatment were apparently healthy, and the molecular analysis carried out on these samples did not detect the pathogen. In T1, T3, and T6, only 6.7% of plants did not show symptoms, and the pathogen was detected in all samples only from T6. The pathogen was also not detected in the healthy plants from T3 and T5. Half of the apparently healthy plants in T9 tested positive for Foc R1 detection by amplifying the marker for the SIX6b gene, so, these plants were asymptomatic even when the pathogen had colonized the inner tissue of the rhizome (Supplementary Table S2).

3.1.4. Effect of Treatments on Physiological and Plant Growth Variables

Plant Physiology Response

Although four measurements of the variables were recorded during the trial period (around 70 days), one every 20 days, the description of the results obtained at 49 days after transplant are shown here, since the statistical differences were wider, preserving the trend observed over time.

The results of photosynthesis-related variables varied among the replications of the experiment, mainly in plants under the effect of *Trichoderma*-based biofungicides (Table 4). For instance, plants treated with the *Trichoderma*-based biofungicides T2 and T9 showed the highest values of CO_2-Net Assimilation Rate (13.94 and 15.63 µmol m^{-2} s^{-1}, respectively) in the first replicate of the experiment, but only T9 was statistically different to T10 [(negative control) 5.72 µmol m^{-2} s^{-1}]. In the second replicate, the response trend of T2 was preserved, reaching the highest values of photosynthesis (8.30 µmol m^{-2} s^{-1}, on average), which was significantly different from the negative control (2.01 µmol m^{-2} s^{-1}). In contrast, in the third replication of the experiment, plants under T2 and T9 treatments showed values of 1.00 and 1.51 µmol m^{-2} s^{-1}, respectively which were significantly lower than the values recorded in the negative control T10 (10.69 µmol m^{-2} s^{-1}) (Table 4).

Table 4. Effect of biofungicides on physiological responses in banana plants from the three replicates of the entire experiment (R1 to R3).

	Net Assimilation Rate (A) (µmol m^{-2} s^{-1})			**Transpiration Rate (E) (mol m^{-2} s^{-1})**			**PSII Photochemical Yield Y(II)**			**Quantum Efficiency of PSII (Fv/Fm)**		
	49 Days after Transplant											
	R1	**R2**	**R3**	**R1**	**R2**	**R3**	**R1**	**R2**	**R3**	**R1**	**R2**	**R3**
T1	9.74 abc	5.54 abc	4.95 abcde	0.004 bcd	0.000 a	0.002 a	0.83 a	0.13 ab	0.08 a	0.56 a	0.80 a	0.55 a
T2	13.94 ab	8.30 ab	1.00 e	0.005 abc	0.001 a	0.001 a	0.14 a	0.11 ab	0.06 ab	0.83 a	0.77 a	0.54 a
T3	8.39 abc	0.00 c	3.12 bcde	0.003 bcd	0.000 a	0.001 a	0.10 a	0.00 c	0.08 a	0.81 a	0.00 b	0.55 a
T4	10.42 abc	5.58 abc	2.75 cde	0.004 bcd	0.000 a	0.001 a	0.14 a	0.07 bc	0.13 a	0.84 a	0.53 a	0.83 a
T5	11.03 abc	4.08 abc	9.96 abc	0.005 abcd	0.001 a	0.003 a	0.13 a	0.18 a	0.13 a	0.83 a	0.79 a	0.82 a
T6	6.69 bc	4.67 abc	0.00 e	0.002 cd	0.000 a	0.000 a	0.13 a	0.15 ab	0.00 b	0.75 a	0.81 a	0.00 b
T7	13.43 abc	10.47 a	10.45 abc	0.006 abc	0.000 a	0.005 a	0.14 a	0.15 ab	0.12 a	0.83 a	0.81 a	0.80 a
T8	4.89 c	4.96 abc	8.71 abcd	0.001 d	0.000 a	0.004 a	0.08 a	0.12 ab	0.12 a	0.55 a	0.80 a	0.82 a
T9	15.63 a	5.51 abc	1.51 de	0.007 ab	0.000 a	0.001 a	0.15 a	0.11 ab	0.08 a	0.83 a	0.55 a	0.51 a
T10	5.72 bc	2.01 bc	10.69 ab	0.001 d	0.000 a	0.004 a	0.09 a	0.07 bc	0.12 a	0.78 a	0.52 a	0.82 a
T12	16.38 a	8.29 ab	11.05 a	0.008 a	0.000 a	0.005 a	0.14 a	0.15 ab	0.13 a	0.84 a	0.84 a	0.82 a
	70 days after transplant											
	R1	R2	R3	R1	R2	R3	R1	R2	R3	R1	R2	R3
T1	4.43 a	0.00 b	0.00 c	0.001 a	0.000 b	0.001 a	0.05 ab	0.00 c	0.03 cd	0.56 ab	0.00 c	0.25 bc
T2	3.78 a	3.59 ab	2.32 ab	0.000 a	0.001 ab	0.001 a	0.05 ab	0.04 abc	0.09 abc	0.50 ab	0.27 bc	0.54 ab
T3	5.37 a	0.00 b	0.33 bc	0.001 a	0.000 b	0.001 a	0.04 ab	0.00 c	0.07 bcd	0.53 ab	0.00 c	0.45 abc
T4	3.19 a	2.63 ab	0.83 bc	0.000 a	0.001 b	0.001 a	0.04 ab	0.07 abc	0.11 abc	0.48 ab	0.38 abc	0.77 a
T5	4.40 a	0.00 b	1.43 bc	0.001 a	0.000 b	0.001 a	0.04 ab	0.00 c	0.12 ab	0.51 ab	0.00 c	0.78 a
T6	0.15 a	0.32 b	0.00 c	0.000 a	0.000 b	0.000 a	0.01 ab	0.02 bc	0.00 d	0.27 ab	0.24 bc	0.00 c
T7	0.72 a	7.56 a	0.69 bc	0.000 a	0.003 a	0.002 a	0.07 a	0.11 a	0.08 abcd	0.57 ab	0.79 a	0.53 ab
T8	0.00 a	3.89 ab	0.61 bc	0.000 a	0.002 ab	0.001 a	0.00 b	0.08 abc	0.14 ab	0.00 b	0.35 abc	0.80 a
T9	6.07 a	3.05 ab	2.22 abc	0.001 a	0.001 ab	0.001 a	0.04 ab	0.06 abc	0.00 d	0.56 ab	0.51 ab	0.00 c
T10	0.00 a	0.00 b	0.50 bc	0.000 a	0.000 b	0.001 a	0.00 b	0.00 c	0.07 bcd	0.00 b	0.00 c	0.51 abc
T12	4.06 a	4.58 ab	3.95 a	0.000 a	0.001 ab	0.001 a	0.07 a	0.11 ab	0.16 a	0.85 a	0.81 a	0.80 a

Each value represents the average from three sampled plants ($n = 3$). Treatments sharing the same letter are not significantly different according to Fisher's minimum significant difference (LSD, $\alpha = 0.05$).

On the other hand, the response of the photosynthesis-related variables in plants treated with the *Bacillus*-based biofungicides T5 to T8 was consistent among the three replicates of the experiment and allowed us to determine that T7 showed promising results in the maintenance of the functioning of the photosynthetic apparatus in banana plants growing in the presence of Foc R1. Thus, the net assimilation rate values in plants treated with T7 were close, 13.43, 10.47, and 10.45 µmol m^{-2} s^{-1} in the first, second, and third replicate, respectively, and no significant differences were found compared with the absolute control T12 (plants free of both Foc R1 inoculum and biocontrol treatments) (Table 4). Plants treated with the *Bacillus*-based biofungicide T5 also showed high values of net assimilation rate, 11.03 and 9.96 µmol m^{-2} s^{-1} in R1 and R3, respectively (Table 4).

The response of plant transpiration was reciprocal to photosynthesis; high variation among *Trichoderma*-based treatments was also observed (Table 4). Thus, while the T2 and T9 treatments showed the highest transpiration values (0.005 and 0.007 mol m^{-2} s^{-1}, respectively) in the first replicate of the experiment, low transpiration values were shown in the third replicate (0.0005 mol m^{-2} s^{-1} for both treatments), reciprocally with the decreasing of its photosynthetic activity. With respect to the second replicate, the transpiration values were like in the third replicate, reporting low transpiration in plants treated with *Trichoderma*, especially in T9. The highest transpiration rate was observed in plants free of both Foc R1 and biofungicides (T12) and was reciprocally associated with the highest values of photosynthetic rate (Table 4).

Plants under treatment T7 showed transpiration values between 0.0003 and 0.006 mol m^{-2} s^{-1} with no significant differences compared with the absolute control (T12), likely due to a lower rate of stomatal closure in these plants and, likewise, better gas exchange and easier transportation of water throughout the xylem.

Similar to the results of photosynthesis, plants treated with *Bacillus*-based T5 treatment showed high consistency among the replicates and reached transpiration values of 0.005, 0.001, and 0.003 mol m^{-2} s^{-1} in R1, R2 and R3, respectively, close to those values recorded in plants under the T7 treatment, and no significant differences were detected compared with the absolute control (T12), with values of 0.008, 0.00002, and 0.0051 mol m^{-2} s^{-1} in replicates 1, 2, and 3, respectively (Table 4**).**

At 49 days, the photochemical yield response of the PSII (Y(II)) was not significantly different among treatments in general, with the exception of T5, which showed a response that was significantly higher than the response in the negative control (T10), and T3, which showed the lowest value in the second replicate of the experiment (Table 4). The maximum quantum efficiency of photosystem II was also similar among the treatments in all three replicates of the experiment, with the exception of the response in T3 and T6 in the second and third replicates, respectively, with the significantly lowest values (Table 4).

Although no significant differences were found between most of the biocontrol treatments and the negative control (T10) for the electron transfer rate (ETR), the T4, T5, and T7 treatments showed the highest values (Supplementary Figure S5).

The chlorophyll indices were stable until 42 days, after which values began to fall in treatments T8 and T10 (first replicate); T1, T9, and T10 (second replicate); and T1, T6, and T9, (third replicate) (Figure 4). Meanwhile, T7, T2, and T9 treatments (first replicate); T7 and T2 (second replicate), and T5 and T3 (third replicate) showed stable values of chlorophyll indices until the end of the experiment, which demonstrates the buffering capacity of chlorophyll degradation associated with affectations caused by the pathogen on the banana plant.

Figure 4 supports what was described above in terms of SPAD index or chlorophyll reduction at 49 days. For instance, the lowest affectations in chlorophyll content were shown by the T2, T7, and T9 treatments, with 20.0, 12.4, and 26.1%, respectively. In contrast, the chlorophyll content was reduced by 75.4% in plants from the negative control (T10) of the first replicate of the experiment, and considerably in the second and the third replicates. The T7 treatment showed the lowest rate of chlorophyll degradation (16.5%) in the second replicate of the experiment. The T5 treatment mitigated chlorophyll damage in the third replication, with a negative value of SPAD unit reduction (−7.2%).

Spearman ordinal correlation analysis was performed to determine whether the behavior of the physiological variables is explained to a significant extent by the disease progression. Although the correlation adjustments were not highly significant, it was noteworthy that increasing disease in plants tends to correlate negatively with values of the chlorophyll fluorescence-related variables (ETR, Fv/Fm, and Y(II)) (Supplementary Table S3).

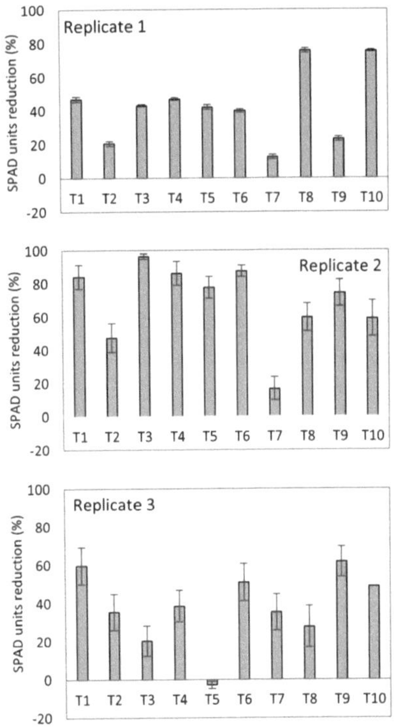

Figure 4. Chlorophyll index reduction in banana plants at 70 days after transplantation in soil artificially inoculated with Foc R1 and treated with biofungicides. Bars on the columns represent the standard deviation of the mean ($n = 3$).

Plant Growth Response

It was highlighted that plant height in the T2, T5, and T7 treatments was consistently less affected along the three replicates of the experiment as compared with the other treatments and with the negative control at 70 days (Figure 5). The plants from the negative control showed a decrease in the pseudostem diameter at 49 days, evidencing the dehydration caused by the disease (Supplementary Table S3). The lowest percentages of diameter reduction were observed in plants under the T2 and T7 treatments, as compared with the negative control (Figure 6), as is consistent with the other variables measured in this study. Thus, this suggests a protective effect by the microorganisms in biofungicides T2 and T7 on the physiological mechanisms of the banana plants facing the presence of Foc R1.

3.1.5. In Vitro Antagonism against Foc Race 1

All the treatments significantly reduced the diametric growth of the Foc R1 colony. However, a higher inhibition was observed for the treatments containing *Trichoderma* as the active ingredient (T1–T4 and T9), with inhibitions from 78.2 to 85.1%, with the treatments T3, T4, and T9 being those with the highest effectiveness in the group of *Trichoderma*-based bioproducts. On the other hand, the highest amount of inhibition by the bacterial-based products was shown by T5 and T8. The lowest inhibition amount was observed with T6 at 48.5% (Figure 7, Supplementary Figure S6, Table S4).

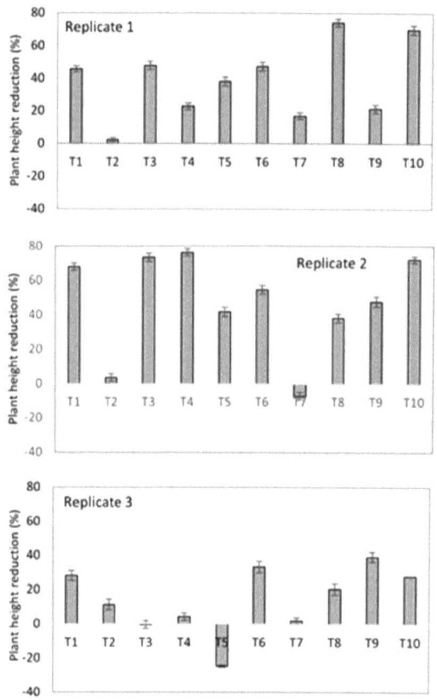

Figure 5. Plant height reduction of banana plants grown in soil artificially inoculated with Foc R1 at 70 days after transplant. Bars on the columns represent the standard deviation of the mean (n = 3).

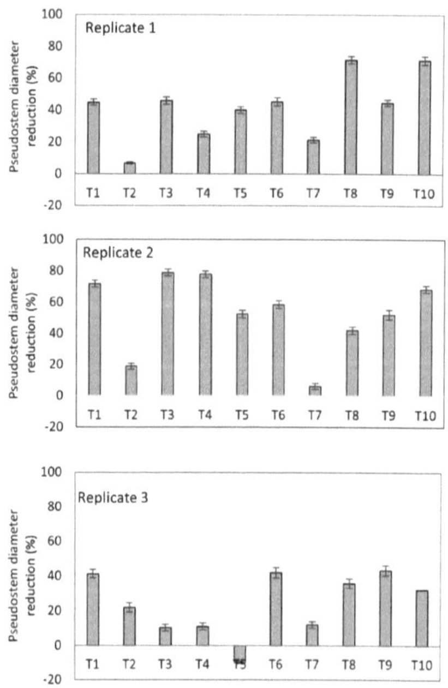

Figure 6. Pseudostem diameter reduction of banana plants at 70 days after transplantation in soil artificially inoculated with Foc R1. Bars on the columns represent the standard deviation of the mean (n = 3).

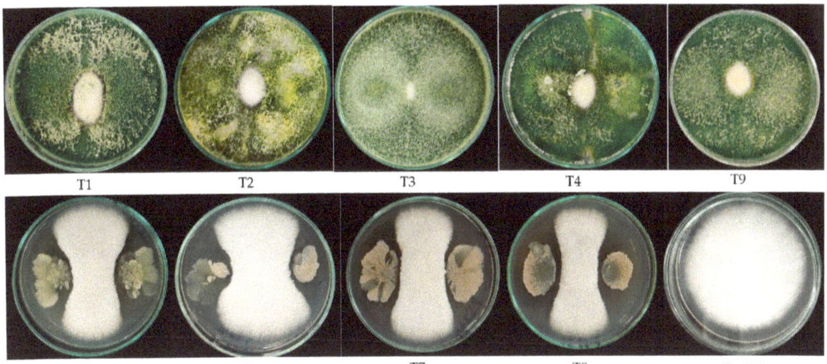

Figure 7. In vitro antagonism test of the microbial fungicides against Foc R1. The inhibition of Foc growth was calculated on the basis of the diameter of the colony measured one week after incubation at 25 °C.

4. Discussion

Biological control and the study of resistance in the host have been the most relevant strategies for the management of Foc in banana plants, with endophytic microorganisms in general, and strains from *Trichoderma*, *Pseudomonas*, and *Bacillus* genera in particular, as the most studied biocontrol agents [30].

The reduction in Foc growth by antagonists has been studied through in vitro dual confrontation tests and has been focused mainly on strains of different *Trichoderma* species. Evaluated strains of *T. viride*, *T. harzianum*, and *T. hamatum* inhibited Foc by from 80 to 84% [62]. A similar value of antagonism was found in the present work with the formulation based on *T. koningiopsis* Th003, which showed 85% inhibition, the other biofungicides reduced Foc growth from 78 to 82%. It was observed that strains of *Trichoderma* grew faster than Foc, suggesting an inhibitory effect through competition for space. On the other hand, an inhibition halo was observed in the interaction zone between Foc and *Bacillus* strains, suggesting the displaying of an antibiosis effect against Foc.

With respect to the effect of bioproduct application on plant physiology, the gas exchange variables allow us to analyze the real-time state of the plant related to the net assimilation of carbon (A) and transpiration (E), which can be affected by Foc R1 colonization in the roots and the subsequent vascular bundle occlusion since they limit the water and nutrients intake capacity stimulating a drought condition, so the plant responds early to the stomata closing [63]. In this work, this phenomenon was observed in the initial phases of the disease infection, without being reflected in damage to the chloroplasts, as was described by Dong et al. [64].

With the disease progression, the plants under treatment (T10 and T8) with the highest level of disease showed stomatal closure and, therefore, a lower rate of CO_2 (A) assimilation. However, Fv/Fm values were not as variable between treatments as expected. In this sense, authors, such as Lawson et al. [65] and Tambussi et al. [66], demonstrated that the antenna-PSII complex integrity is not significantly affected early during water stress. The chlorophyll fluorescence-related variables are a good indicator of the plant stress responses since they measure the state of photosystem II, which, in the case of stress, dissipates excess light energy in different ways (fluorescence or heat) instead of harnessing it for photosynthesis [67]. The effective quantum yield of PSII (Y(II)) shows the probability of a photon being absorbed to drive the photochemical process. This process depends on the photosystem state and the leaf temperature [68].

In general, it is assumed that the normal Fv/Fm value is around 0.82, which indicates an optimal condition in the plant, and the values falling below 0.60 would indicate a loss of photosynthetic function [69]. At the end of the three replicates of the experiment, treatments that could maintain values above 0.70, despite infection damages, showed a stable trend in their response to Foc R1 infection, indicating that biofungicides may be associated with physiological condition maintenance and avoiding the oxidative damage in thylakoidal membranes, which results in photosynthetic efficiency decreasing, as mentioned by Cheng et al. [70].

The reductions in variables such as ETR indicate that the electrons captured during the photosynthesis light phase cease to be destined towards the dark phase, and, as a consequence, they must be dissipated in the form of heat, fluorescence, or by the photorespiratory response of severe stress in plants caused by pathogen presence [65]. In this work, T4, T5, and T7 maintained higher values of ETR indicating stability on thylakoidal electron flow conservation, unlike other treatments, where data distribution allowed us to observe low values, indicating a decrease in the ETR as an indicator of photosystem II thylakoid damage [71].

At this point, the non-stomatal limitations could be related to the production of reactive oxygen species (ROS) and the oxidative cascades that lead to chlorophyll degradation and chloroplast component damage [72,73], which is consistent with the results of the chlorophyll index variable (SPAD units) in the present study, which was significantly lower in plants with severe disease symptoms. The chlorophyll index (SPAD) measures the greenness of the leaves, allowing a correlation to be made between SPAD units and chlorophyll concentration. Accordingly, the damage caused by Foc can be related to chlorophyll degradation as a consequence of severe damage in the photosystems caused by mycotoxins, such as fusaric acid, or oxidative stress [74,75].

In nature, there are species of fungi and bacteria capable of colonizing plant tissues and inhibiting the affectation caused by plant pathogens; some can be higher than the pathogens, but others fail due to their low survival rate in the soil [76]. In this study, we evaluated nine treatments of bioproducts as biological control agents with active ingredients, five based on *Trichoderma* and four based on *Bacillus*.

As a consequence of physiological parameter alterations, plant growth can be severely affected, depending on damage degree and extent. The analysis of plant growth parameters, such as height and pseudostem diameter, showed differences among treatments, which were also associated with disease evolution and the inhibition caused according to the evaluated treatments. Reduction percentages lower than 10% and 20% for treatments T2 (*T. harzianum*) and T7 (consortium of *B. amyloliquefaciens*, *A. radiobacter*, and *B. pumilus*), respectively, were observed with respect to plant height, as well as a smaller reduction in pseudostem diameter in treatments T4 (*T. harzianum, T. koningii, T. viride*) and T7.

Different studies have shown that *Bacillus* species act as disease suppressors, with antifungal characteristics and plant growth promotion traits [58]. The strain of *Bacillus licheniformis* CSR-D4 significantly reduced the incidence of wilt disease caused by *F. oxysporum* f. sp. *cubense* Tropical Race 4 (Foc TR4) in banana plants evaluated under in vitro and in vivo conditions. Bioactive metabolites, such as iturin C, bacillomycin, and fengycin, produced by CSR-D4 inhibited the invasion of the pathogen, proving promising for use as an effective biocontrol agent in the management of Foc TR4 [12].

In the case of fungi, strains of *T. harzianum* have been studied to control *Fusarium* wilt in young banana plants treated with different *Glomus* species [77]. Growth parameters such as plant height and pseudostem diameter showed a significant increase when individual applications of *T. harzianum* and *Glomus* were made as compared with the control. The plant growth-promoting effect exerted by the species of *Trichoderma* is attributed to the production of some phytohormones, such as auxins, phosphorus solubilizing enzymes, and siderophores [78]. In plants inoculated with Foc R4T, the species of *Glomus* showed a positive effect related to seedling growth, and, in turn, healthy seedlings were obtained at the end of the trial. Combined treatments of *T. harzianum* and *Glomus* spp. resulted in a delay in the progression of the disease. However, a phenomenon of competition occurred when the application of the two biocontrol agents was carried out simultaneously.

Bubici et al. [30] reported efficacies in different studies on Foc bananas in field conditions with *Pseudomonas* (79%) and *Trichoderma* spp. (70%), and lower efficacies for *Bacillus* sp.

(69%), mycorrhizae, and nonpathogenic strains of *Fusarium* (42–55%) finding that efficacies between pot experiments and field experiments showed similar efficacies on average (pot 65% and field 70%). However, Thangavelu and Gopi [53] reported isolates of *Trichoderma* sp. that decreased the disease and increased the growth parameters compared with non-treated plants inoculated with Foc R1, and the effectiveness was increased up to 100% when two isolates of *Trichoderma* sp. and *T. asperellum* were combined, and the growth parameters were increased up to 250%. In the present work, we found the highest effectiveness by the treatment of the consortium of two strains of *Bacillus* spp. and *A. radiobacter*, with efficacy values up to 74%, agreeing with the reports in other studies. However, in the case of *Trichoderma*-based bioproducts the highest efficacy was 44.3%, this is lower than the previous reports for *Trichoderma* spp. Considering these results, it is important to continue with studies to evaluate doses, volumes, and frequencies of application for promising biofungicides and the search for other microorganisms that could increase the effectiveness, as well as the construction, of consortia of microorganisms with different modes of action.

There has been evidence for more than 30 years that *A. radiobacter* is an effective biological control agent with the strain K84. Coincidentally, this is one of the components of the active ingredient of the most promissory treatment in this research. K84 is classified as non-pathogenic for its lack of capacity to form galls, and its mode of action against *Agrobacterium tumefaciens* relies on the production of one highly specific anti-agrobacterial antibiotic, Agrocin 84, and also ALS 84, an antibiotic type of compound related to siderophore production [79,80]. Although its mechanism of action against other types of phytopathogens is still unclear, it has been reported to control *Erwinia* sp. and *Pseudomonas* sp. by the action of antibiotic compounds [81]. Additionally, there are reports of old data that already described the inhibition of Foc growth by *A. radiobacter* under in vitro conditions [82].

Some of the microorganisms evaluated and reported as biological control agents with promising results for the control of wilting caused by Foc include genera of fungi as well as bacteria that have shown indirect modes of action, like induction of systemic resistance in plants, production of fungitoxic metabolites, stimulation from competition for essential nutrients [47], as well as direct modes of action, such as antibiosis, mycoparasitism, and production of cell wall degrading enzymes (CWDEs) [83]. Studying and identifying the mechanisms triggered by the interaction between the biological control agents and the plant pathogens could be a step toward creating the auspicious conditions for their success or improving the strategies of biological control [84].

In the present study, we found that the biofungicides that showed higher inhibition of the Foc vegetative growth were T3, T9, T4, and T8, and that a reduction in the viable population of the pathogen in the soil was exerted by T3, T1, T4, and T9. These conditions, where the direct mechanisms of action could be displayed against the pathogen, were not the best treatments for reducing the disease in the in planta experiment, where T7 and T2 were the most effective treatments for controlling Fusarium wilt. Additionally, T7 biofungicide prevented the inner colonization of plant tissues by the pathogen in the 80% of plants inoculated with Foc, suggesting that the most effective control of the disease could be associated with other mechanisms different than direct ones, such as induction of resistance in the host.

Akila et al. [52] found a correlation between the decrease in the incidence of the disease caused by Foc R1 and the induction of the defense-related enzymes peroxidase (PO) and polyphenol oxidase (PPO) when combining botanical extracts and BCA *P. fluorescens* and *B. subtilis*. Li et al. [55] reported that the application of an alkaline fertilizer and biocontrol fungi had a synergistic effect in the reduction of disease related to the peroxidase (PO), catalase, and superoxide dismutase (SOD) activity associated with the antioxidant system of the plant. Lin et al. [85] found that when the plant defense system is activated by BCA it can improve its ability to resist the colonization of the pathogen, similar to the observed effect of the T7 treatment in the present study. However, these hypotheses related to the modes of action of T7 members deserve further studies to elucidate biocontrol performance, since it was the most promising treatment against Foc R1 in this study. In addition, currently, we are validating the results obtained in the present study with the Fusarium wilt caused by Foc R4T under biosecurity conditions.

5. Conclusions

In this study, the treatments that showed high efficacy in the inhibition of diametral growth and the reduction of the viable population of the pathogen in the soil were not the best at reducing the disease under the in planta assays, suggesting that the control of the disease was related to indirect mechanisms of action, like induction of systemic resistance in the plant. Although available bioproducts in the market were identified with the potential to reduce the disease and to mitigate the biotic stress in the plant by the presence of Foc R1, the effect of different cost-effective doses and the frequency of applications of biocontrol agents deserve further research. Continuing with the search for more efficient microorganisms and the evaluation of microbial consortia with different modes of action that could improve the control of the disease, and characterizing the mechanisms of action by the most effective treatments, are also important issues to be studied, with the aim of understanding the interactions between the host, the pathogen, and the biocontrol agent governing the effectiveness against Foc infections.

Supplementary Materials: The following supporting information can be downloaded at the following location: https://www.mdpi.com/article/10.3390/jof10060419/s1, Table S1: Severity scale of the Fusarium wilt of banana; Figure S1: Colony Forming Units (CFU) of the active ingredient per gram or per milliliter of each bioproduct according to the quality control compared to the information in the label; Figure S2: Progress curves of the disease intensity index (DII) for 72 to 77 days in each replicate of the entire experiment; Figure S3: Verification by electrophoresis in agarose gel (1.4%) of PCR amplification of the internal control BrepI (specific for repetitive elements of musaceous) on plant DNA from banana plants (replicate 2); Figure S4: Verification by electrophoresis in agarose gel (1.4%) of PCR amplification of the marker SIX6b (specific for Foc R1) on plant DNA from banana plants (replicate 3); Table S2: Detection of Foc R1 in plant material from the replicate 1 of the in vivo experiment; Figure S5: Electron transport rate (ETR) recorded in banana plants; Table S3: Spearman's correlation between disease severity index (SI) and physiological and growth variables Figure S6: Growth of Foc race 1 colony on PDA medium in presence of biopesticide inoculum; Table S4: Inhibition of diametral growth of the colony of Foc R1.

Author Contributions: Conceptualization, M.B.-V., G.A.R.-Y., L.F.I.-G. and C.A.M.-V.; methodology, L.F.I.-G. and C.A.M.-V.; formal analysis, C.A.M.-V., A.d.P.V.-N. and S.L.C.-G.; investigation, L.F.I.-G., S.L.C.-G., C.A.M.-V., A.d.P.V.-N., R.Y.Q.-M., D.M.B.-D. and M.R.G.-M.; writing—original draft preparation, L.F.I.-G. and C.A.M.-V.; writing—review and editing, C.A.M.-V.; visualization, L.F.I.-G. and C.A.M.-V.; supervision, M.B.-V.; project administration, L.F.I.-G.; funding acquisition, M.B.-V. All authors have read and agreed to the published version of the manuscript.

Funding: This research was funded by the earmarked fund to Agrosavia from the National Plant Protection Organization (ICA—Instituto Colombiano Agropecuario) (Agreement 10 (2021) and 11 (2022), derived from 021, 2018), and the earmarked fund to Agrosavia and Fontagro: ATN/RF-18761-RG (1002280): Strengthening capacities for the prevention and management of Fusarium wilt in Latin America and the Caribbean, and the APC was funded by Derived agreement No. 1 Lines of epidemiology and models for pest mitigation and diagnosis of the inter-administrative framework agreement GGC-017–2023 signed between ICA and AGROSAVIA.

Institutional Review Board Statement: Not applicable.

Informed Consent Statement: Not applicable.

Data Availability Statement: The datasets generated during and/or analyzed during the current study are available from the corresponding author upon reasonable request.

Acknowledgments: This research was carried out with the collaboration of Fontagro, BID Invest, and Agrovid S.A.S. Special thanks are given to the Colombian Agricultural Institute (ICA) for the technical support and sanitary surveillance of the farms under quarantine and financial support, to the Colombian Corporation for Agricultural Research (Agrosavia) for the technical and financial support, and to AGROVID group for the collaboration during the visits of the researchers, samplings carried out in field, and the provision of field information.

Conflicts of Interest: The authors declare that they have no conflicts of interest with third parties or with Agrosavia. The funders had no role in the design of the study; in the collection, analyses, or interpretation of data; in the writing of the manuscript; or in the decision to publish the results.

References

1. Voora, V.; Larrea, C.; Bermúdez, S. Global Market Report: Bananas. International Institute for Sustainable Development. 2020. Available online: https://www.iisd.org/publications/report/global-market-report-bananas (accessed on 7 September 2023).
2. Altendorf, S. Banana Fusarium Wilt Tropical Race 4: A mounting threat to global banana markets? In *Food Outlook—Biannual Report on Global Food Markets*; FAO: Rome, Italy, 2019. Available online: https://www.fao.org/3/ca6911en/CA6911EN_TR4EN.pdf (accessed on 20 December 2023).
3. Altendorf, S. Bananas and Major Tropical Fruits in Latin America and the Caribbean: The Significance of the Region to World Supply. In *Food Outlook—Biannual Report on Global Food Markets*; FAO: Rome, Italy, 2019.
4. Dita, M.A.; Garming, H.; Van den Bergh, I.; Staver, C.; Lescot, T. Banana in Latin America and the Caribbean: Current state, challenges and perspectives. *Acta Hortic.* **2013**, *986*, 365–380. [CrossRef]
5. López-Zapata, S.P.; Castaño-Zapata, J. Manejo integrado del mal de Panamá *Fusarium oxysporum* Schlechtend.: Fr. sp. *cubense* (E.F. SM.) W.C. Snyder & H.N. Hansen: Una revisión. *Rev. U.D.C.A. Actual. Divulg. Científica* **2019**, *22*, e1240. [CrossRef]
6. Kema, G.H.; Drenth, A.; Dita, M.; Jansen, K.; Vellema, S.; Stoorvogel, J.J. Fusarium wilt of banana, a recurring threat to global banana production. *Front. Plant Sci.* **2021**, *11*, 628888. [CrossRef] [PubMed]
7. Dita, M.; Barquero, M.; Heck, D.; Mizubuti, E.S.; Staver, C.P. Fusarium wilt of banana: Current knowledge on epidemiology and research needs toward sustainable disease management. *Front. Plant Sci.* **2018**, *9*, 1468. [CrossRef] [PubMed]
8. Pegg, K.G.; Coetes, L.M.; O'Neill, W.T.; Turner, D.W. The epidemiology of Fusarium Wilt of Banana. *Front. Plant Sci.* **2019**, *10*, 1395. [CrossRef] [PubMed]
9. Fones, H.N.; Bebber, D.P.; Chaloner, T.M.; Kay, W.T.; Steinberg, G.; Gurr, S.J. Threats to global food security from emerging fungal and oomycete crop pathogens. *Nat. Food.* **2020**, *1*, 332–342. [CrossRef]
10. García-Bastidas, F.A.; Quintero-Vargas, J.C.; Ayala-Vasquez, M.; Schermer, T.; Seidl, M.F.; Santos-Paiva, M.; Noguera, A.M.; Aguilera-Galvez, C.; Wittenberg, A.; Hofstede, R.; et al. First report of Fusarium wilt Tropical Race 4 in Cavendish bananas caused by *Fusarium odoratissimum* in Colombia. *Plant Dis.* **2019**, *104*, 994. [CrossRef]
11. Rodríguez-Yzquierdo, G.; Olivares, B.O.; Silva-Escobar, O.; González-Ulloa, A.; Soto-Suarez, M.; Betancourt-Vásquez, M. Mapping of the susceptibility of colombian *Musaceae* lands to a deadly disease: *Fusarium oxysporum* f. sp. *cubense* Tropical Race 4. *Horticulturae* **2023**, *9*, 757. [CrossRef]
12. Yadav, K.; Damodaran, T.; Dutt, K.; Singh, A.; Muthukumar, M.; Rajan, S.; Gopal, R.; Sharma, P.C. Effective biocontrol of banana fusarium wilt tropical race 4 by a bacillus rhizobacteria strain with antagonistic secondary metabolites. *Rhizosphere* **2021**, *18*, 100341. [CrossRef]
13. Carmona, S.L.; Villarreal, A.; Burbano-David, D.; Gómez, M.; Torres-Rojas, E.; Soto-Suárez, M. Protection of tomato plants against *Fusarium oxysporum* f. sp. lycopersici induced by chitosan. *Rev. Colomb. Cienc. Hortic.* **2020**, *15*, e12822. [CrossRef]
14. Zakaria, M.A.T.; Sakimin, S.Z.; Ismail, M.R.; Ahmad, K.; Kasim, S.; Baghdadi, A. Biostimulant Activity of Silicate Compounds and Antagonistic Bacteria on Physiological Growth Enhancement and Resistance of Banana to Fusarium Wilt Disease. *Plants* **2023**, *12*, 1124. [CrossRef]
15. Dita, M.; Teixeira, L.A.J.; O'Neill, W.; Pattison, A.B.; Weinert, M.P.; Li, C.Y.; Zheng, S.J.; Staver, C.; Thangavelu, R.; Viljoen, A. Current state of Fusarium wilt of banana in the subtropics. *Acta Hortic.* **2020**, *1272*, 45–56. [CrossRef]
16. Izquierdo-García, L.F.; Carmona, S.L.; Zuluaga, P.; Rodríguez, G.; Dita, M.; Betancourt, M.; Soto-Suárez, M. Efficacy of disinfectants against *Fusarium oxysporum* f. sp. *cubense* Tropical Race 4 Isolated from La Guajira, Colombia. *J. Fungi* **2021**, *7*, 297. [CrossRef] [PubMed]
17. Dale, J.; James, A.; Paul, J.Y.; Khanna, H.; Smith, M.; Peraza-Echeverria, S.; Garcia-Bastidas, F.; Kema, G.; Waterhouse, P.; Mengersen, K.; et al. Transgenic Cavendish bananas with resistance to Fusarium wilt tropical race 4. *Nat. Commun.* **2017**, *8*, 1496. [CrossRef] [PubMed]
18. Mintoff, S.J.L.; Nguyen, T.V.; Kelly, C.; Cullen, S.; Hearnden, M.; Williams, R.; Daniells, J.W.; Tran-Nguyen, L.T.T. Banana cultivar field screening for resistance to *Fusarium oxysporum* f. sp. *cubense* Tropical Race 4 in the Northern territory. *J. Fungi.* **2021**, *7*, 627. [CrossRef] [PubMed]
19. Shen, Z.; Xue, C.; Penton, C.R.; Thomashow, L.S.; Zhang, N.; Wang, B.; Ruan, Y.; Li, R.; Shen, Q. Suppression of banana Panama disease induced by soil microbiome reconstruction through an integrated agricultural strategy. *Soil Biol. Biochem.* **2019**, *128*, 164–174. [CrossRef]
20. Jamil, F.N.; Hashim, A.M.; Yusof, M.T.; Saidi, N.B. Analysis of soil bacterial communities and physicochemical properties associated with Fusarium wilt disease of banana in Malaysia. *Sci. Rep.* **2022**, *12*, 999. [CrossRef] [PubMed]

21. Dror, B.; Amutuhaire, H.; Frenkel, O.; Jurkevitch, E.; Cytryn, E. Identification of bacterial populations and functional mechanisms potentially involved in biochar-facilitated antagonism of the soilborne pathogen *Fusarium oxysporum*. *Phytobiomes J.* **2022**, *6*, 139–150. [CrossRef]
22. Gazolla, C.; Britto, B.; Freitas, J.; Beneduzi, A.; Eichelberger, G.; Kayser, L. Soil-plant-microbiota interaction to enhance plant growth. *Rev. Bras. Cienc. Solo* **2022**, *46*, e0210098. [CrossRef]
23. Zhu, Z.; Wu, G.; Deng, R.; Hu, X.; Tan, H.; Chen, Y.; Tian, Z.; Li, J. Spatiotemporal biocontrol and rhizosphere microbiome analysis of Fusarium wilt of banana. *Commun. Biol.* **2023**, *6*, 27. [CrossRef]
24. Haddad, F.; Rocha, L.S.; Soares, A.C.F.; Martins, I.P.S.; Teixeira, L.A.J.; Staver, C.; Dita, M. Management of Fusarium wilt of bananas in Minas Gerais, Brazil. *Acta Hortic.* **2018**, *1196*, 137–145. [CrossRef]
25. Nowembabazi, A.; Tautaya, G.; Tinzaara, W.; Karamura, E. Effect of integrated potassium nutrition on Fusarium wilt tolerance in apple banana. *Afr. J. Plant Sci.* **2021**, *15*, 257–265. [CrossRef]
26. Yang, J.; Ren, X.; Liu, M.; Fan, P.; Ruan, Y.; Zhao, Y.; Wang, B.; Li, R. Suppressing soil-borne *Fusarium* pathogens of bananas by planting different cultivars of pineapples, with comparisons of the resulting bacterial and fungal communities. *Appl. Soil Ecol.* **2022**, *169*, 104211. [CrossRef]
27. Teixeira, L.; Nomura, E.; Damatto, E.; Vieira, H.; Staver, C.; Dita, M. Effectiveness of soil management practices on Fusarium wilt of banana in the Ribeira Valley, Brazil. *Trop. Plant Pathol.* **2022**, *47*, 411–420. [CrossRef]
28. Fira, D.; Dimkić, I.; Berić, T.; Lozo, J.; Stanković, S. Biological control of plant pathogens by *Bacillus* species. *J. Biotechnol.* **2018**, *285*, 44–55. [CrossRef] [PubMed]
29. Pattison, A.; Limbaga, C.; Gervacio, T.; Notarte, A.; Juruena, M.; Dennis, P.; Lindsay, S. Molina. Integrated Management of Fusarium wilt of Banana in the Philippines and Australia. Final Report. Australian Centre for International Agricultural Research—ACIAR. 2020; 91p. Available online: https://www.aciar.gov.au/publication/integrated-management-fusarium-wilt-bananas-philippines-and-australia-final-report (accessed on 20 November 2023).
30. Bubici, G.; Kaushal, M.; Prigigallo, M.I.; Gómez-Lama, C.; Mercado-Blanco, J. Biological control agents against Fusarium wilt of banana. *Front. Microbiol.* **2019**, *10*, 616. [CrossRef] [PubMed]
31. Thambugala, K.M.; Daranagama, D.A.; Phillips, A.J.L.; Kannangara, S.D.; Promputtha, I. Fungi vs. Fungi in Biocontrol: An Overview of Fungal Antagonists Applied Against Fungal Plant Pathogens. *Front. Cell. Infect. Microbiol.* **2020**, *10*, 604923. [CrossRef] [PubMed]
32. Wang, Q.; Zhou, L.; Jin, H.; Cong, B.; Yang, H.; Wang, S. Investigating the responses of microbial communities to banana Fusarium wilt in suppressive and conductive soils based on soil particle size differentiation. *Agronomy* **2022**, *12*, 229. [CrossRef]
33. Sharma, A.; Gupta, B.; Verma, S.; Pal, J.; Akanksha, M.; Chauhan, P. Unveiling the biocontrol potential of *Trichoderma*. *Eur. J. Plant Pathol.* **2023**, *167*, 569–591. [CrossRef]
34. Tyśkiewicz, R.; Nowak, A.; Ozimek, E.; Jaroszuk-Ściseł, J. *Trichoderma*: The current status of its application in agriculture for the biocontrol of fungal phytopathogens and stimulation of plant growth. *Int. J. Mol. Sci.* **2022**, *23*, 2329. [CrossRef]
35. Pérez-García, A.; Romero, D.; de Vicente, A. Plant protection and growth stimulation by microorganisms: Biotechnological applications of Bacilli in agriculture. *Curr. Opin. Biotech.* **2011**, *22*, 187–193. [CrossRef] [PubMed]
36. Monaci, L.; Quintieri, L.; Caputo, L.; Visconti, A.; Baruzzi, F. Rapid profiling of antimicrobial compounds characterizing *B. subtilis* TR50 cell-free filtrate by high performance liquid chromatography coupled to high-resolution Orbitrap™ mass spectrometry. *Rapid Commun. Mass Spectrom.* **2016**, *30*, 45–53. [CrossRef] [PubMed]
37. Fan, B.; Blom, J.; Klenk, H.-P.; Borriss, R. *Bacillus amyloliquefaciens*, *Bacillus velezensis*, and *Bacillus siamensis* Form an "Operational Group *B. amyloliquefaciens*" within the *B. subtilis* Species Complex. *Front. Microbiol.* **2017**, *8*, 22. [CrossRef] [PubMed]
38. Bais, H.P.; Fall, R.; Vivanco, J.M. Biocontrol of *Bacillus subtilis* against infection of Arabidopsis roots by *Pseudomonas syringae* is facilitated by biofilm formation and surfactin production. *J. Plant Physiol.* **2004**, *134*, 307–319. [CrossRef] [PubMed]
39. Chen, X.H.; Scholz, R.; Borriss, M.; Junge, H.; Mögel, G.; Kunz, S.; Borriss, R. Difficidin and bacilysin produced by plant-associated *Bacillus amyloliquefaciens* are efficient in controlling fire blight disease. *J. Biotechnol.* **2009**, *140*, 38–44. [CrossRef] [PubMed]
40. Cawoy, H.; Debois, D.; Franzil, L.; De Pauw, E.; Thonart, P.; Ongena, M. Lipopeptides as main ingredients for inhibition of fungal phytopathogens by *Bacillus subtilis/amyloliquefaciens*. *Microb. Biotechnol.* **2014**, *8*, 281–295. [CrossRef]
41. Malfanova, N.; Franzil, L.; Lugtenberg, B.; Chebotar, V.; Ongena, M. Cyclic lipopeptide profile of the plant-beneficial endophytic bacterium *Bacillus subtilis* HC8. *Arch. Microbiol.* **2012**, *194*, 893–899. [CrossRef] [PubMed]
42. Moreno-Velandia, C.A.; Ongena, M.; Cotes, A.M. Effects of Fengycins and Iturins on *Fusarium oxysporum* f. sp. *physali* and root colonization by *Bacillus velezensis* Bs006 protect golden berry against vascular wilt. *Phytopathology* **2021**, *111*, 2227–2237. [CrossRef]
43. Cawoy, H.; Mariutto, M.; Henry, G.; Fisher, C.; Vasilyeva, N.; Thonart, P.; Domes, J.; Ongena, M. Plant defense stimulation by natural isolates of *Bacillus* depends on efficient surfactin production. *Mol. Plant Microb. Interact.* **2014**, *27*, 87–100. [CrossRef]
44. Pertot, I.; Puopolo, G.; Hosni, T.; Pedrotti, L.; Jourdan, E.; Ongena, M. Limited impact of abiotic stress on surfactin production in planta and on disease resistance induced by *Bacillus amyloliquefaciens* S499 in tomato and bean. *FEMS Microbiol. Ecol.* **2013**, *86*, 505–519. [CrossRef]
45. Ongena, M.; Jourdan, E.; Adam, A.; Paquot, M.; Brans, A.; Joris, B.; Arpigny, J.-L.; Thonart, P. Surfactin and fengycin lipopeptides of *Bacillus subtilis* as elicitors of induced systemic resistance in plants. *Environ. Microbiol.* **2007**, *9*, 1084–1090. [CrossRef]
46. Chen, K.; Tian, Z.; He, H.; Long, C.; Jiang, F. *Bacillus* species as potential biocontrol agents against citrus diseases. *Biol. Control* **2020**, *151*, 104419. [CrossRef]

47. Khan, N.; Maymon, M.; Hirsch, A.M. Combating Fusarium Infection Using *Bacillus*-Based Antimicrobials. *Microorganisms* **2017**, *5*, 75. [CrossRef]
48. Miljaković, D.; Marinković, J.; Balešević-Tubić, S. The Significance of *Bacillus* spp. in disease suppression and growth promotion of field and vegetable crops. *Microorganisms* **2020**, *8*, 1037. [CrossRef]
49. Ntushelo, K.; Ledwaba, L.K.; Rauwane, M.E.; Adebo, O.A.; Njobeh, P.B. The mode of action of *Bacillus* species against *Fusarium graminearum*, tools for investigation, and future prospects. *Toxins* **2019**, *11*, 606. [CrossRef] [PubMed]
50. Saxena, A.K.; Kumar, M.; Chakdar, C.; Anuroopa, N.; Bagyaraj, D.J. *Bacillus* species in soil as a natural resource for plant health and nutrition. *J. Appl. Microbiol.* **2019**, *6*, 1583–1594. [CrossRef]
51. Shafi, J.; Tian, H.; Ji, M. *Bacillus* species as versatile weapons for plant pathogens: A review. *Biotechnol. Biotec. Equip.* **2017**, *31*, 446–459. [CrossRef]
52. Akila, R.; Rajendran, L.; Harish, S.; Saveetha, K.; Raguchander, T.; Samiyappan, R. Combined application of botanical formulations and biocontrol agents for the management of *Fusarium oxysporum* f. sp. *cubense* (Foc) causing Fusarium wilt in banana. *Biol. Control* **2011**, *57*, 175–183. [CrossRef]
53. Thangavelu, R.; Gopi, M. Combined application of native *Trichoderma* isolates possessing multiple functions for the control of Fusarium wilt disease in banana cv. Grand Naine. *Biocontrol Sci. Technol.* **2015**, *25*, 1147–1164. [CrossRef]
54. Wong, N.B.; Saidi, G.; Vadamalai, C.Y.; Zulperi, D. Effect of bioformulations on the biocontrol efficacy, microbial viability and storage stability of a consortium of biocontrol agents against *Fusarium* wilt of banana. *J. Appl. Microbiol.* **2019**, *127*, 544–555. [CrossRef]
55. Li, Y.; Jiang, S.; Jiang, J.; Gao, C.; Qi, X.; Zhang, L.; Sun, S.; Dai, Y.; Fan, X. Synchronized efficacy and mechanism of alkaline fertilizer and biocontrol fungi for *Fusarium oxysporum* f. sp. *cubense* Tropical Race 4. *J. Fungi* **2022**, *8*, 261. [CrossRef] [PubMed]
56. ICA. 2023. Bioinsumos Registrados a 1 de Diciembre de 2023. Available online: https://www.ica.gov.co/areas/agricola/servicios/fertilizantes-y-bio-insumos-agricolas/listado-de-bioinsumos/2023/6-bd_productos-bioinsumos_19-de-abril-de-2023-1.aspx (accessed on 29 December 2023).
57. Dita, M.; Teixeira, L.; Li, C.; Zheng, S.; O'Neill, W.; Daniels, J.; Pérez-Vicente, L.; Carreel, F.; Roussel, V.; Carlier, J.; et al. *Practical Guidelines for Early Screening and Field Evaluation of Banana against Fusarium wilt, Pseudocercospora Leaf Spots and Drought*; Bioversity International: Montpellier, France, 2021; 83p, Available online: https://hdl.handle.net/10568/111159 (accessed on 20 January 2023).
58. Mateus-Cagua, D.; Rodríguez-Yzquierdo, G. Effect of biostimulants on dry matter accumulation and gas exchange in plantain plants (Musa AAB). *Rev. Colomb. Cienc. Hortícolas* **2019**, *13*, 151–160. [CrossRef]
59. Carvalhais, L.C.; Henderson, J.; Rincon-Florez, V.A.; O'Dwyer, C.; Czislowski, E.; Aitken, E.A.; Drenth, A. Molecular diagnostics of banana Fusarium Wilt targeting secreted-in-xylem genes. *Front. Plant Sci.* **2019**, *10*, 547. [CrossRef] [PubMed]
60. Daire, X.; Clair, D.; Reinert, W.; Boudon-Padieu, E. Detection and differentiation of grapevine yellows phytoplasmas belonging to the elm yellows group and to the stolbur subgroup by PCR amplification of non-ribosomal DNA. *Eur. J. Plant Pathol.* **1997**, *103*, 507–514. [CrossRef]
61. Mansoor, S.; Qazi, J.; Amin, I.; Khatri, A.; Khan, I.A.; Raza, S.; Zafar, Y.; Briddon, R.W. A PCR-based method, with internal control, for the detection of banana bunchy top virus in banana. *Mol. Biotechnol.* **2005**, *30*, 167–169. [CrossRef] [PubMed]
62. Nayak, D.; Mishra, M.K.; Pradhan, B.; Sharma, K.K. Evaluation of some bio-control agents in in vitro control of *Fusarium oxysporum* f. sp. *cubense*, an incitant of banana panama wilt. *J. Pharmacogn. Phytochem.* **2020**, *9*, 751–753. Available online: https://www.phytojournal.com/archives/2020.v9.i3.11365/evaluation-of-some-bio-control-agents-in-in-vitro-control-of-fusarium-oxysporum-f-sp-cubense-an-incitant-of-banana-panama-wilt (accessed on 20 December 2023).
63. Segarra, G.; Casanova, E.; Avilés, M.; Trillas, I. *Trichoderma asperellum* strain T34 controls Fusarium wilt disease in tomato plants in soilless culture through competition for iron. *Microb. Ecol.* **2010**, *59*, 141–149. [CrossRef] [PubMed]
64. Dong, X.; Wang, M.; Ling, N.; Shen, Q.; Guo, S. Potential role of photosynthesis-related factors in banana metabolism and defense against *Fusarium oxysporum* f. sp. cubense. *Environ. Exp. Bot.* **2016**, *129*, 4–12. [CrossRef]
65. Lawson, T.; Oxborough, K.; Morison, J.I.L.; Baker, N.R. Responses of photosynthetic electron transport in stomatal guard cells and mesophyll cells in intact leaves to light, CO_2 and humidity. *Plant Physiol.* **2002**, *128*, 52–62. [CrossRef]
66. Tambussi, E.A.; Casadesus, J.; Munné-Bosch, S.; Araus, J.L. Photoprotection in water stressed plants of durum wheat (*Triticum turgidum* var. *durum*): Changes in chlorophyll fluorescence, spectral signature and photosynthetic pigments. *Funct. Plant Biol.* **2002**, *29*, 35–44. [CrossRef]
67. Moustaka, J.; Moustakas, M. Early-stage detection of biotic and abiotic stress on plants by chlorophyll fluorescence imaging analysis. *Biosensors* **2023**, *13*, 796. [CrossRef] [PubMed]
68. Pérez-Bueno, M.L.; Pineda, M.; Barón, M. Phenotyping plant responses to biotic stress by chlorophyll fluorescence imaging. *Front. Plant Sci.* **2019**, *10*, 1135. [CrossRef]
69. Goltsev, V.N.; Kalaji, H.M.; Paunov, M.; Bąba, W.; Horaczek, T.; Mojski, J.; Kociel, H.; Allakhverdiev, S.I. Variable chlorophyll fluorescence and its use for assessing physiological condition of plant photosynthetic apparatus. *Russ. J. Plant Physl.* **2016**, *63*, 869–893. [CrossRef]
70. Cheng, C.; Li, D.; Qi, Q.; Sun, X.; Anue, M.R.; David, B.M.; Zhang, Y.; Hao, X.; Zhang, Z.; Lai, Z. The root endophytic fungus *Serendipita indica* improves resistance of banana to *Fusarium oxysporum* f. sp. *cubense* tropical race 4. *Eur. J. Plant Pathol.* **2020**, *156*, 87–100. [CrossRef]

71. Ibrahimova, U.; Zivcak, M.; Gasparovic, K.; Rastogi, A.; Allakhverdiev, S.I.; Yang, X.; Brestic, M. Electron and proton transport in wheat exposed to salt stress: Is the increase of the thylakoid membrane proton conductivity responsible for decreasing the photosynthetic activity in sensitive genotypes? *Photosynth. Res.* **2021**, *150*, 195–211. [CrossRef]
72. Noctor, G.; Veljovic-Jovanovic, S.; Driscoll, S.; Novitskaya, L.; Foyer, C. Drought and oxidative load in the leaves of C3 plants: A predominant role for photorespiration. *Ann. Bot.* **2002**, *89*, 841–850. [CrossRef] [PubMed]
73. Basso, C.; Rodríguez-Yzquierdo, G.; Rivero, G.; León, R.; Barrios, M.; Díaz, G. Respuesta del cultivo de maracuyá (*Passiflora edulis* Sims) a condiciones de estrés por inundación. *Bioagro* **2019**, *31*, 185–192. Available online: https://dialnet.unirioja.es/servlet/articulo?codigo=7146779 (accessed on 15 December 2023).
74. Bueno, A.; Castro, G.; Rêgo, M.; Batista, T.; Filippi, M.; da Silva, G. *Trichoderma* reduces scald and protects the photosynthetic apparatus in rice plants. *Biocontrol Sci. Technol.* **2017**, *27*, 449–460. [CrossRef]
75. Wong, C.; Zulperi, D.; Saidi, N.B.; Vadamalai, G.A. Consortium of *Pseudomonas aeruginosa* and *Trichoderma harzianum* for improving growth and induced biochemical changes in Fusarium wilt infected bananas. *Trop. Life Sci. Res.* **2021**, *32*, 23–45. [CrossRef]
76. Sajeena, A.; Nair, D.S.; Sreepavan, K. Non-pathogenic *Fusarium oxysporum* as a biocontrol agent. *Indian Phytopathol.* **2020**, *73*, 177–183. [CrossRef]
77. Castillo, A.G.; Puig, C.G.; Cumagun, C.J.R. Non-synergistic effect of *Trichoderma harzianum* and *Glomus* spp. in reducing infection of Fusarium wilt in banana. *Pathogens* **2019**, *8*, 43. [CrossRef]
78. Doni, F.; Isahak, A.; Zain, C.R.C.M.; Ariffin, S.M.; Mohamad, W.N.W.; Yusoff, W.M.W. Formulation of *Trichoderma* sp. SL2 inoculants using different carriers for soil treatment in rice seedling growth. *SpringerPlus* **2014**, *3*, 532. [CrossRef] [PubMed]
79. Penyalver, R.; Vicedo, B.; López, M.M. Use of the genetically engineered *Agrobacterium* strain K1026 for biological control of crown gall. *Eur. J. Plant Pathol.* **2000**, *106*, 801–810. [CrossRef]
80. Vicedo, B.; Peñalver, R.; Asins, M.J.; López, M.M. Biological control of *Agrobacterium tumefaciens*, colonization, and pAgK84 transfer with *Agrobacterium radiobacter* K84 and the Tra mutant strain K1026. *Appl. Environ. Microbiol.* **1993**, *59*, 309–315. [CrossRef] [PubMed]
81. Peñalver, R.; Vicedo, B.; Salcedo, C.I.; López, M.M. *Agrobacterium radiobacter* strains K84, K1026 and K84 Agr- produce an antibiotic-like substance, active in vitro against *A. tumefaciens* and phytopathogenic *Erwinia* and *Pseudomonas* spp. *Biocontrol Sci. Technol.* **1994**, *4*, 259–267. [CrossRef]
82. Marshall, K.C.; Alexander, M. Competition between soil bacteria and *Fusarium*. *Plant Soil.* **1960**, *12*, 143–153. [CrossRef]
83. Köhl, J.; Kolnaar, R.; Ravensberg, W.J. Mode Action of Microbial Biological Control Agents Against Plant Diseases: Relevance Beyond Efficacy. *Front. Plant Sci.* **2019**, *10*, 845. [CrossRef]
84. El-Wakeil, N.; Saleh, M.; Abu-hashim, M. (Eds.) *Cottage Industry of Biocontrol Agents and Their Applications. Practical Aspects to Deal Biologically with Pests and Stresses Facing Strategic Crops*; Springer: Cham, Switzerland, 2020; 466p. [CrossRef]
85. Lin, P.; Zhang, M.Y.; Wang, M.Y.; Li, Y.Q.; Liu, J.F.; Chen, Y.L. Inoculation with arbuscular mycorrhizal fungi modulates defense-related genes expression in banana seedlings susceptible to wilt disease. *Plant Signal Behav.* **2021**, *16*, e1884782. [CrossRef]

Journal of *Fungi*

Article

Transcriptome Sequencing and Analysis of *Trichoderma polysporum* Infection in *Avena fatua* L. Leaves before and after Infection

Haixia Zhu [1,2,3,*] and Yushan He [1,2,3]

1 Academy of Agriculture and Forestry Sciences, Qinghai University, Xining 810016, China; yushanhe312@163.com
2 State Key Laboratory of Plateau Ecology and Agriculture, Qinghai University, Xining 810016, China
3 Key Laboratory of Agricultural Integrated Pest Management of Qinghai Province, Xining 810016, China
* Correspondence: zhuhaixia0101@163.com

Abstract: Biological control is a scientific management method used in modern agricultural production, and microbially derived biopesticides are one effective method with which to control weeds in agricultural fields. In order to determine the key genes for weed control by *Trichoderma polysporum*, transcriptome sequencing was carried out by high-throughput sequencing technology, and the strains of *T. polysporum* HZ-31 infesting *Avena fatua* L. at 24, 48, and 72 h were used as the experimental group, with 0 h as the control group. A total of 690,713,176 clean reads were obtained, and the sequencing results for each experimental group and the control group (0 h) were analyzed. In total, 3464 differentially expressed genes were found after 24 h of infection with the pathogen, including 1283 down-regulated genes and 2181 up-regulated genes. After 48 h of infection, the number of differentially expressed genes was 3885, of which 2242 were up-regulated and 1643 were down-regulated. The number of differentially expressed genes after 72 h of infection was the highest among all the groups, with 4594 differentially expressed genes, of which 2648 were up-regulated and 1946 were down-regulated. The up-regulated genes were analyzed by GO and KEGG, and the results showed that the up-regulated differentially expressed genes were mainly enriched in the biosynthesis of phenylalanine, tyrosine, and tryptophan; the degradation of aromatic compounds; methane metabolism; and other pathways. Among them, the *PHA2, GDH, ADH2,* and *AROF* genes were significantly enriched in the above-mentioned pathways, so they were hypothesized to play an important role in the synthesis of the herbicidally active substances of *T. polysporum* HZ-31. The results of this study can provide a theoretical basis for further studies on the pathogenicity of *T. polysporum* to *A. fatua* L., and accelerate the development and utilization of new and efficient bioherbicides.

Keywords: transcriptome sequencing; *T. polysporum* HZ-31; *Avena fatua* L.; key genes

Citation: Zhu, H.; He, Y. Transcriptome Sequencing and Analysis of *Trichoderma polysporum* Infection in *Avena fatua* L. Leaves before and after Infection. *J. Fungi* **2024**, *10*, 346. https://doi.org/10.3390/jof10050346

Academic Editors: Katrina Maria Ramonell and Ofir Degani

Received: 29 March 2024
Revised: 4 May 2024
Accepted: 8 May 2024
Published: 13 May 2024

Correction Statement: This article has been republished with a minor change. The change does not affect the scientific content of the article and further details are available within the backmatter of the website version of this article.

1. Introduction

Avena fatua L. is considered one of the most serious weeds in the world [1], and it poses a serious economic threat to crop yields due to its unique seed traits, including high competitiveness, staggered germination, chemosensory potential, and the ability to persist in the soil seed bank [2]. *A. fatua* L. is the second-most-important grass weed in Australia, costing cereal growers more than AUD 10,000 per year in lost income [3]. In China, 60% of the wheat-sown area in Qinghai Province (about 8.7 million hectares of farmland) is damaged by *A. fatua* L. [4]. However, the use of chemical herbicides to control it faces numerous obstacles, the most difficult of which is the development of resistance, with *A. fatua* L. ranking second among the 10 weeds most likely to develop resistance. The selection of the wrong type of herbicide, prolonged use of a single type of herbicide, application of an insufficient dose, late application, and the use of generics can increase the risk of selecting resistant biotypes [5]. Therefore, there is an urgent need for new, safe, non-polluting, and efficient bioherbicides to deal with *A. fatua* L. control and resistance development. Microorganisms and their metabolites have herbicidal activity. Microbial herbicides developed using microbial metabolites, especially plant pathogenic toxins, usually have the characteristics of safety, providing environmental protection, high efficiency, and multiple target sites, which can achieve green and efficient control and complicate the development of weed resistance.

Its rapid growth and ability to metabolize a wide range of substrates make *Trichoderma* a major component of the soil flora in diverse ecosystems, such as agricultural fields, pastures, grasslands, forests, saline wetlands, deserts, and polar regions [6]. It has an effective antagonistic mechanism with which to inhibit the growth of plant pathogenic fungi, weeds, or nematodes, and is a source of biological control agents in agriculture. *Trichoderma polysporum* is a biocontrol fungus of the *Trichoderma* with anti-stress, antibacterial and anti-grass activities, and studies have shown that it possesses biocontrol activity against a variety of fungi, such as *Armilaria gallica* [7], *Pseudogymnoascus destructans* [8], *Alternaria panax* [9], and *Ceratocystis paradoxa* [10]. Previous studies in our laboratory have found that *T.polysporum* can also be highly effective for the inhibition of weeds, such as *A.fatua* L., *Chenopodium album* L., and *Polygonum lapathifolium*, and has the potential to be developed as a biological control agent for weeds [11]. Because *T.polysporum* can also produce secondary metabolites, such as cyclosporin [12], xylomycin [13], and anthraquinones [14], the secondary metabolites have been extracted and identified in our laboratory. It has been found that it can produce various active substances, such as 1, 8-propanediol-o-xylene and 2, 3-dihydroxypropyl propionate, which have an inhibitory effect on weeds [15]. Therefore, the above studies have indicated that *T.polysporum* is a fungal resource of high value to be utilized in the biological control of weeds.

Although *T. polysporum* HZ-31 has the potential to be exploited as a biocontrol agent, further research is needed to improve the efficiency of its biocontrol and to discover new genes and active secondary substances. Transcriptomics, the global analysis of gene expression at the RNA level, is a key element of functional genomics that will greatly contribute to our understanding of gene function in the post-genomic era [16]. Therefore, in this study, transcriptome sequencing was performed at different time points after the inoculation of *A. fatua* L. with *T. polysporum* HZ-31, and the sequencing data were comparatively analyzed to reveal the key pathways and genes involved in the infection of *A. fatua* L. by *T. polysporum* HZ-31. The results are expected to provide candidate resources for future research on key gene functions.

2. Materials and Methods

2.1. Test Strains

T. polysporum strain HZ-31 was isolated from the stem base of a diseased *Cirsium arvense var. setosum* by the Key Laboratory of Comprehensive Management of Agricultural Pests in Qinghai Province, and it was identified and preserved.

2.2. Infection of A. fatua L. with T. polysporum HZ-31 and Culture Treatments

A small amount of the mycelium of *T. polysporum* strain HZ-31, which was maintained in the Key Laboratory of Comprehensive Management of Agricultural Pests in Qinghai Province, was placed in the center of PDA medium and cultured in an incubator at 25 °C for 4–5 days. The cultured strain was then inoculated into new PDA medium by punching out the cake with a 5 mm circular punch, and then used for the subsequent experiments when it had grown for 5 days. The *A. fatua* L. leaves were washed three times in sterile water and placed in a petri dish with filter paper. Moist cotton wool was used to cover the breaks in the *A. fatua* L. leaves, and sterile water was added to the petri dish until the filter paper was slightly wet for the subsequent experiments. The edges of the cultured *T. polysporum* HZ-31 strains were punched, and the fungus cakes were randomly picked and inoculated on the surface of the *A. fatua* L. leaves, such that the side of the cakes with mycelium was guaranteed to be in contact with the *A. fatua* L. leaves, and they were placed in an incubator at 25 °C and sampled at 24 h, 48 h, and 72 h. When sampling, the cultured cakes were placed into 2 mL centrifugal tubes after removing the residues of the medium and *A. fatua* L. leaves as much as possible. Ten pieces were extracted from each centrifugal tube and clearly labeled, and the uninoculated cakes were used as the control. Each type of control and treatment were repeated three times, and then frozen in liquid nitrogen and stored at −80 °C.

2.3. RNA Extraction and Quality Testing of Samples

The total RNA was extracted using the TRNzol Universal Reagent extraction kit according to the extraction instructions. Qubit2.0 was used to measure the RNA concentration, and agarose gel was used to detect RNA integrity and genomic contamination.

2.4. Library Construction and Sequencing

The transcriptome library construction, sequencing, and assembly of the extracted samples were completed by Sangon Bioengineering (Shanghai) Co., Shanghai, China.

2.5. Analysis of Transcriptome Sequencing Information

2.5.1. Data Quality Control

The raw data obtained from sequencing contains sequences with junctions and low quality. To ensure the quality of the information analysis, the raw data must be filtered to obtain the clean data. Data processing using Trimmomatic includes the removal of sequences with N bases; the removal of splice sequences in the reads; the removal of low-quality bases (Q-value < 20) starting from the 3′ to 5′ direction of the reads; the removal of low-quality bases (Q-value < 20) starting from the 5′ to 3′ direction to start removing the low-quality bases (Q-value < 20); removing bases with quality values below 20 in the tails of the reads using the sliding window method (window size of 5 bp); and removing the reads themselves and their paired reads, for lengths of reads less than 35 nt, to obtain the clean data.

2.5.2. Reference Sequence Comparison Analysis

The clean data obtained from sequencing were compared with the reference genome of *T. polysporum* HZ-31 PRJNA941260 (https://www.ncbi.nlm.nih.gov/datasets/genome/?bioproject=PRJNA941260, accessed on 16 June 2023) using HISAT2 2.1.0, and the results of the comparison were statistically analyzed by RSeQC2.6.1.

2.5.3. Expression Statistics

The gene expression levels were estimated by counting the sequenced sequences (reads) localized to the genomic regions or exonic regions of the genes. *TPM* (Transcripts Per Million) is a measure of the proportion of a particular transcript in a pool of RNAs. *TPM* accounts for both the sequencing depth and the length of the gene as well as the effect of the sample on the counts of the reads, and it is a commonly used method for estimating gene expression levels. The *TPM* formula is as follows:

$$TPM_i = \frac{X_i}{L_i} * \frac{1}{\sum_j \frac{X_j}{L_j}} * 10^6$$

$$X_i = total\ exon\ fragment/reads\ L_i = \frac{exon\ length}{KB}$$

2.5.4. Differential Expression Analysis

In order to screen the differential genes in the treatment and control groups of *A. fatua* L. infected with *T. polysporum* HZ 31, DESeq 1.26.0 was used to analyze the differential expression of the genes in each sample, and the screening conditions were as follows: DEGs with a q-value < 0.05 and a multiplicity of differences | fold change | > 2. The genes meeting these two criteria were considered to be significantly differentially expressed.

2.5.5. Differential Gene Enrichment Analysis and Functional Annotation

GO (Gene Ontology) and KEGG (Kyoto Encyclopedia of Genes and Genomes) functional enrichment analyses were performed using clusterProfiler 3.0.5. GO enrichment was performed using topGO 2.24.0 alone because GO is a directed acyclic structure. In general, when the corrected *p*-value (Q-value) was < 0.05, the function was considered to be significantly enriched.

2.5.6. Validation of Transcriptome Results

The samples used for qRT-PCR were from the same batch as the transcriptome sequencing. The cDNA first-strand synthesis was performed using a FastKing gDNA Dispelling RT SuperMix kit from Tengen (Beijing, China). Based on the analysis of the *T. polysporum* HZ-31 transcriptome sequencing results, 20 genes related to the biosynthesis of phenylalanine, tyrosine, and tryptophan; methane metabolism; the degradation of aromatic compounds; and the phenylalanine metabolism pathway were selected from the batch, and primers were designed and subjected to qRT-PCR experiments. The internal reference for qRT-PCR is Actin, a commonly used internal reference gene in fungi. The qRT-PCR reaction system was as follows: 2 × SuperReal Color PreMix, 25 μL; PCR Forward Primer (10 μM), 1 μL; PCR Reverse Primer (10 μM), 1 μL; cDNA, 1 μL; 50 × ROX Reference Dye, 0.5 μL; ddH_2O, 21.5 μL. The reaction was carried out in a qRT-PCR instrument using the qRT-PCR reaction program: predenaturation at 95 °C for 15 min, 1×; denatured at 95 °C for 10 s; and annealed at 60 °C for 30 s, 40×. The primers are shown in Table 1.

Table 1. Candidate gene primer sequences.

Gene ID	Forward Primer (5′→3′)	Reverse Primer (5′→3′)	Size (bp)
actin	CCTCCTCTTCCTTGCCAGCATTG	CCTCCTCTTCCTTGCCAGCATTG	111
Unigene2109	TGGCGGTGGTATTGGAGGTCTG	TCCTTGTATGCTGGTGCTTGTTCG	102
Unigene6375	ATTCTGAACGACCGTGAACTCTTGG	TGGTGTGGAAGCATCGCATGTG	96
Unigene8535	GCTGACTGGCAATGCTCTATGAC	CAATGGGCTGGGCGGTCTC	139
Unigene5854	CAATGGGCTGGGCGGTCTC	CAGCGTTCGGAGTGTTGATGG	139
Unigene3028	TCTCGCTCGCTTGATGGAAGG	ATGAGGAGTTGGAGGTCGTATCG	119
Unigene4053	CTTGGCTTAGAATCCCTCGTTGAC	TGATGGTGCCGCCGTTCTC	136
Unigene5271	GATGGCTTCTCCTCTGCTATGTG	TCTGAACTCCAATATGTCCCAAACC	134
Unigene6412	CGAAGATGCTGCTGCGTTAGG	CGAACAGTATCATACACGGCTAGG	90
Unigene6713	TGTTGCGAATGAGCGAGATGC	CCCTTGTAGTTTGTTCCTCCGTTAG	99
Unigene339	GGTTCTCTCTGGCGGTTTCTTG	CCTCGTTAGCATGTGCGTAGTC	80
Unigene2759	TGGCGGACTTGGTGGATGAG	TGCGTGATATTGATAGAGGCTTGC	84
Unigene3429	TGCGAACTGGACTGGACTCTG	GTTGATCTCGGTGACGGCTTTC	110
Unigene5483	TCGCCTATGCCGTGGTTCC	GGTTCGTCCATTCGCCAGATG	150
Unigene7084	GGACCATGAGACCTGATAGCATTG	AGTCGCTGGCTCGGTTACG	137
Unigene259	CAATGCCGCCGATTCACAGG	GGACCTTGTAATAGTTGCCGTATGG	92
Unigene8087	AGGTGTTGGCGAGGAGTATCAG	TCAAGCGTCTCTTCAGTCTTTAGTG	81
Unigene2730	GCCATTGAATGCCTTGCTTGAC	ATGTCCGCCGCCTTGGTAG	102
Unigene753	TCGCTCGGTGTCGCCATC	AAGCAGTCTTCGTTACCTGTTGTG	116
Unigene6048	TCGCCTCGGAGCAGATTGTC	GTTCATCAGCCATCGCAGGTAG	143
Unigene10110	GGCGGCGGCATTGTATTCG	GCGGATTGCTGCTGGTCATAG	125

3. Results and Analysis

3.1. Transcriptome Data Assembly Results

In order to screen the genes related to the prevention of infection of *A. fatua* L. by *T. polysporum* HZ-31, samples of the mycelium of *T. polysporum* HZ-31 were collected at 0, 24, 48, and 72 h of infestation of *A. fatua* L. for transcriptome sequencing, and quality control was carried out on the sample data after obtaining the raw data. As shown in Table 2, an average of 59,053,547 raw reads were measured, and the samples were filtered for the raw data to obtain an average of 57,559,431-retained clean reads. All the samples with Q20 above 98% (greater than 95%) and Q30 above 95% (greater than 80%) indicated high data quality, and the GC content was above 50%, which proved that the sequencing was not obviously biased. Therefore, the data quality was satisfactory.

Table 2. Sequencing data quality.

Sample	Raw Reads	Clean Reads	Clean Bases	Q20 (%)	Q30 (%)	GC Content (%)
CK-1	64,467,532	62,859,772	9,198,358,055	99.00%	96.17%	52.39%
CK-2	61,689,040	60,084,542	8,787,122,185	98.98%	96.16%	52.54%
CK-3	55,680,734	54,362,330	7,879,606,421	99.07%	96.47%	53.05%
T24-1	56,045,126	54,310,362	7,791,837,914	98.85%	95.69%	52.06%
T24-2	55,269,126	54,006,616	7,771,363,231	99.03%	96.29%	52.61%
T24-3	61,983,022	60,503,534	8,682,351,056	99.04%	96.32%	52.66%
T48-1	57,216,160	55,900,958	8,163,997,746	99.03%	96.31%	51.87%
T48-2	49,386,632	48,061,810	7,007,420,562	98.97%	96.09%	51.68%
T48-3	51,854,456	50,528,528	7,354,822,322	99.01%	96.19%	51.68%
T72-1	77,526,690	75,760,238	10,964,415,778	99.06%	96.41%	52.34%
T72-2	59,115,428	57,610,338	8,362,733,753	99.01%	96.24%	52.51%
T72-3	58,408,616	56,724,148	8,171,046,956	98.93%	95.97%	51.80%

Note: CK is the control group and T is the treatment group. Sample: indicates the sample name. Clean reads: data after QC. Q20 means that the sequencing error rate of this base is 0.01, and Q30 means that the sequencing error rate of this base is 0.001; therefore, the higher the ratio of Q20 and Q30, the better the data quality.

3.2. Reference *T. polysporum* Genome Comparison Results

Using the *T. polysporum* genome measured in this article as the reference genome, the comparison results are shown in Table 3. The total reads are the sum of read1 and read2, which is known as the clean reads. The reads on the genome compared to the clean reads are more than 95%, so more than 95% of the reads in a unique position are compared, and less than 1% of the reads in multiple positions are compared. Both read1 and read2 are compared to the genome, with more than 47% and more than 92% respectively of the reads simultaneously compared to the genome at both ends.

Table 3. Reads in the reference genome alignment results.

Sample	Clean Reads	Total Mapped	Multiple Mapped	Uniquely Mapped	Read1 Mapped	Read2 Mapped	Proper Mapped
CK-1	56,568,506 (100.00%)	56,110,923 (99.19%)	163,743 (0.29%)	55,947,180 (98.90%)	27,979,165 (49.46%)	27,968,015 (49.44%)	54,648,680 (96.61%)
CK-2	59,334,990 (100.00%)	58,694,852 (98.92%)	238,005 (0.40%)	58,456,847 (98.52%)	29,220,950 (49.25%)	29,235,897 (49.27%)	57,088,794 (96.21%)
CK-3	54,102,082 (100.00%)	53,611,134 (99.09%)	176,271 (0.33%)	53,434,863 (98.77%)	26,719,566 (49.39%)	26,715,297 (49.38%)	51,866,660 (95.87%)
T24-1	50,455,296 (100.00%)	49,908,731 (98.92%)	147,119 (0.29%)	49,761,612 (98.63%)	24,899,621 (49.35%)	24,861,991 (49.28%)	48476952 (96.08%)
T24-2	53,627,174 (100.00%)	52,963,908 (98.76%)	179,316 (0.33%)	52,784,592 (98.43%)	26,400,523 (49.23%)	26,384,069 (49.20%)	50,871,496 (94.86%)
T24-3	59,771,456 (100.00%)	59,042,883 (98.78%)	209,930 (0.35%)	58,832,953 (98.43%)	29,428,530 (49.24%)	29,404,423 (49.19%)	56,860,444 (95.13%)
T48-1	52,454,282 (100.00%)	51,511,562 (98.20%)	238,042 (0.45%)	51,273,520 (97.75%)	25,640,147 (48.88%)	25,633,373 (48.87%)	49,681,412 (94.71%)
T48-2	34,189,154 (100.00%)	33,535,178 (98.09%)	114,753 (0.34%)	33,420,425 (97.75%)	16,714,982 (48.89%)	16,705,443 (48.86%)	32,285,094 (94.43%)
T48-3	29,116,484 (100.00%)	27,919,427 (95.89%)	94,113 (0.32%)	27,825,314 (95.57%)	13,918,967 (47.80%)	13,906,347 (47.76%)	26,793,860 (92.02%)
T72-1	75,273,746 (100.00%)	74,106,054 (98.45%)	345,027 (0.46%)	73,761,027 (97.99%)	36,888,884 (49.01%)	36,872,143 (48.98%)	71,239,844 (94.64%)
T72-2	57,056,046 (100.00%)	56,079,845 (98.29%)	281,043 (0.49%)	55,798,802 (97.80%)	27,908,963 (48.91%)	27,889,839 (48.88%)	53,954,714 (94.56%)
T72-3	54,409,280 (100.00%)	53,175,243 (97.73%)	320,973 (0.59%)	52,854,270 (97.14%)	26,441,123 (48.60%)	26,413,147 (48.55%)	51,321,778 (94.33%)

3.3. Expression Analysis

As shown in Figure 1, the highest, lowest, and median log2(*TPM*) values of gene expression in the different samples can be known according to the box line plot of gene expression. The results show that the discrete degree of gene expression of *T. polysporum* HZ-31 is more concentrated, the degree of difference is higher in the treatment group than in the control group, and the median values of log2(*TPM*) are all concentrated around 4. The moderately expressed genes account for more of the total, and the gene expression of the treated group is higher than that of the control group.

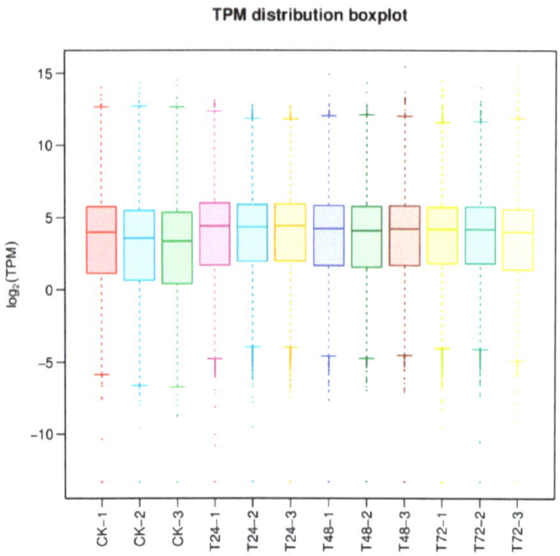

Figure 1. Results of gene expression analysis. The horizontal axis is the sample name and the vertical axis is the log(*TPM*) value. The box plot for each region is for five statistics (top to bottom for maximum, upper quartile, median, lower quartile, and minimum, respectively). Different colors represent different samples.

3.4. Differential Gene Analysis

The differentially expressed genes (DEGs) in *A. fatua* L. infected with *T. polysporum* HZ-31 were screened by the FDR-corrected *p*-value $\leq$ 0.05 and the absolute value of the log_2 ratio > 2. The sequencing results for each experimental group and for the control group were analyzed. When compared with the control group, there were 3464 differentially expressed genes, including 1283 down-regulated genes and 2181 up-regulated genes after 24 h of infection by the pathogen (Figure 2A,D). After 48 h of infection, there were 3885 differentially expressed genes, of which 2242 were up-regulated and 1643 were down-regulated (Figure 2B,D). The number of differentially expressed genes after 72 h of infestation was the highest among all the groups, with 4594 differentially expressed genes, of which 2648 were up-regulated and 1946 were down-regulated (Figure 2C,D).

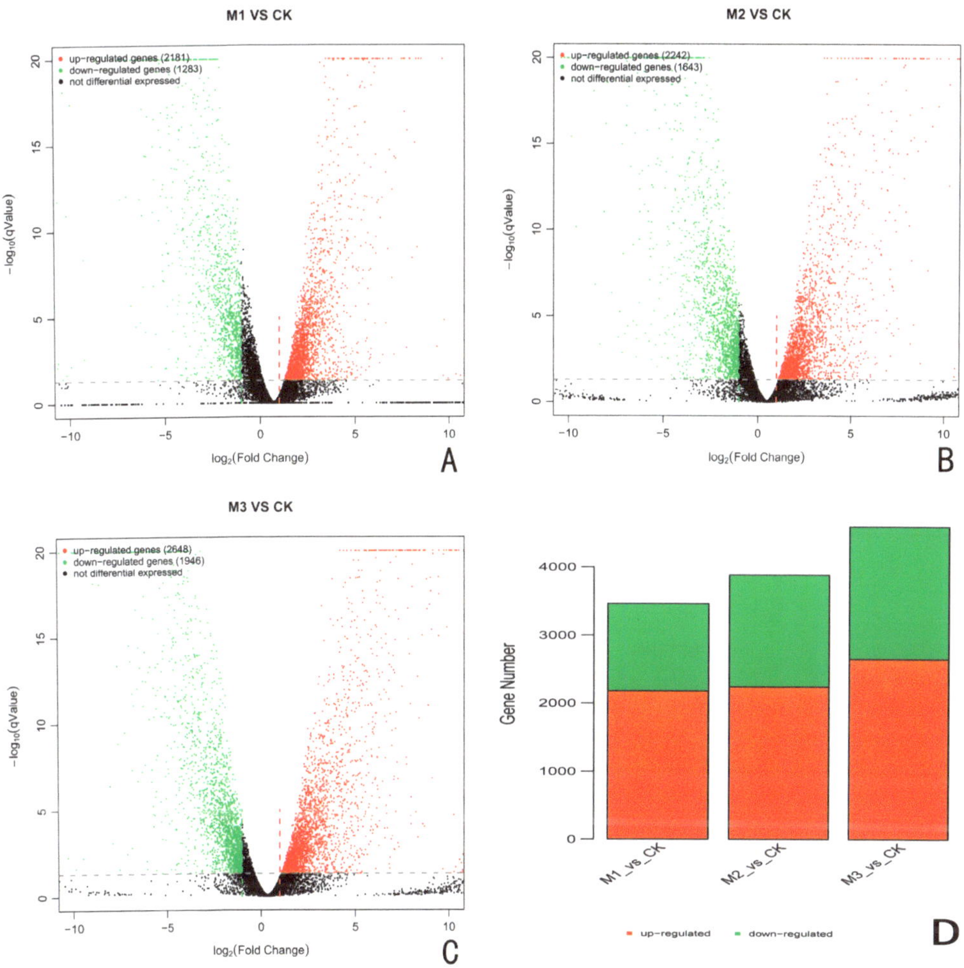

Figure 2. Results of differential gene analysis. (**A**) CK vs. 24 h, (**B**) CK vs. 48 h, (**C**) CK. vs. 72 h. M stands for group, M1 stands for 24 h, M2 stands for 48 h, and M3 stands for 72 h. The horizontal axis is the average log(TPM) of the two groups of samples, that is, (log(A) + log(B))/2, and the vertical axis is log(fold change), that is, the log(B/A) value. Each dot represents a gene, Green dots to the left of the green dashed line and above the gray dashed line indicate down-regulated genes, and red dots to the right of the red dashed line and above the gray dashed line indicate up-regulated genes, and black representing the non-differential genes. (**D**) Statistical map of the differentially expressed genes. The horizontal axis is the difference comparison name, and the vertical axis is the number of down-regulated differential genes. Where the green is down, the red is up.

3.5. Venn Analysis

A Venn analysis of the DEGs after the three pathogen-infection time points and comparative analysis showed that 980 genes were co-occurring DEGs in all three groups at all time points, which accounted for a smaller portion of the total. In total, 3960 differentially expressed genes were present (Figure 3).

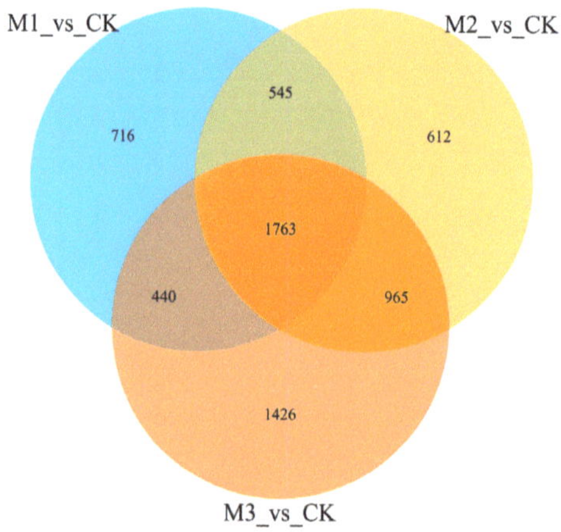

Figure 3. Results of Venn analysis. M stands for group, M1 stands for 24 h, M2 stands for 48 h, and M3 stands for 72 h. Different comparison groups are represented by different colors, and the numbers in the figure represent the number of specific or common differentially expressed genes. The overlapping region represents the number of differentially expressed genes shared by the different comparison groups, while the non-overlapping region represents the number of differentially expressed genes unique to the different comparison groups.

3.6. *Enrichment Analysis of Up-Regulated Differentially Expressed Genes*

3.6.1. GO Functional Enrichment Analysis of Up-Regulated Differentially Expressed Genes

Using the GO database, the above up-regulated DEGs of the CK vs. 24 h, CK vs. 48 h, and CK vs. 72 h were analyzed by GO functional enrichment. The results indicated that the up-regulated DEGs at 24 h, 48 h, and 72 h were distributed among 43, 48, and 45 categories of the three major functional annotations, respectively. A total of 1320 up-regulated differentially differentiated genes were annotated for CK vs. 24 h (Figure 4), with 476 biological processes (BPs), 410 molecular functions (MFs), and 434 cellular components (CCs), accounting for 36.06%, 31.06%, and 32.88%, respectively. These up-regulated differential genes were significantly enriched in the organic cyclic compound metabolic process, nucleic acid metabolic process, heterocycle metabolic process, cellular aromatic compound metabolic process, and catalytic activity, with the highest number of genes.

A total of 1542 up-regulated differential genes, including 558 biological processes, 463 molecular functions, and 521 cellular components were annotated for CK vs. 48 h (Figure 5), accounting for 36.19%, 30.03% and 33.79%, respectively. These up-regulated differential genes were significantly enriched in the nitrogen compound metabolic process, gene expression, and RNA metabolic process, with the highest number of genes.

A total of 1767 up-regulated differential genes, including 635 biological processes, 545 molecular functions, and 587 cellular components were annotated by CK vs. 72 h (Figure 6), which account for 35.94%, 30.84%, and 33.22%, respectively. These up-regulated differential genes were significantly enriched in the organic cyclic compound metabolic process, nucleic acid metabolic process, heterocycle metabolic process, gene expression, cellular nitrogen compound metabolic process, cellular aromatic compound metabolic process, and RNA metabolic process, with the highest number of genes.

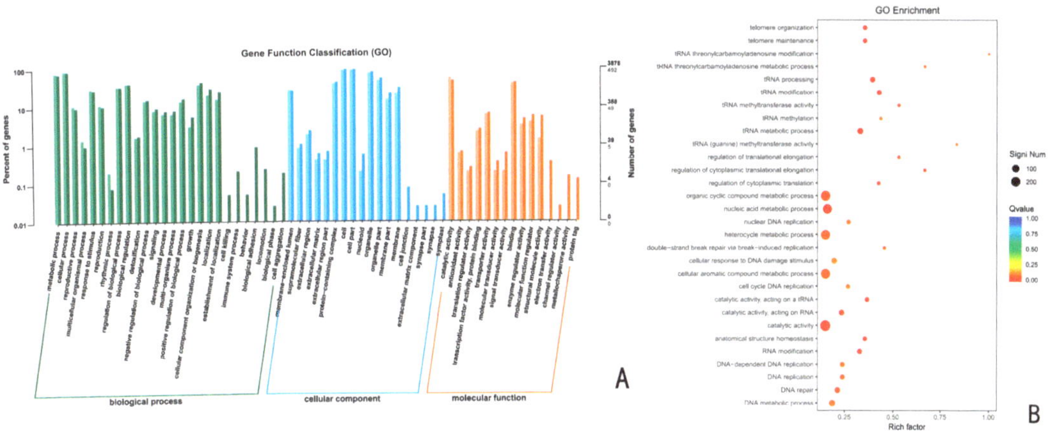

Figure 4. GO functional enrichment analysis of *Trichoderma polysporum* HZ-31 differential genes after treatment with CK vs 24 h. (**A**) The horizontal axis of the differential gene GO annotation classification bar is the functional classification, and different colors represent different classifications. The values on the bar chart and the vertical axis are lighter colored for the differential genes and darker colored for all the genes. The right vertical axis is the number of genes in the classification, and the left vertical axis is the number of genes annotated to the function (differential genes/all genes). (**B**) GO enrichment bubble charts. The vertical axis represents the functional annotation information, and the horizontal axis represents the Rich factor corresponding to the function (the number of differential genes annotated to the function divided by the number of genes annotated to the function). The size of the Q value is represented by the color of the dot. The smaller the Q value is, the closer the color is to red. The number of differential genes contained in each function is expressed by the size of the dots. (We only selected the top 30 GOs with the highest degree of enrichment.) Figures 4–6 are the same.

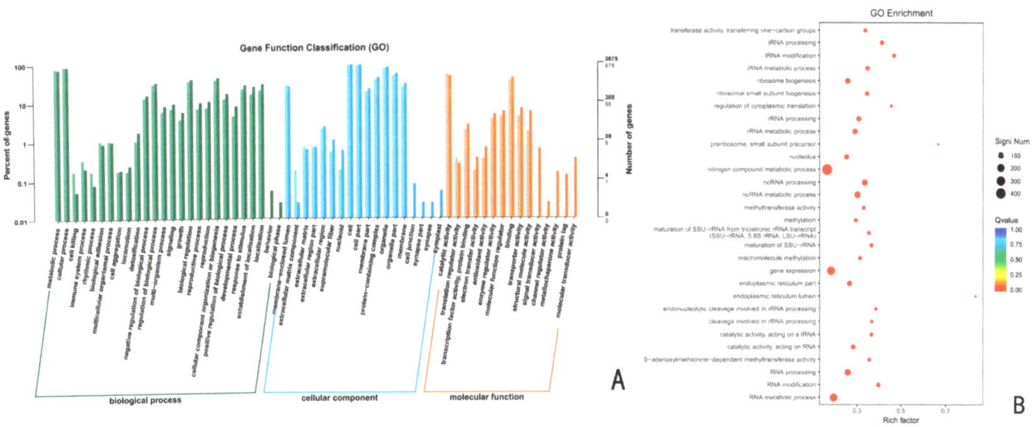

Figure 5. (**A**) GO functional enrichment analysis of *Trichoderma polysporum* HZ-31 differential genes after treatment with CK vs. 48 h. (**B**) GO enrichment bubble charts.

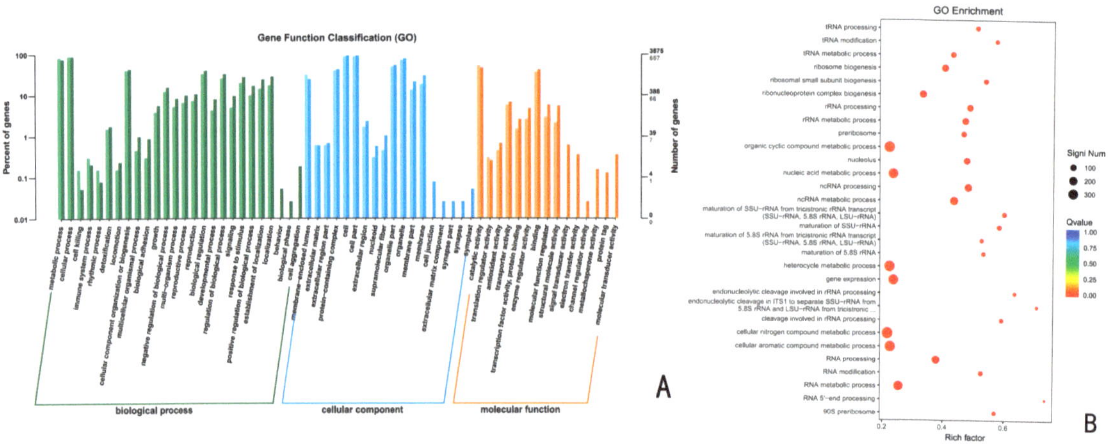

Figure 6. (**A**) GO functional enrichment analysis of *Trichoderma polysporum* HZ-31 differential genes after treatment with CK vs. 72 h. (**B**) GO enrichment bubble charts.

3.6.2. KEGG Functional Enrichment Analysis of Up-Regulated Differentially Expressed Genes

The top 30 KEGG pathways that were significantly enriched in the up-regulated expressed genes compared to the CK at the three different time periods were analyzed after the infection of *A. fatua* L. with *T. polysporum* HZ-31 for 24 h, 48 h, and 72 h (Figure 7). The results show that phenylalanine, tyrosine, and tryptophan biosynthesis; naphthalene degradation; methane metabolism; dioxin degradation; the degradation of aromatic compounds; chlorocyclohexane and chlorobenzene degradation; and the carbapenem biosynthesis pathways appeared in all three groups, suggesting that the above pathways play an important role in the synthesis of herbicidal secondary metabolites by *T. polysporum* HZ-31. Among these pathways, phenylalanine, tyrosine, and tryptophan biosynthesis; the degradation of aromatic compounds; and methane metabolism were associated with the herbicidally active compounds of *T. polysporum* HZ-31, such as P-hydroxyphenyl-2,3-dihydroxypropyl ether, O-hydroxy-3-carbonyl-1-O-hydroxy-3-carbonyl-1-hydroxyphenylpropanol, 1,8-propanediol-o-xylene, and the biosynthesis of P-hydroxyphenyl-2,3-dihydroxypropyl ether.

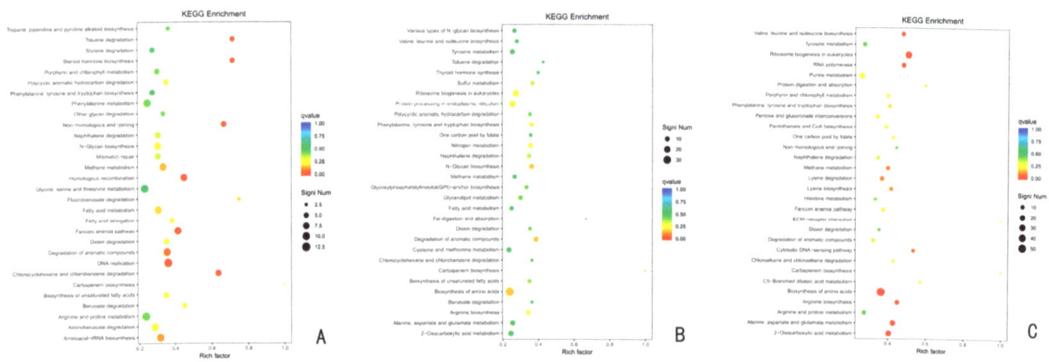

Figure 7. KEGG functional enrichment analysis of *Trichoderma polysporum* HZ-31 differential genes under the different treatments. (**A**) CK vs. 24 h, (**B**) CK vs. 48 h, and (**C**) CK vs. 72 h. The vertical axis

represents the functional annotation information, and the horizontal axis represents the Rich factor corresponding to the function (the number of differential genes annotated to the function divided by the number of genes annotated to the function). The size of the Q value is represented by the color of the dot. The smaller the Q value is, the closer the color is to red. The number of differential genes contained in each function is expressed by the size of the dots. (We only selected the top 30 KEGGs with the highest degree of enrichment).

3.6.3. Metabolic Pathway Analysis

The phenylalanine biosynthesis pathway was significantly enriched in all three periods after the infection of *A. fatua* L. with *T. polysporum* HZ-31. The phenylalanine biosynthesis pathway was annotated to a multitude of genes in the different periods after the infection of *A. fatua* L. with *T. polysporum* HZ-31 (Figure 8A); the genes that were significantly up-regulated were *TRP* (8866_g), *PHA2* (2454_g), *AATR1* (6048_g), *DHQA* (2492_g), *AATR1* (8535_g), *3DHQ* (2490_g), *AATC* (7327_g), *TRPE* (10110_g), *TRPG* (9206_g), *AROF* (753_g), and *AROM* (9839_g), encoding, respectively, tryptophan synthase, putative prephenate dehydratase, aromatic amino acid aminotransferase, quinate dehydrogenase, aromatic amino acid aminotransferase, catabolic 3-dehydroquinase, aspartate aminotransferase, anthranilate synthase component 1, anthranilate synthase component 2, 3-deoxyheptanose-7-phosphate synthetase and pentafunctional *AROM* polypeptide.

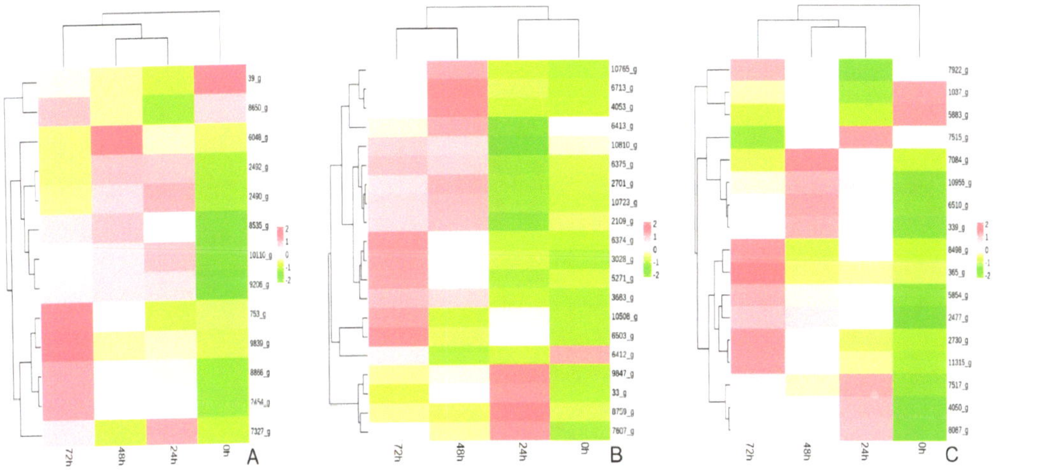

Figure 8. Analysis of the key metabolic pathways of *T. polysporum* HZ-31 under different treatments: (**A**) phenylalanine biosynthesis, (**B**) aromatic compound metabolism, and (**C**) methane metabolism. Each row represents a gene, each column represents a sample, and the color indicates the size of the gene's expression in the sample, with red representing a higher expression of the gene in the sample and green representing a lower expression.

The metabolic pathways of the aromatic compounds were also significantly enriched in all three periods after the infection of *A. fatua* L. by *T. polysporum* HZ-31 (Figure 8B). The significantly up-regulated genes were *CP53* (3028_g), *YJU6* (6374_g), *NHG2* (2109_g), *GNL* (6713_g), *NHG1* (6375_g), *ADH2* (5271_g), and *3HBHI* (10810_g), encoding, respectively, benzoate 4-monooxygenase, the putative transcriptional regulatory protein, salicylate hydroxylase, gluconolactonase, salicylate hydroxylase, and alcohol dehydrogenase 2,3-hydroxybenzoate 6-hydroxylase.

In the methane metabolic pathway (Figure 8C), the genes that were significantly up-regulated during the three periods were *FDH* (5854_g), *DAS* (10855_g), *GDH* (7084_g), *DAS* (6510_g), *ALF* (11315_g), *ALOX* (8087_g), and *PGDH* (4050_g), encoding formate dehydrogenase, dihydroxyacetone synthase, glycerate dehydrogenase, dihydroxyacetone synthase, glycerate dehydrogenase, dihydroxyacetone synthase, and glycerate dehydrogenase, respectively.

3.7. Validation of Transcriptome Results

In this experiment, the differential genes were analyzed by qRT-PCR. The FPKM values of the transcriptome sequencing and the relative expressions of the genes that were determined by real-time fluorescence quantification were plotted and analyzed on the vertical axis, and the horizontal axis indicated the different treatment time points. The results are shown in Figure 8. The RNA-seq results for 20 genes were basically consistent with the expression trends of the qRT-PCR sequencing results (Figure 9), which prove that the transcriptome sequencing results are reliable.

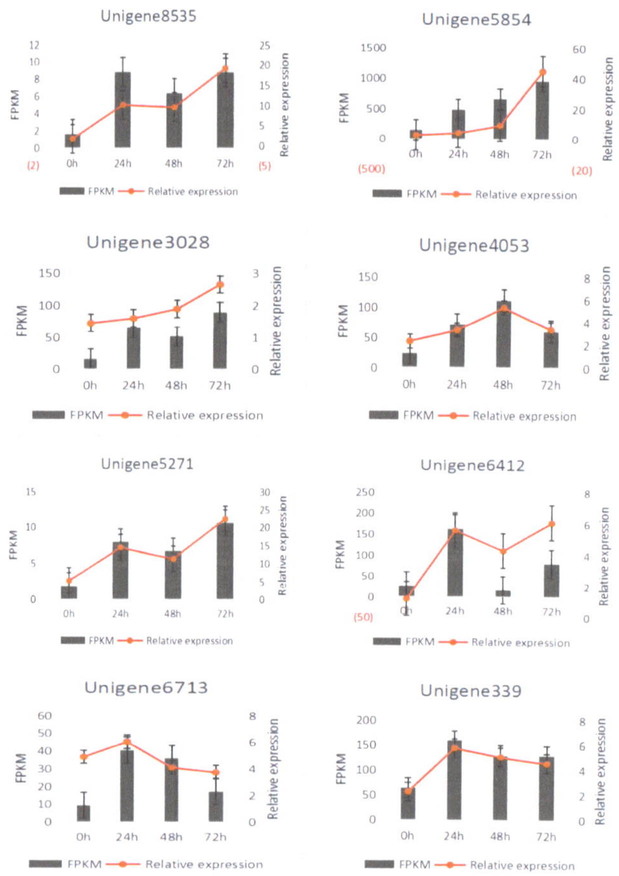

Figure 9. *Cont.*

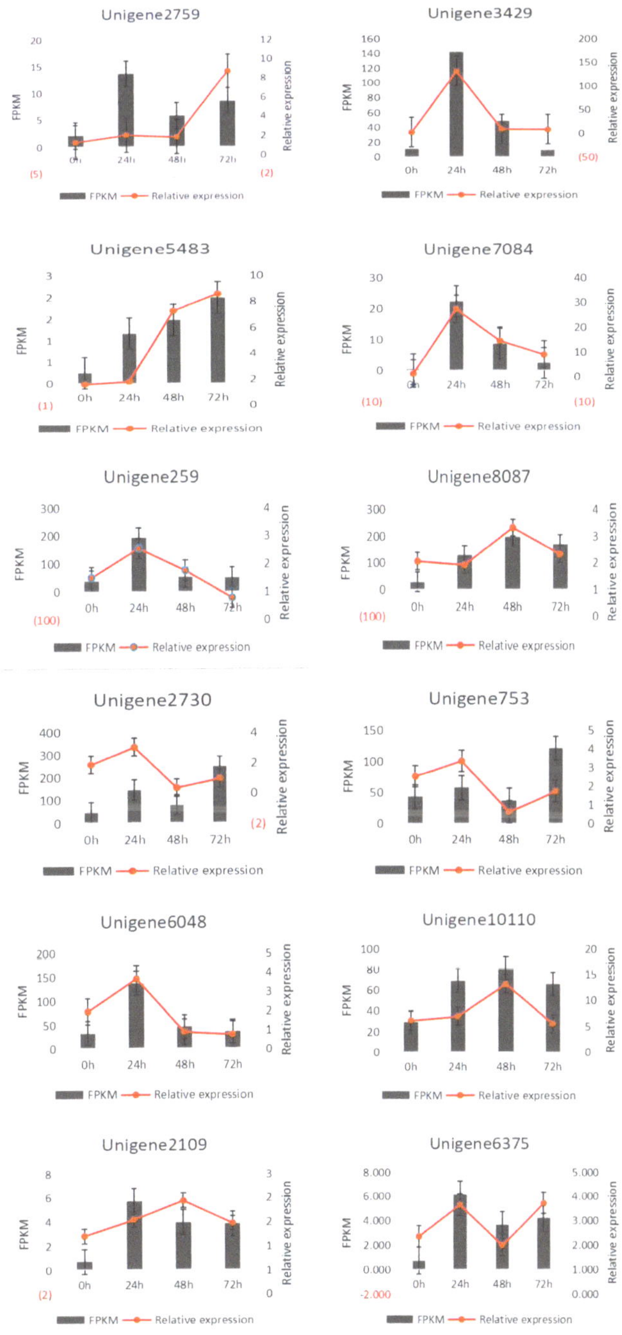

Figure 9. qRT-PCR of the differential genes. The title of the chart represents the i.d. of the different differential genes. The left ordinate represents the FPKM values of the differential genes in the transcriptome. The right ordinate represents the relative expression of the differential genes in the qRT-PCR.

4. Discussion

The ultimate goal of this study was to understand the differential expression of pathogenic fungal genes and their involvement in metabolic pathways during the infestation of *A. feua* L. leaves by *T. polysporum* HZ-31 by transcriptome sequencing, and to explore the differentially expressed genes related to the pathogenicity of the pathogenic fungi. The results show that there were 3464, 3885, and 4594 differential genes during the three different time periods when *T. polysporum* HZ-31 infected *A. feua* L., showing an increasing trend and an increasing number of up-regulated genes. This is consistent with the results of Wang et al.'s study, which showed that *Trichoderma harzianum* increased the number of up-regulated differential genes over time during parasitism [17]. The trend of the up-regulated differential genes indicates that the infectious activity of *T. polysporum* HZ-31 on *A. feua* L. may increase with time within 72 h after infection. After the up-regulated genes were enriched with the GO function, the number of differential genes enriched that fell into the category of biological processes was more than that for the number of genes in the categories of cell components and molecular functions, indicating that *T. polysporum* has a higher activity of genes related to biological processes during infection, including the GO classification of metabolic processes, cellular processes, biological regulation, etc. The results of the GO enrichment analysis were similar to the transcriptome analysis of *Fusarium solani*-infected sweet potato studied by Luo et al. [18].

To date, at least 545 fungal phytotoxic secondary metabolites have been reported, including 207 polyketides, 46 phenols and phenolic acids, 135 terpenoids, 146 nitrogenous metabolites, and 11 other metabolites. Among them, aromatic polyketides and sesquiterpenoids are the major phytotoxic compounds [19]. The metabolism of phenylalanine, tyrosine, and tryptophan (also known as the three aromatic amino acids) are involved in fungal growth and are known precursors of several mycotoxins. These amino acids can be synthesized and utilized by fungi to produce potentially damaging mycotoxins. In this study, *A. feua* L. was infected with *T. polysporum* HZ-31. During the three periods, the phenylalanine biosynthesis pathway was significantly enriched, indicating that phenylalanine plays an important role in the pathogenicity of *T. polysporum* HZ-31, and it may be a precursor to the synthesis of *T. polysporum* HZ-31 herbicidal active substances. This is consistent with the results of Santiago et al.'s study, which showed that the metabolic process of *Ganoderma boninense* is mainly enriched in the biosynthesis pathway of phenylalanine, tyrosine and tryptophan, and that this pathway plays an important role in the growth and development of *Ganoderma boninense* and toxin synthesis [20]. The theoretical pathway for the microbial production of phenylalanine from glucose involves glucose first generating phosphoenolpyruvic acid and erythrose 4-phosphate via the glycolysis pathway and the pentose phosphate pathway, respectively, which then enters the mangiferic acid pathway to generate mangiferic acid, and phenylalanine is finally generated from mangiferic acid [21]. Among the enzymes, 3-deoxyheptanose-7-phosphate synthetase encoded by *AROF*, and five functional fusion proteins encoded by *AROM* are the first key enzymes of the mangiferic acid synthesis pathway. Their up-regulated expression promotes the biosynthesis of chorismic acid, and the putative *PHA2* gene encoding prephenate dehydratase catalyzes the production of prephenic acid from chorismic acid, which in turn promotes phenol biosynthesis [22]. The proteins encoded by the *AROF*, *AROM*, and *PHA2* genes showed up-regulation in this study, and the expression of the *AROF* and *PHA2* genes in the phenylalanine biosynthesis pathway in the treatment group showed an up-regulated trend compared with the control group, so it is speculated that the *AROF* and *PHA2* genes are key genes in the synthesis of the herbicidal active substances of *T. polysporum* HZ-31.

The metabolic pathway of aromatic compounds in this study was also significantly enriched in the three time periods of *A. feua* L. infestation by *T. polysporum* HZ-31, which is consistent with the result that the KEGG pathway, which was predominantly enriched to be the metabolic pathway of aromatic compounds at the 6th d of apple infestation by the *Valsa mali var.* mali, as shown by Jin et al. [23]. The gene *ADH2*, encoding alcohol dehydrogenase ADHS, is enriched in the metabolic pathway of the aromatic compounds, and the expression level is always up-regulated compared with the control group. Alcohol dehydrogenase ADHS is widely found in plants, animals, microorganisms, and other organisms [24]. In recent years, ADHS has been found to play an important role in the self-growth of many pathogens and their pathogenicity to plants. Smidt [25] found that the transcription levels of the *BcADH1* gene were significantly up-regulated during the tomato–Botrytis cinerea strain B0510 interaction and in the early stages of infection. The virulence of the mutant on tomato leaves after knockout of the gene was significantly different from that of the wild type. These results provide strong evidence for the importance of *ADH1* genes for fungal development, including their environmental adaptation and reaching full pathogenicity [26]. Therefore, it is speculated that the *ADH2* gene encoding alcohol dehydrogenase in this study is involved in the prevention of *A. feua* L. infection by *T. polysporum* HZ-31. The reason for its continued up-regulation during the three periods of *T. polysporum* HZ-31 infection is that it may regulate the growth and development and pathogenicity of *T. polysporum* HZ-31.

Previous studies in our laboratory have shown that the *GDH* gene in the methane metabolic pathway is located in the secondary metabolite synthesis gene cluster encoding enniatin. Structurally, enniatin is a cyclohexadecapeptide composed of alternating residues of three N-methylamino acids (usually valine, leucine, and isoleucine) and three hydroxyl acids (usually hydroxyisovaleric acid). It is a well-known antibacterial, repellent, antifungal, herbicidal, and insecticidal compound. [27]. In order to determine whether the strain virulence was affected by fusarium enniatin, potato tuber tissues were measured. The results showed that 7 strains that produced enniatin and 16 strains that did not produce enniatin induced tissue necrosis of potato tubers. As a result, the production of enniatin by the strains that synthesize it may influence its pathogenicity [28]. In this study, the expression of the *GDH* gene was up-regulated during the infection of *A. feua* L. by *T. polysporum* HZ-31, and the expression level was higher after 24 h of infection. Therefore, it is speculated that the *GDH* gene might be involved in the synthesis of the secondary metabolite of *T. polysporum* HZ-31, thus regulating its pathogenicity.

In this study, we first analyzed the gene expression characteristics of *T. polysporum* HZ-31 during three different periods after infection of *A. fatua* L. at the transcriptome level, and found the key disease-causing genes to be *AROF*, *ADH2*, *PHA2*, and *GDH*, which are related to the synthesis of the secondary metabolites of the pathogen. The biological functions of these genes will be further investigated in the future in order to utilize the herbicidal potential of *T. polymorpha* HZ-31 to develop new and efficient bioherbicides that are suitable for the control of *A. fatua* L. in the Tibetan Plateau region.

Author Contributions: Y.H. and H.Z. conceived of this study, collected samples, and performed the lab work. Y.H. and H.Z. analyzed the results and wrote the manuscript. All authors have read and agreed to the published version of the manuscript.

Funding: This research was supported by the Project of the Science and Technology Department of Qinghai Province with the grant number 2024-ZJ-928.

Institutional Review Board Statement: Not applicable.

Informed Consent Statement: Not applicable.

Data Availability Statement: The data that support the findings of this study are available from the corresponding author upon reasonable request.

Acknowledgments: We are grateful to the anonymous reviewers for their valuable suggestions and comments.

Conflicts of Interest: The authors declare no conflict of interest.

References

1. Iannucci, A.; Fragasso, M.; Platani, C.; Papa, R. Plant growth and phenolic compounds in the rhizosphere soil of wild oat (*Avena fatua* L.). *Front. Plant Sci.* **2013**, *4*, 509. [CrossRef] [PubMed]
2. Dahiya, A.; Sharma, R.; Sindhu, S.; Sindhu, S.S. Resource partitioning in the rhizosphere by inoculated *Bacillus* spp. towards growth stimulation of wheat and suppression of wild oat (*Avena fatua* L.) weed. *Physiol. Mol. Biol. Plants* **2019**, *25*, 1483–1495. [CrossRef]
3. Chauhan, B.S. The world's first glyphosate-resistant case of *Avena fatua* L. and *Avena sterilis* ssp. ludoviciana (Durieu) Gillet & Magne and alternative herbicide options for their control. *PLoS ONE* **2022**, *17*, e0262494.
4. Li, W.; Shen, S.; Chen, H. Bio-herbicidal potential of wheat rhizosphere bacteria on *Avena fatua* L. grass. *Bioengineered* **2021**, *1*, 516–526. [CrossRef] [PubMed]
5. Tafoya-Razo, J.A.; Mora-Munguía, S.A.; Torres-García, J.R. Diversity of Herbicide-Resistance Mechanisms of *Avena fatua* L. to Acetyl-CoA Carboxylase-Inhibiting Herbicides in the Bajio, Mexico. *Plants* **2022**, *11*, 1644. [CrossRef] [PubMed]
6. Kamo, M.; Tojo, M.; Yamazaki, Y.; Itabashi, T.; Takeda, H.; Wakana, D.; Hosoe, T. Isolation of growth inhibitors of the snow rot pathogen *Pythium iwayamai* from an arctic strain of *Trichoderma polysporum*. *J. Antibiot.* **2016**, *69*, 451–455. [CrossRef] [PubMed]
7. Zhang, T.; Chaturvedi, V.; Chaturvedi, S. Novel *Trichoderma polysporum* Strain for the Biocontrol of *Pseudogymnoascus destructans*, the Fungal Etiologic Agent of Bat White Nose Syndrome. *PLoS ONE* **2015**, *10*, e0141316. [CrossRef] [PubMed]
8. Singh, A.; Lasek-Nesselquist, E.; Chaturvedi, V.; Chaturvedi, S. *Trichoderma polysporum* selectively inhibits white-nose syndrome fungal pathogen *Pseudogymnoascus destructans* amidst soil microbes. *Microbiome* **2018**, *6*, 139. [CrossRef] [PubMed]
9. Park, Y.H.; Kim, Y.; Mishra, R.C.; Bae, H. Fungal endophytes inhabiting mountain-cultivated ginseng (*Panax ginseng* Meyer): Diversity and biocontrol activity against ginseng pathogens. *Sci. Rep.* **2017**, *7*, 16221. [CrossRef]
10. Eziashi, E.; Uma, N.U.; Adekunle, A.A.; Airede, C.E.; Odigie, E.E. Evaluation of lyophilized and non lyophilized toxins from *Trichoderma* species for the control of *Ceratocystis paradoxa*. *Afr. J. Agric. Res.* **2010**, 5, 1733–1738.
11. Zhu, H.X.; Ma, Y.Q.; Guo, Q.Y.; Xu, B.L. Biological weed control using *Trichoderma polysporum* strain HZ-31. *Crop Prot.* **2020**, *134*, 105161. [CrossRef]
12. Rüegger, A.; Kuhn, M.; Lichti, H.; Loosli, H.R.; Huguenin, R.; Quiquerez, C.; von Wartburg, A. Cyclosporin A, ein immunsuppressiv wirksamer Peptidmetabolit aus *Trichoderma polysporum* (Link ex Pers.) Rifai [Cyclosporin A, a Peptide Metabolite from *Trichoderma polysporum* (Link ex Pers.) Rifai, with a remarkable immunosuppressive activity. *Helv. Chim. Acta.* **1976**, *59*, 1075–1092. [CrossRef] [PubMed]
13. Fujita, T.; Iida, A.; Uesato, S.; Takaishi, Y.; Shingu, T.; Saito, M.; Morita, M. Structural elucidation of trichosporin-B-Ia, IIIa, IIId and V from *Trichoderma polysporum*. *J. Antibiot.* **1988**, *41*, 814–818. [CrossRef] [PubMed]
14. Qi, J.; Zhao, P.; Zhao, L.; Jia, A.; Liu, C.; Zhang, L.; Xia, X. Anthraquinone Derivatives from a Sea Cucumber-Derived *Trichoderma* sp. Fungus with Antibacterial Activities. *Chem. Nat. Compd.* **2020**, *56*, 112–114. [CrossRef]
15. Zhu, H.; Chen, H.Y.; Ma, Y.Q.; Guo, Q.Y. Identification and extraction of herbicidal compounds from metabolites of *Trichoderma polysporum* HZ-31. *Weed Sci.* **2023**, *71*, 39–49. [CrossRef]
16. Bhadauria, V.; Popescu, L.; Zhao, W.S.; Peng, Y.L. Fungal transcriptomics. *Microbiol. Res.* **2007**, *162*, 285–298. [CrossRef] [PubMed]
17. Xu, D.; Xue, M.; Shen, Z.; Jia, X.; Hou, X.; Lai, D.; Zhou, L. Phytotoxic Secondary Metabolites from Fungi. *Toxins* **2021**, *13*, 261. [CrossRef] [PubMed]
18. Wang, Y.; Wang, J.; Zhu, X.; Wang, W. Genome and transcriptome sequencing of *Trichoderma harzianum* T4, an important biocontrol fungus of *Rhizoctonia solani*, reveals genes related to mycoparasitism. *Can. J. Microbiol.* **2024**, *70*, 86–101. [CrossRef] [PubMed]
19. Luo, Q.C.; Tang, W.; Ma, J.K.; Chen, J.W.; Yang, D.J.; Gao, F.Y.; Sun, H.J.; Xie, Y.P.; Zhang, C.J. Transcriptome analysis of *Fusarium solani* infected sweet potato. *Acta Agric. Zhejiangensis* **2023**, *35*, 1097–1107.
20. Santiago, K.A.A.; Wong, W.C.; Goh, Y.K.; Tey, S.H.; Ting, A.S.Y. Pathogenicity of monokaryotic and dikaryotic mycelia of *Ganoderma boninense* revealed via LC-MS-based metabolomics. *Sci. Rep.* **2024**, *14*, 5330. [CrossRef]
21. Zhu, M. Cloning and Functional Verification of Key Genes for L-Phenylalanine Biosynthesis in Rose. Master's Thesis, Yangzhou University, Yangzhou, China, 2020.
22. Wang, B. Metabolic Engineering of Engineered Strains of Phenol Biosynthesis. Master's Thesis, Qufu Normal University, Qufu, China, 2016.
23. Jin, J.Y.; Diao, Y.F.; Yu, C.M.; Xiong, X.; Zhao, T.; He, B.L.; Liu, H.X. Transcriptome analysis of the process of infectation of apple branches by Fusarium fusarium. *J. Shandong Agric. Univ. (Nat. Sci. Ed.)* **2021**, *52*, 187–193.
24. Reid, M.F.; Fewson, C.A. Molecular characterization of microbial alcohol dehydrogenases. *Crit. Rev. Microbiol.* **1994**, *20*, 13–56. [CrossRef] [PubMed]
25. de Smidt, O.; du Preez, J.C.; Albertyn, J. The alcohol dehydrogenases of Saccharomyces cerevisiae: A comprehensive review. *FEMS Yeast Res.* **2008**, *8*, 967–978. [CrossRef] [PubMed]
26. DafaAlla, T.E.I.M.; Abdalla, M.; El-Arabey, A.A.; Eltayb, W.A.; Mohapatra, R.K. Botrytis cinerea alcohol dehydrogenase mediates fungal development, environmental adaptation and pathogenicity. *J. Biomol. Struct. Dyn.* **2022**, *40*, 12426–12438. [CrossRef] [PubMed]

27. Prosperini, A.; Berrada, H.; Ruiz, M.J.; Caloni, F.; Coccini, T.; Spicer, L.J.; Perego, M.C.; Lafranconi, A. A Review of the Mycotoxin Enniatin, B. *Front. Public Health* **2017**, *5*, 304. [CrossRef]
28. Herrmann, M.; Zocher, R.; Haese, A. Enniatin production by fusarium strains and its effect on potato tuber tissue. *Appl. Environ. Microbiol.* **1996**, *62*, 393–398. [CrossRef]

Journal of *Fungi*

MDPI

Article

Biological Control of Root Rot of Strawberry by *Bacillus amyloliquefaciens* Strains CMS5 and CMR12

Ruixian Yang [1,*], Ping Liu [1], Wenyu Ye [2,3,*], Yuquan Chen [1], Daowei Wei [1], Cuicui Qiao [1], Bingyi Zhou [1] and Jingyao Xiao [1]

[1] School of Environmental Engineering and Chemistry, Luoyang Institute of Science and Technology, Luoyang 471002, China; lylp76lp@163.com (P.L.); luomuling520@gmail.com (Y.C.); weidaowei05@163.com (D.W.); qiaocuicui2004@outlook.com (C.Q.); zby2680641714@outlook.com (B.Z.); 13091296356@139.com (J.X.)

[2] China National Engineering Research Center of JUNCAO Technology, College of JunCao Science and Ecology (College of Carbon Neutrality), Fujian Agriculture & Forestry University, Fuzhou 350002, China

[3] Technology Innovation Center for Monitoring and Restoration Engineering of Ecological Fragile Zone in Southeast China, Ministry of Natural Resources, Fuzhou 350002, China

* Correspondence: fairy19790805@163.com (R.Y.); wenyuye08@163.com (W.Y.)

Abstract: Strawberry root rot caused by *Fusarium solani* is one of the main diseases of strawberries and significantly impacts the yield and quality of strawberry fruit. Biological control is becoming an alternative method for the control of plant diseases to replace or decrease the application of traditional chemical fungicides. To obtain antagonistic bacteria with a high biocontrol effect on strawberry root rot, over 72 rhizosphere bacteria were isolated from the strawberry rhizosphere soil and screened for their antifungal activity against *F. solani* by dual culture assay. Among them, strains CMS5 and CMR12 showed the strongest inhibitory activity against *F. solani* (inhibition rate 57.78% and 65.93%, respectively) and exhibited broad-spectrum antifungal activity. According to the phylogenetic tree based on 16S rDNA and *gyrB* genes, CMS5 and CMR12 were identified as *Bacillus amyloliquefaciens*. Lipopeptide genes involved in surfactin, iturin, and fengycin biosynthesis were detected in the DNA genomes of CMS5 and CMR12 by PCR amplification. The genes related to the three major lipopeptide metabolites existed in the DNA genome of strains CMS5 and CMR12, and the lipopeptides could inhibit the mycelial growth of *F. solani* and resulted in distorted hyphae. The inhibitory rates of lipopeptides of CMS5 and CMR12 on the spore germination of *F. solani* were 61.00% and 42.67%, respectively. The plant-growth-promoting (PGP) traits in vitro screening showed that CMS5 and CMR12 have the ability to fix nitrogen and secreted indoleacetic acid (IAA). In the potting test, the control efficiency of CMS5, CMR12 and CMS5+CMR12 against strawberry root rot were 65.3%, 67.94% and 88.00%, respectively. Furthermore, CMS5 and CMR12 enhanced the resistance of strawberry to *F. solani* by increasing the activities of defense enzymes MDA, CAT and SOD. Moreover, CMS5 and CMR12 significantly promoted the growth of strawberry seedlings such as root length, seedling length and seedling fresh weight. This study revealed that *B. amyloliquefaciens* CMS5 and CMR12 have high potential to be used as biocontrol agents to control strawberry root rot.

Keywords: *Fragaria×ananassa*; *Fusarium solani*; *Bacillus amyloliquefaciens*; antifungal lipopeptides; biological control; growth-promoting effect

Citation: Yang, R.; Liu, P.; Ye, W.; Chen, Y.; Wei, D.; Qiao, C.; Zhou, B.; Xiao, J. Biological Control of Root Rot of Strawberry by *Bacillus amyloliquefaciens* Strains CMS5 and CMR12. *J. Fungi* **2024**, *10*, 410. https://doi.org/10.3390/jof10060410

Academic Editor: Ofir Degani

Received: 4 May 2024
Revised: 2 June 2024
Accepted: 4 June 2024
Published: 6 June 2024

1. Introduction

Strawberry (*Fragari×ananassa*) is the most economically important cultivated small berry plant in the world [1]. China is currently the world's largest producer of strawberry, accounting for about one-third of the global output [2]. Nevertheless, soil-borne plant pathogens are found in most strawberry-planted soils. Among soil-borne fungal diseases, root rot is considered the most important and destructive disease, leading to serious economic losses in the strawberry industry [3]. The main symptoms of root rot disease include the gradual blackening and decay of the plant root system with consequent suppression in vigor and yield of the strawberry plant [4]. Many fungal species can cause strawberry root rot, such as *Colletotrichum* spp., *Fusarium* spp., *Verticillium* spp., *Pestalotiopsis* spp., *Neopestalotiopsis* spp., *Rhizoctonia* spp., *Phytophthora* spp., *Pythium* spp., *Ilyonectria* spp., and *Dactylonectria* spp. [5–9]. Among those pathogens, *Fusarium* spp. are the major agents of strawberry root rot in all the main strawberry-growing areas of the world, and *F. solani* is considered to be the major pathogen in strawberry nurseries in many parts of China, such as Liaoning province, Jiangsu province, Guizhou province, Shandong province, and Xinjiang autonomous region [10–12], and it has also been reported as a strawberry pathogen in Iran, Spain, Italy, and Mexico along with other pathogens [13–15]. The control of strawberry root rot diseases is very challenging because several fungi usually work together, giving rise to its occurrence [16]. Several approaches are currently applied to control strawberry root rot, such as the breeding and selection of resistant cultivars, crop rotation, cultivation patterns, improved drainage and chemical plant protection, individually or combined [17]. The most common approach for the prevention and control of strawberry root rot is to use chemical fungicides, but strawberry fruits are consumed directly, and fungicide applications bring a range of issues such as the accumulation of toxin residue on the fruits, increased regulation, and negative impacts on the environment and human consumers, considering the increasing demand for sustainable agriculture [18]. Therefore, it is important to find a safer and more efficient biocontrol agent to control strawberry root rot.

Biological control is an efficient and environmentally friendly way to prevent strawberry root rot. Currently, many species of antagonistic microorganisms, such as *Streptomyces* spp., *Trichoderma* spp., and *Bacillus* spp., have been widely isolated and developed as agents to prevent strawberry root rot [19,20]. *Bacillus* spp., known as plant-growth-promoting rhizobacteria (PGPR), have received considerable attention as biocontrol agents [21]. Many *Bacillus* strains have been used to promote plant growth and control soil-borne plant pathogens [22,23]. *B. amyloliquefaciens*, the type species of *Bacillus* genus, isolated from the rhizosphere, has exhibited notable antifungal properties and the ability to promote plant growth [24]. *B. amyloliquefaciens* suppress the pathogens through various mechanisms, such as the production of antibiosis, competition for places and nutrients, and induction of plant defenses [25].

Inducing disease resistance (ISR) is the one of key mechanisms of *B. amyloliquefaciens* in its biocontrol effects. *B. amyloliquefaciens* enhances plant resistance to pathogens by activating ISR system of plant, which induces the production of defensive enzymes and disrupts normal physiological metabolism of the cells, thereby enhancing the ability to resist plant pathogens [26]. Currently, the defensive enzymes related to plant resistance against pathogen invasion mainly include peroxidase (POD), superoxide dismutase (SOD), catalase (CAT), polyphenol oxidase (PPO), and phenylalanine ammonia-lyase (PAL) [27]. Additionally, some studies have shown that *B. amyloliquefaciens* can promote plant growth by secreting plant growth hormones, non-symbiotic nitrogen fixation, solubilizing insoluble soil phosphates, producing extracellular polysaccharides, volatile organic compounds, and iron carriers [28].

Although previous studies have extensively explored the use of *Bacillus* species in controlling strawberry root rot diseases, the focus has mainly been on the control the pathogens including *R. solani*, *F. oxysporum*, and *C. gloeosporioides*. Studies on the screening of *Bacillus* strains is still relatively lacking regarding strawberry root rot caused by *F. solani*. Consequently, there remains a necessity to isolate and screen biocontrol strains against the pathogen *F. solani*. To achieve this goal, the objectives of this study were as follows: (1) to isolate, screen and identify rhizobacteria for controlling strawberry root rot caused by *F. solani*; (2) to evaluate the potential of *B. amyloliquefaciens* CMS5 and CMR12 for the reduction in strawberry root rot in vitro and in vivo; (3) to explore biocontrol mechanisms of strains CMS5 and CMR12 against *F. solani*; and (4) to assess growth promotion in strawberry seedling with strains CMS5 and CMR12 through pot experiments. This study revealed that *B. amyloliquefaciens* CMS5 and CMR12 can effectively control the occurrence of strawberry root rot, and had an obvious promotion effect on strawberry seedling growth, indicating that *B. amyloliquefaciens* CMS5 and CMR12 are promising biocontrol microorganisms for controlling strawberry root rot caused by *F. solani*.

2. Materials and Methods

2.1. Plant Pathogen and Plant Materials

In this study, *Fusarium solani* was isolated from strawberry cultivars "Hongyan" by using the tissue isolation method [29]. *F. solani* was identified based on the sequence of internal transcription intervals of ribosomal DNA (which were 99.9% identical to GenBank accession number MN461539). *Neopestalotiopsis clavispora* was isolated from strawberry fruit blight in March 2022. *F. oxysporum* was isolated from tree peony root rot in September 2020. *Alternaria alternata* was isolated from peony leaf spot in May 2022. All the plant pathogens were stored in the Laboratory of Microbiology in the Luoyang Institute of Science and Technology (Luoyang, Henan province, China). One-year-old strawberry seedlings of the commercial cultivar "Hongyan" or control effects test and growth-promoting effects were provided by the "Shilixiang" strawberry seedling cultivation facility (Luoyang, Henan province, China).

2.2. Isolation of Bacteria from Strawberry Rhizosphere Soil

Soil samples were collected from the healthy strawberry rhizosphere soil of the "Shilixiang" strawberry planting field (112°57′14.51″ E, 34°79′42.23″ N) in Luoyang, Henan province. Five samples were taken, with approximately 1 kg soil per sample collected from the vicinity of healthy strawberry plants within a 10 cm diameter of the tap root and up to 20 cm deep. After collection, the soil from each plant was thoroughly mixed. The mixed soil was then placed in plastic bags and transported to the laboratory.

The isolation for rhizosphere strains was slightly modified from Ding et al. [30]. Briefly, the soil samples were diluted from 10^{-1}- to 10^{-5}-fold with sterile distilled water, and a 100 μL dilution (10^{-5}-fold) was spread onto tryptic soy agar (TSA) plates and placed in 28 °C incubator for 2 days. Different single colonies were inoculated to fresh TSA. The pure strains were then stored in 20% glycerol at −80 °C.

2.3. Screening of Antagonistic Strains

Antimicrobial activity against plant pathogens *F. solani*, *N. clavispora*, *F. oxysporum*, and *A. alternata* was estimated by dual culture assay [31]. A fungal disc of 5 mm diameter was placed on the center of PDA plate, while the tested bacteria were parallel at 25 mm away from the center on both sides. PDA plates inoculated with plant pathogens only were used as control. The plates were kept at 28 °C for 7 days. All experiments were carried out three times with three repetitions each. The inhibition zone was recorded and the mycelial growth inhibition rate was calculated using the following formula:

$$\text{Inhibition rate (\%)} = (A - B)/A \times 100$$

where A is the diameter of the fungal colonies grown in the control, and B is the diameter of the fungal colonies grown with the bacterial culture.

2.4. Identification of Strains CMS5 and CMR12

Strains CMS5 and CMR12 were cultured on NA medium (beef extract 10 g/L, peptone 5 g/L, NaCl 5 g/L, agar 15 g/L) at 28 °C for 24 h, and then identified preliminarily by Gram-staining and morphological characteristics. The Gram reaction was performed, and morphological characteristics were observed under a microscope (Olympus, Shinjuku City, Japan). Strains CMS5 and CMR12 were further identified on the basis of 16S rDNA and *gyrB* sequences analysis. Strains CMS5 and CMR12 were inoculated into NB and cultured at 28 °C with 150 rpm shaking overnight. Genomic DNA was isolated using the TIANamp Bacterial DNA Kit (Tiangen Biotech Co., Beijing, China). The extracted DNA was used as a template to amplify 16S rDNA gene and gyrase gene (*gyrB*) [32,33]. The amplified primers and PCR conditions are presented in Table 1. PCR was performed using the TIANGEN Golden Easy PCR kit (Tiangen Biotech Co., Beijing, China). The PCR products were subjected to direct automated sequencing using fluorescent terminators on an ABI 377 Prism Sequencer (Sangon Biotech Co., Shanghai, China). The sequences were confirmed with a BLAST (Basic Local Alignment Search Tool) search of the NCBI (National Center for Biotechnology Information) database (https://www.ncbi.nlm.nih.gov/, accessed on 5 March 2024), and phylogenetic trees were constructed using the neighbor-joining (NJ) method, with 1000 bootstrap replications in the MEGA 10.0 package.

Table 1. Amplification sites, primer sequences and PCR conditions of antagonistic bacteria.

Gene	Primer	Primer Sequence (5′-3′)	PCR Conditions	Reference
16S rRNA	27F	AGAGTTTGATCCTGGCTCAG	94 °C for 5 min (94 °C for 30 s, 55 °C for 30 s, and 72 °C for 1 min) × 35 cycles, 72 °C for 10 min	[32]
	1492R	TACGGCTACCTTGTTACGACTT		
gyrB	UP-1S	GAAGTCATCATGACCGTTCTGCA		[33]
	UP-2Sr	AGCAGGGTACGGATGTGCGAGCC		

2.5. Characterization of Antifungal Lipopeptides Substances of Strains CMS5 and CMR12

Three primer pairs were used for the detection of genes (*fenA*, *srfAA*, *ituD*) of CMS5 and CMR12 involved in lipopeptide synthesis [34,35]. The primers and PCR conditions used for amplification are presented in Table 2. PCR was performed and sequenced using the same method as in Section 2.4. The sequences were confirmed with a BLAST (Basic Local Alignment Search Tool) search of the NCBI (National Center for Biotechnology Information) database (https://www.ncbi.nlm.nih.gov/, accessed on 10 March 2024).

Table 2. Primer sequence and PCR conditions for lipopeptide biosynthesis genes of antagonistic bacteria CMS5 and CMR12.

Gene	Primer	Primer Sequence (5′-3′)	Lipopeptide	PCR Conditions	Reference
fenA	FenAa	AAGAGATTCAGTAAGTGGCCCATCCAG	Fengycin	94 °C for 5 min (94 °C for 30 s, 55 °C for 30 s, and 72 °C for 1 min) × 35 cycles, 72 °C for 10 min	[34]
	FenAb	CGCCCTTTGGGAAGAGGTGC			
srfAA	Srfkn-1	AGCCGTCCTGTCTGACGACG	Surfactin		[35]
	Srfkn-2	TCTGCTGCCATACCGCATAGTC			
ituD	ItuD-f	ATGAACAATCTTGCCTTTTTA	Iturin		
	ItuD-r	TTATTTAAATTCCCCAATT			

The production of lipopeptides of strains CMS5 and CMR12 on Landy medium was examined following a predescribed protocol of Kim et al. [36]. The antifungal activity of crude extracts of lipopeptides produced by CMS5 and CMR12 was evaluated against *F. solani* by Oxford cup method [37]. A volume of 100 μL of crude extracts was dropped into an Oxford cup placed 15 mm from the edge of a Petri plate. A plug (5 mm diameter) of *F. solani* was placed in the center. A volume of 100 μL of methanol was used as control. The plates were then incubated for 7 days at 28 °C, and the inhibition zone was recorded.

2.6. Effects of Lipopeptides of Strains CMS5 and CMR12 on the Pathogenic Fungal Mycelial Morphology and Spore Germination

The inhibitory effect of lipopeptide substances of strains CMS5 and CMR12 on mycelial growth was slightly modified from Pan et al. [38]. Briefly, 0.2 mL from spore suspension of *F. solani* with 1×10^6 spores/mL was inoculated into 8 mL of PDB medium and incubated at 28 °C with shaking at 140 rpm for 60 h. The hyphae of *F. solani* were collected by centrifugation (10,000 rpm, 5 min) and washed with PBS buffer. Then, 0.8 mL of lipopeptides (concentration 100 mg/L) of strains CMS5 and CMR12 was added. The mixture was cultured at 28 °C with shaking at 140 rpm for 24 h. The mycelial morphology of *F. solani* was observed one time per 4 h under a microscope (Olympus). The sterile distilled water was served as a control. Each treatment was repeated three times.

For spore preparation, an equal volume of lipopeptides (concentration 100 mg/L) of strains CMS5 and CMR12 was mixed with spore suspension of *F. solani* (concentration 1×10^6 spores/mL). A 10 μL drop of the mixture was placed on a concave glass slide, covered with a cover slip, and placed in a sterilized petri dish. The setup was incubated at 28 °C with moisture maintained, and spore germination was observed one time per 4 h under a microscope (Olympus). For each treatment, 100 spores were observed, and observations continued for up to 28 h. The sterile distilled water was served as a control. Each treatment was repeated three times. The germination rate and inhibition rate of pathogenic fungal spores were calculated using the following formula:

$$\text{Germination rate (\%)} = (C/D) \times 100;$$

$$\text{Inhibition rate (\%)} = (E - F)/E \times 100.$$

where C is the number of germinated spores, and D is the total observed spores. E is the conidial germination rate in the control, and F is the conidial germination rate in the treatment.

2.7. Determination of Plant Growth-Promoting Traits of Strains CMS5 and CMR12

Strains CMS5 and CMR12 were individually inoculated on nitrogen-free Ashby medium [39] and CAS agar medium [40], and incubated at 28 °C for 3 days. The abilities to fix nitrogen and secrete iron carriers were evaluated according to the growth condition of the strains in Ashby medium and CAS agar medium [41]. For phosphate solubilization, strains CMS5 and CMR12 were inoculated on the PKO inorganic phosphate medium and incubated at 28 °C for 5 days. The phosphate solubilization capabilities were assessed by the clear zone method [42]. IAA production of strains CMS5 and CMR12 was determined using Salkowski's reagent [43].

2.8. Control Effects of Strains CMS5 and CMR12 on Strawberry Root and Defensive Enzyme Activity

Strains CMS5 and CMR12 were inoculated into 5 mL of NB medium and cultured to stationary phase. Then, 100 μL of CMS5 and CMR12 NB culture was inoculated into 100 mL PDB medium and cultivated at 28 °C for 40 h at 180 rpm. The concentration of CMS5 and CMR12 PDB culture was adjusted to 1×10^8 cfu/mL. *F. solani* was inoculated at the center of PDA plates at 28 °C for 7 d. Conidial suspensions (1×10^8 spores/mL) of pathogen were prepared and then stored at 4 °C for later use. One-year-old strawberry seedlings of the commercial cultivar "Hongyan" were used. The experiment included five treatments: (1) strawberry seedling+water (T1); (2) strawberry seedling+*F. solani* (T2); (3) strawberry seedling+*F. solani*+CMS5 (T3); (4) strawberry seedling+*F. solani*+CMR12 (T4); and (5) strawberry seedling+*F. solani*+CMS5+CMR12 (T5). After strawberry seedling acclimation for 20 d, 5 mL *F. solani* conidia (1×10^8 spores/mL) was inoculated with strawberry roots. After 5 days of *F. solani* treatment, 20 mL CMS5 and CMR12 PDB culture were used to inoculate the soil around each seedling through a root irrigation method. Each treatment included 5 pots with three repetitions each. All the treatments were incubated for 60 days at 25 °C and 80% relative humidity. The disease severity of the strawberry seedlings was assessed using a scoring system of 0–5, from the report of Vestberg et al. [44]. The germination rate and inhibition rate of pathogenic fungal spores were calculated using the following formula:

$$\text{Disease Index} = \sum (\text{disease level} \times \text{number of plants at that level})/(\text{total number of plants} \times \text{highest disease level}) \times 100;$$

$$\text{Control efficiency (\%)} = (\text{control disease index} - \text{treatment disease index})/\text{control disease index} \times 100.$$

To determine the induced resistance to *F. solani* by strains CMS5 and CMR12, strawberry seedlings leaves were randomly collected at 0, 36, 72, 120, and 168 h after five treatments (T1, T2, T3, T4 and T5) to detect the activity of superoxide dismutase (SOD), malondialdehyde (MDA) and catalase (CAT). MDA content in strawberry leaves was determined using the thiobarbituric acid method [45]. CAT activity was measured using ultraviolet spectrophotometry, and SOD activity was determined using the nitroblue tetrazolium method [46]. All experiments were conducted in triplicate.

2.9. Growth-Promoting Effects of Strains CMS5 and CMR12 on Strawberry Seedlings

The same method used in Section 2.8 was used in this experiment, which consisted of four treatments: (1) strawberry seedling+water (T1); (2) strawberry seedling+CMS5 (T2); (3) strawberry seedling+CMR12 (T3); and (4) strawberry seedling+CMS5+CMR12 (T4). Each treatment included 5 pots with three repetitions each. All the treatments were incubated 60 days at 25 °C and 80% relative humidity. The strawberry seedlings were carefully excavated, and their height, root length, and fresh weight were measured.

2.10. Data Statistics and Analysis

Data obtained from the experiments were processed using a one-way analysis of variance (ANOVA) was performed using DPS 7.05 statistical software. Duncan's new multiple range test was used to assess the significant differences, and the significance level was set at $p < 0.05$.

3. Results

3.1. Isolation and Screening of Rhizosphere Soil Bacteria

The total 72 bacterial strains were isolated from strawberry rhizosphere soil. Their antagonistic activity against *Fusarium solani* was measured using plate confrontation method. The results showed that two strains CMS5 and CMR12 exhibited strong antagonistic activity against *F. solani*. The inhibition zones of CMS5 and CMR12 against *F. solani* were 5.00 ± 0.00 mm and 6.00 ± 0.10 mm, respectively. The inhibition rates of CMS5 and CMR12 against *F. solani* were 57.78% and 65.93%, respectively (Table 3, Figure 1). Meanwhile, CMS5 and CMR12 also exhibited inhibitory effects against *F. oxysporum*, *Neopestalotiopsis clavispora*, and *Alternaria alternata*. The strain CMS5 displayed inhibition zones of 3.70 mm, 2.30 mm, and 10.10 mm against these pathogens, and the inhibition rates to the three pathogens were 65.92%, 59.26%, and 70.37%, respectively. The strain CMR12 had inhibition zones of 4.30 mm, 4.00 mm, and 11.30 mm against the three pathogens, with inhibition rates of 70.37%, 61.48%, and 71.85%, respectively(Table 3, Figure 1).

Table 3. Inhibitory effect of the rhizosphere bacteria CMS5 and CMR12 against four different plant pathogens on PDA plates.

Pathogenic Strains	CMS5		CMR12	
	Inhibition Zones (mm)	**Inhibitory Rate (%)**	**Inhibition Zones (mm)**	**Inhibitory Rate (%)**
F. solani	5.00 ± 0.00 a	57.78 ± 0.04 a	6.00 ± 0.10 a	65.93 ± 0.01 a
F. oxysporum	3.70 ± 0.12 a	65.92 ± 0.01 b	4.30 ± 0.06 a	70.37 ± 0.03 a
N. clavispora	2.23 ± 0.06 b	59.26 ± 0.01 b	4.00 ± 0.10 a	61.48 ± 0.01 a
A. alternata	10.01 ± 0.12 a	70.37 ± 0.03 a	11.3 ± 0.06 a	71.85 ± 0.01 a

Note: Data were mean ± SD of triplicates. Means were tested with Duncan's multiple range test for the last column. Means followed by the same letter are not significantly different ($p < 0.05$) for the same column. Lowercase letters in the same column indicate differences between CMS5 and CMR12 against for each pathogenic strain.

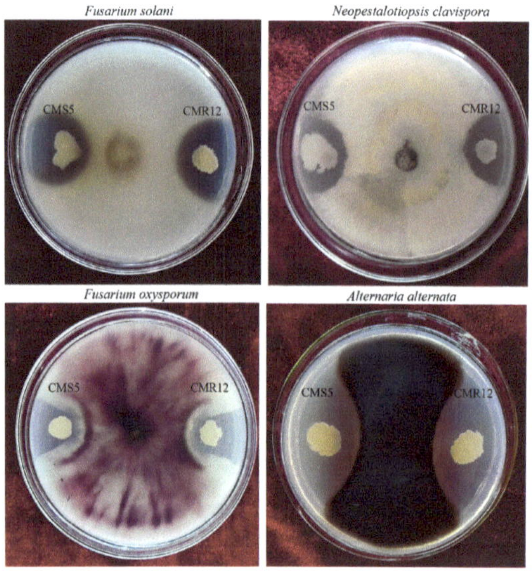

Figure 1. Inhibitory effect of the rhizosphere bacteria against four different plant pathogens on PDA plates.

3.2. Identification of Strains CMS5 and CMR12

The strains CMS5 and CMR12 formed circular, milky white colonies with neat edges, a moist surface, slight convexity, and opacity on NA medium. CMS5 and CMR12 were Gram-positive and rod-shaped, rounded at both ends. Based on 16S rDNA sequence alignment, strains CMS5 and CMR12 obtained 1450 bp and 1443 bp sequence fragments, with GenBank accession numbers PP338070 and PP338071, respectively. The *gyrB* genes sequence fragments of strains CMS5 and CMR12 were 1162 bp and 1152 bp, and the GenBank accession numbers were PP355738 and PP355739, respectively. Phylogenetic trees constructed based on the 16S rDNA and *gyrB* gene sequences showed that strains CMS5, CMR12 and *Bacillus amyloliquefaciens* were in the same branch (Figure 2A,B). Based on the combined morphological and molecular characteristics, strains CMS5 and CMR12 were identified as *B. amyloliquefaciens*.

3.3. Antifungal Activity of Lipopeptide Substances of Strains CMS5 and CMR12

Gene fragments of *ituD*, *fenA*, and *srfAA* were obtained from the DNA of strains CMS5 and CMR12. To further confirm the genes corresponding to the antimicrobial substances, gene-specific bands were recovered and cloned for sequencing, and sequence analysis was performed using BLASTX (https://www.ncbi.nlm.nih.gov/blast.cgi, accessed on 10 March 2024) in GenBank. It was found that the sequences of the *ituD*, *fenA*, and *srfAA* gene fragments from strains CMS5 and CMR12 exhibited 99% similarity to the sequences of the non-ribosomal peptide synthetase proteins from *B. amyloliquefaciens*, specifically the iturin A, bacillomycin D synthetase, fengycin synthetase, and surfactin synthetase proteins, respectively (Table 4). The results indicated that the genes related to the three major metabolites (fengycin, surfactin and iturin) exist in the genomes of strains CMS5 and CMR12. The lipopeptide extract of strains CMS5 and CMR12 had an obvious inhibitory effect on *F. solani* in vitro. The diameters of inhibition zones induced by lipopeptide extract were 7.50 ± 0.20 mm and 10.50 ± 0.45 mm, respectively (Figure 3).

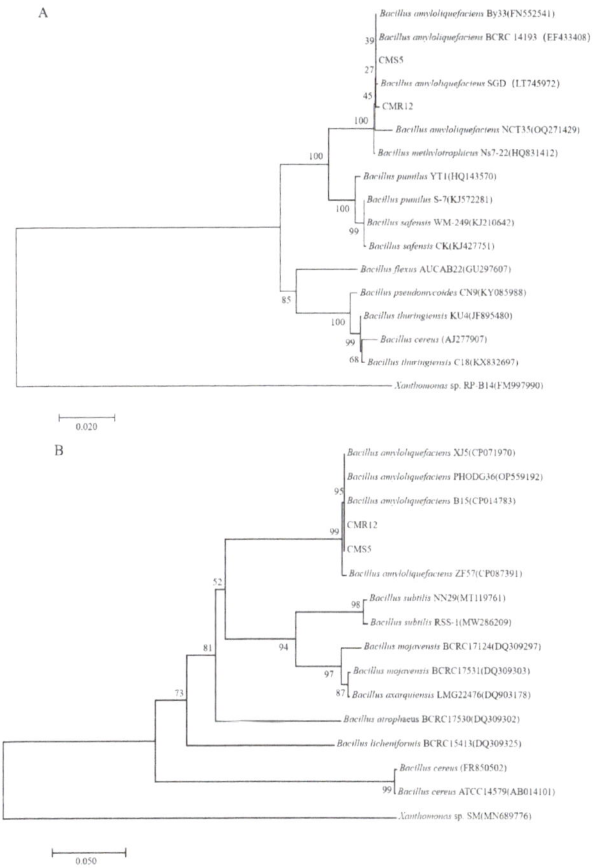

Figure 2. Phylogenetic trees based on the 16S rDNA sequence (**A**) and *gyrB* sequence (**B**) of CMS5 and CMR12 and their homologous sequences. Phylogenetic trees were constructed by the neighbor-joining method of MEGA10.0 with bootstrap values based on 1000 replications. *Xanthomonas* sp. RP-B14 and *Xanthomonas* sp. SM were chosen as the outgroup. Gene accession numbers of bacterial strains are indicated in parentheses. The scale bar represents the number of substitutions per base position.

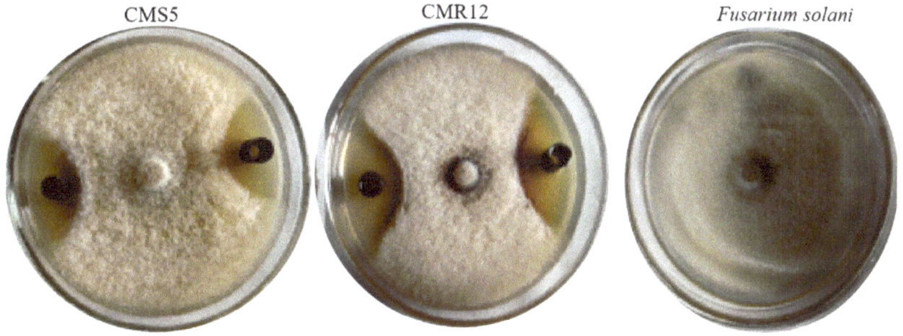

Figure 3. Inhibition effect of lipopeptides of strains CMS5 and CMR12 against *F. solani* on PDA plates.

Table 4. Mimilar protein sequences of lipopeptide synthetase genes of strains CMS5 and CMR12 generated from NCBI using BLASTX.

Strain	Genes	Product Size (bp)	Similar Protein Sequences	Accession Numbers	Identity (%)
CMS5	*ituD*	1122	bacillomycin D biosynthesis malonyl-CoA transacylase BamD (*B. amyloliquefaciens*)	WP_061860597	99%
	fenA	1365	non-ribosomal peptide synthetase (*B. amyloliquefaciens*)	WP_060674287	99%
	srfAA	1464	surfactin non-ribosomal peptide synthetase SrfAA (*B. amyloliquefaciens*)	WP_223255265	99%
CMR12	*ituD*	1121	bacillomycin D biosynthesis malonyl-CoA transacylase BamD (*B. amyloliquefaciens*)	WP_094246888	99%
	fenA	1369	non-ribosomal peptide synthetase (*B. amyloliquefaciens*)	WP_243939152	99%
	srfAA	1463	surfactin non-ribosomal peptide synthetase SrfAA (*B. amyloliquefaciens*)	WP_241457877	99%

3.4. Inhibitory Effects of Lipopeptides of Strains CMS5 and CMR12 on the Pathogenic Fungal Mycelial Morphology and Spore Germination

As shown in Figure 4, 12 h after treatment with the lipopeptides of strains CMS5 and CMR12, the mycelia of *F. solani* exhibited partial staining. Some hyphal tips were swollen, and there was an increase in hyphal septum thickness. After 16 h, nearly all the mycelia of *F. solani* were stained, with a large number of hyphal tips swollen and a few hyphae breaking. After 20 h, the permeability of the treated hyphae further increased, extensive staining of the hyphal was observed, and numerous hyphae were broken. After 24 h, the mycelia of *F. solani* treated with the lipopeptides fragmented completely. In contrast, the mycelial morphology of *F. solani* treated with distilled water remained smooth and intact, and produced conidia. The results indicate that the lipopeptides of strains CMS5 and CMR12 can cause deformities and breakage of the mycelia of of *F. solani*, increasing the permeability of hyphal cells' membranes.

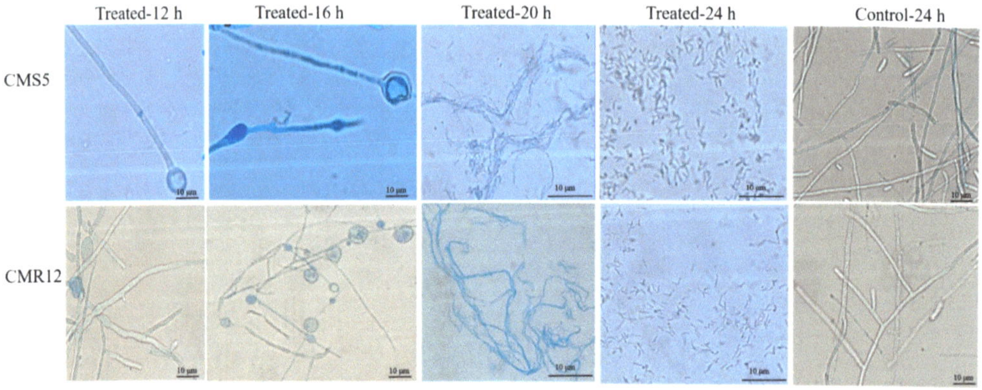

Figure 4. Inhibitory effects of lipopeptide substances of strains CMS5 and CMR12 on the mycelial morphology of *F. solani*.

To further understand the effects of lipopeptides of CMS5 and CMR12 on spores of *F. solani*, the germination and inhibition rates of spores were analyzed. As shown in Figure 5A,B, the spore germination of *F. solani* was inhibited. After 8 h treatment with lipopeptides of CMS5 and CMR12, the spore germination rates were only 9.00% and 4.00%, and the spore germination rate of the control treatment was 39.00%. The inhibition rates of lipopeptides of CMS5 and CMR12 were 76.76% and 90.15%, respectively. After 28 h, all spores of the control group had germinated, whereas the germination rates of lipopeptides of CMS5 and CMR12 were 61.00% and 42.67%, with inhibition rates of 39.00% and 57.00%. Microscopic observation showed that spores treated with lipopeptides were deformed, swelling at the ends or middle. Conversely, the spores germinated normally and formed mycelia in the control. The results indicated that the lipopeptides of CMS5 and CMR12 significantly delay the germination of spores of *F. solani*. Additionally, the lipopeptides of CMS5 and CMR12 also caused spores deformities, thereby inhibiting their germination.

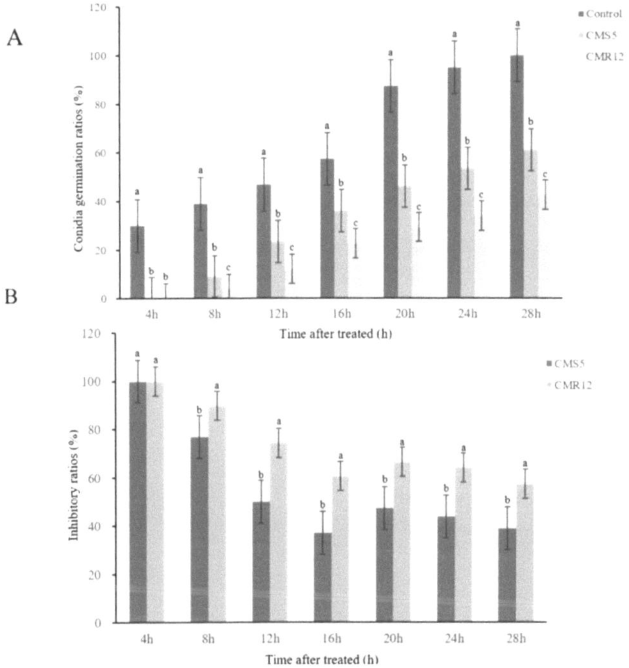

Figure 5. Inhibitory effects of lipopeptides of strains CMS5 and CMR12 on the conidia germination of *F. solani*. (**A**) Conidia germination ratios; (**B**) inhibitory ratios of conidia germination. Bars indicate the standard error of the mean. Columns marked with the same letter are not significantly different according to Duncan's multiple range test at $p < 0.05$.

3.5. Plant Growth-Promoting Traits of Strains CMS5 and CMR12

Strains CMS5 and CMR12 successfully grew on Ashby medium, indicating that they had nitrogen-fixing capabilities (Figure 6A). However, strains CMS5 and CMR12 could not dissolve phosphate and did not produce siderophore. The supernatant of strains CMS5 and CMR12 was mixed with the Salkowski reagent, and the mixed solution turned pink, indicating that CMS5 and CMR12 produced IAA. The concentration of IAA synthesized by strains CMS5 and CMR12 was 25.68 mg/L and 32.97 mg/L (Figure 6B).

3.6. Biocontrol Efficiency of Strains CMS5 and CMR12 against Strawberry Root Rot

The incidence of strawberry root rot on each treatment was investigated after inoculation for 60 days (Table 5). The results showed that strains CMS5 and CMR12 had a strong control effect on root rot of strawberry (Figure 7). As shown in Table 5, the disease index of the control group (T2) inoculated only with the pathogen was 83.30 ± 1.66, whereas the disease index of the CMS5-treated group (T3), CMR12-treated group (T4), and CMS5+CMR12-treated group (T5) were 28.90 ± 1.57, 26.70 ± 0.80, and 10.00 ± 1.20, respectively. The biocontrol efficiency of T3, T4 and T5 group against strawberry root rot were 65.3 ± 0.07%, 67.94 ± 0.10% and 88.00 ± 0.06%, respectively. The strawberry treated with the distilled water group showed good growth and no signs of disease. The results indicated that strains CMS5 and CMR12 effectively controlled the occurrence of potted strawberry root rot, and the CMS5+CMR12-treated group showed significantly better biocontrol efficiency than the individual strain treatments.

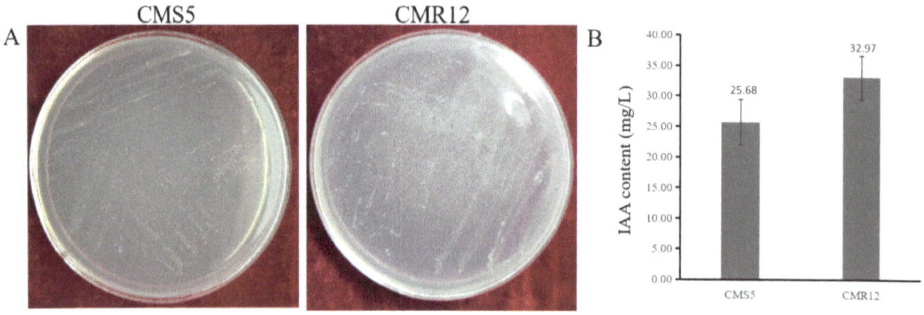

Figure 6. Evaluation of growth-promoting traits of strains CMS5 and CMR12. (**A**) Nitrogen-fixing capabilities of CMS5 and CMR12 on Ashby's medium; (**B**) the concentration of IAA synthesized by strains CMS5 and CMR12 by Salkowski colorimetric method.

Table 5. Biocontrol efficiency of CMS5 and CMR12 against strawberry root rot by different treatments.

Treatments	Disease Index	Biocontrol Efficiency (%)
T1 (Water-treated)	0.00 ± 0.00 d	-
T2 (*F. solani*-treated)	83.30 ± 1.66 a	-
T3 (*F. solani*+CMS5-treated)	28.90 ± 1.57 b	65.30 ± 0.07 b
T4 (*F. solani*+CMR12-treated)	26.70 ± 0.80 b	67.94 ± 0.10 b
T5 (*F. solani*+CMS5+CMR12-treated)	10.00 ± 1.20 c	88.00 ± 0.06 a

Note: Data are mean ± SD of triplicates. Means were tested with Duncan's multiple range test for the last column. Means followed by the same letter are not significantly different ($p < 0.05$) for the same column.

Figure 7. Biocontrol efficiency of CMS5 and CMR12 against strawberry root rot by different treatments.

3.7. Effects of Strains CMS5 and CMR12 on the Activity of Strawberry Defense-Related Enzyme Activities

To determine whether the defense system is activated in response to strains CMS5, CMR12 against *F. solani*, the activity of three defense enzymes (MDA, CAT, and SOD) were selected. Compared with the T2 group (inoculated only with the pathogen), the content of MDA in the T3, T4 and T5 groups (CMS5-treated group, CMR12-treated group, and CMS5+CMR12-treated group) began to decrease 36 h after inoculation. The MDA content was lower in the T5 group than in the T3 and T4 groups, indicating that CMS5 and CMR12 effectively inhibit MDA synthesis in strawberry leaves, with CMS5+CMR12-treated group proving more effective than treatments with individual strains (Figure 8A). The activities of CAT and SOD in the T2 group began to decrease 36 h after inoculation, while the activities of CAT and SOD were significantly higher in the T3, T4 and T5 groups. Seventy-two hours after inoculation, the activities of CAT and SOD in the T5 group were higher than in the T2, T3 and T4 groups (Figure 8B,C). These results showed that strains CMS5 and CMR12 can induce an increase in the defensive enzyme activities of CAT and SOD in strawberries, and CMS5+CMR12-treated group proved more effective than treatments with individual strains. Therefore, strains CMS5 and CMR12 could rapidly activate the defense enzyme system in strawberry plants, enabling strawberries to respond rapidly to *F. solani* infection and improve their disease resistance.

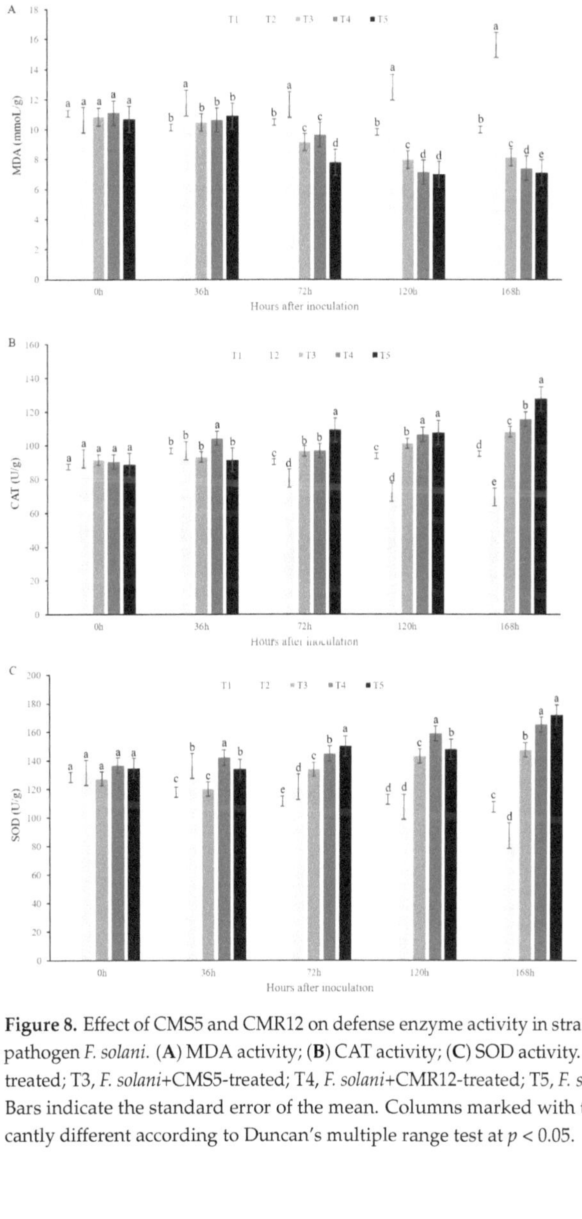

Figure 8. Effect of CMS5 and CMR12 on defense enzyme activity in strawberry leaves inoculated with pathogen *F. solani*. (**A**) MDA activity; (**B**) CAT activity; (**C**) SOD activity. T1, water-treated; T2, *F. solani*-treated; T3, *F. solani*+CMS5-treated; T4, *F. solani*+CMR12-treated; T5, *F. solani*+CMS5+ CMR12-treated. Bars indicate the standard error of the mean. Columns marked with the same letter are not significantly different according to Duncan's multiple range test at $p < 0.05$.

3.8. Growth Promotion Effects of Strains CMS5 and CMR12 on Strawberry Seedlings

In the greenhouse, after the strawberry seedlings were treated with strains CMS5 and CMR12 fermentation broth for 60 days, compared with the control, strains CMS5 and CMR12 significantly enhanced the plant height, root length, and fresh weight of the strawberry seedlings (Table 6, Figure 9). The growth promotion rates for strain CMS5 on plant height, root length, and fresh weight were 8.89%, 34.18%, and 99.84%, respectively. For strain CMR12, the corresponding rates were 42.22%, 18.21%, and 95.62%, respectively. When strains CMS5 and CMR12 were co-inoculated, the growth promotion rates for plant height, root length, and fresh weight reached 25.65%, 26.51%, and 159.16%, respectively.

Table 6. Growth--promoting effects of CMS5 and CMR12 on strawberry seedings.

Treatments	Plant Height (cm)	Root Length (cm)	Total Fresh Weight (g)
T1 (water)	20.86 ± 0.65 b	16.50 ± 0.36 c	14.57 ± 1.15 c
T2 (CMS5)	28.00 ± 0.78 a	17.97 ± 0.76 c	29.12 ± 2.83 b
T3 (CMR12)	24.67 ± 2.32 a	23.47 ± 0.83 a	28.50 ± 1.69 b
T4 (CMS5+CMR12)	26.40 ± 2.94 a	20.73 ± 4.84 ab	37.76 ± 4.78 a

Note: Data are mean ± SD of triplicates. Means were tested with Duncan's multiple range test for the last column. Means followed by the same letter are not significantly different ($p < 0.05$) for the same column.

Figure 9. Growth promoting effects of strains CMS5 and CMR12 on strawberry seedings.

4. Discussion

4.1. Significance of Exploring Biocontrol Resources for Plant Disease

Strawberry root rot is a complex disease that significantly challenges agricultural production, especially in protected cultivation environments where economic losses can be substantial [47]. *Bacillus amyloliquefaciens* is known for producing various secondary metabolites and exhibits broad-spectrum antifungal activities. It is widely used as a biocontrol agent in the management of crop diseases and also acts as a plant-growth-promoting rhizobacterium, significantly enhancing plant growth [48]. Some studies have demonstrated that *B. amyloliquefaciens* play a pivotal role in the control of strawberry root rot. Wu et al. [49] found that strains *B. amyloliquefaciens* PMB04 and PMB05 have demonstrated effective antagonism against *Colletotrichum gloeosporioides*, significantly reducing the incidence of anthracnose on strawberry seedlings. Chen et al. [50] obtained a strain of *B. amyloliquefaciens* CM3, which achieved a control efficacy of 64.86% against root rot caused by *F. oxysporum*, and promoted strawberry growth, indicating that strain CM3 is suitable for development into a microbial formulation. These studies provide essential microbial resources for the biological control of strawberry root rot. However, due to the diverse composition of root rot pathogens, most studies have focused on *B. amyloliquefaciens* against *C. gloeosporioides* and *F. oxysporum* [51,52]. Studies focusing on *F. solani*, a pathogen associated with strawberry root rot, are scarce. This study focuses on *F. solani* as the target pathogen, obtaining two strains of *B. amyloliquefaciens*, CMS5 and CMR12, which have demonstrated significant biocontrol efficacy against strawberry root rot and notably promote the growth of strawberry seedlings. These findings contributed to the development of biocontrol resources for managing strawberry root rot and broadened the potential applications of *B. amyloliquefaciens* in agriculture.

A potential biocontrol strain may have a broad antimicrobial spectrum. Previous studies have shown that most antagonistic *B. amyloliquefaciens* strains have broad-spectrum activity [53]. The type strain of the *Bacillus* species, *B. amyloliquefaciens* FZB42, is commercially used as biocontrol bacterium, being especially efficient against a broad spectrum of pathogens, such as *F. graminearum*, *Rhizoctonia solani*, *Magnaporthe oryzae*, and *Phytophthora nicotianae* [54]. Similarly, in present study, we screened two broad-spectrum bacteria, CMS5 and CMR12, which exhibited obvious inhibitory activity against plant pathogenic *Alternaria alternata*, *F. oxysporum*, and *Neopestalotiopsis clavispora* on PDA medium. Strains CMS5 and CMR12 showed an especially strong effect in inhibiting the mycelial growth of *A. alternata*. The results showed that strains CMS5 and CMR12 may have potential application prospects in controlling other fungal diseases.

4.2. Biocontrol Mechanisms of B. amyloliquefaciens CMS5 and CMR12

Current research indicates that the suppression of various plant diseases by *B. amyloliquefaciens* involves mechanisms including competitive action, antibiotics, inducing plant resistance (ISR), and plant-growth promotion. Antibiotics are a pivotal mechanism in controlling plant diseases through the secretion of lipopeptide substance, polyketide compounds, and antimicrobial proteins [55]. Among these secondary metabolites, lipopeptides are extensively studied as antimicrobial agents and are categorized into three main groups based on their structure: iturins, surfactins, and fengycins [56]. Studies have demonstrated that the inhibitory effects of lipopeptides on pathogens primarily manifest in the suppression of mycelial growth and impact spore germination. For instance, Kang et al. [57] reported that the culture filtrate of *B. amyloliquefaciens* PPL containing fengycin can inhibit plant wilt diseases caused by *F. oxysporum* both in vitro and in vivo by affecting the growth of spores and mycelia. Jiao et al. [58] discovered that bacillomycin D produced by *B. amyloliquefaciens* YN201732 could inhibit spore germination of powdery mildew, effectively controlling tobacco powdery mildew. Additionally, Chowdhury et al. [59] found that the bacillomycin D of *B. amyloliquefaciens* FZB42 could directly inhibit the growth of *R. solani* in lettuce and alter mycelial morphology, conidial cell walls, and cytoplasmic membranes. In this study, lipopeptides of strains CMS5 and CMR12 exhibited strong inhibitory effects on

F. solani causing strawberry root rot. The biocontrol mechanisms primarily involve causing deformities and breakage of *F. solani* mycelia, delaying spore germination, and inducing deformities in the spores. These results aligned with previously reported studies on the biocontrol mechanisms of *B. amyloliquefaciens*. Future work could aim to further purify and identify the lipopeptide substances produced by strains CMS5 and CMR12 that are crucial for antimicrobial activity. Techniques such as scanning electron microscopy, fluorescence microscopy, and comparative transcriptomic analysis could elucidate the antimicrobial mechanisms of high-purity lipopeptide metabolites from CMS5 and CMR12 against *F. solani* at cellular, physiological, biochemical, and molecular biology levels.

ISR is another key mechanism of *B. amyloliquefaciens* in its biocontrol effects. ISR is induced by non-pathogenic strains, attenuated strains, proteins or glycoproteins, extracellular polysaccharides, and lipopolysaccharides, which activate defense enzymes in the plant such as PAL, POD, SOD, CAT, and PPO. These enzymes facilitate the development of ISR, enhancing the ability of plants to resist pathogen invasion [60,61]. SOD and CAT are integral to the intracellular defense enzyme system; the accumulation of reactive oxygen species and free radicals can damage the cell membrane system, affecting plant health. The function of SOD and CAT is to scavenge reactive oxygen species and free radicals to enhance plant resistance to pathogens [62,63]. Changes in the content of MDA are directly related to plant disease resistance. When plants are damaged, the disruption of the reactive oxygen metabolism system leads to an increase in the MDA content in lipids, increases membrane permeability, and reduces plant resistance [64,65]. Pei et al. [66] showed that *B. amyloliquefaciens* Oj-2.16 enhances resistance in tomato plants to *Verticillium* wilt by increasing the activities of defense-related enzymes CAT and SOD. Compared to tomato seedlings inoculated with *V. dahliae*, the treated with strain Oj-2.16 exhibited significantly reduced MDA content. Our study found that after 36 **h** of treatment with strains CMS5 and CMR12, there was an increasing trend of the activities of CAT and SOD, and a decreasing trend of MDA content, whereas in the control group inoculated only with the pathogen, the activities of CAT and SOD decreased, and MDA content increased. This result showed that strains CMS5 and CMR12 increased strawberry resistance to root rot by increasing the activity of some defense. However, the mechanisms of strains CMS5 and CMR12 inducing systemic resistance in strawberries remain to be fully elucidated. Previous studies have suggested that *B. amyloliquefaciens* may contribute to inducing plant systemic resistance through the production of various secondary metabolites. Therefore, there is a need to further explore the effects of the secondary metabolites of CMS5 and CMR12 on inducing systemic resistance in strawberries through gene expression analysis, proteomics and metabolomics analysis.

4.3. Practical Applications of *B. amyloliquefaciens* CMS5 and CMR12

This study demonstrated that *B. amyloliquefaciens* CMS5 and CMR12 exhibit antimicrobial effects and induce systemic resistance against *F. solani*, which causes strawberry root rot. Under greenhouse conditions, strains CMS5 and CMR12 effectively control the occurrence of strawberry root rot. However, the biocontrol mechanisms at various interaction stages between the plant and the pathogen still require further investigation to provide a theoretical basis for the application of *B. amyloliquefaciens* CMS5 and CMR12 under strawberry field conditions. Consequently, further exploration of interactions among *B. amyloliquefaciens*, the root rot pathogen, and the host will be a primary focus of future research. Additionally, investigating the synergistic effects of CMS5 and CMR12 with other strains, chemical pesticides and fertilizers, and assessing their effectiveness under field conditions to develop integrated antimicrobial and fertilizing solutions, will also be a key point of future work.

5. Conclusions

In summary, *B. amyloliquefaciens* CMS5 and CMR12 were obtained among 72 rhizosphere bacteria strains as potent biocontrol agent against *F. solani*, the pathogenic fungi causing strawberry root rot. The results of the pot experiment demonstrated that *B. amyloliquefaciens* CMS5 and CMR12 effectively inhibited root rot and significantly enhanced the growth of strawberry seedlings. These findings indicate that *B. amyloliquefaciens* CMS5 and CMR12 have great potential as a biocontrol resource for preventing and controlling strawberry root rot, making it a promising candidate for future development.

Author Contributions: Conceptualization, R.Y.; methodology, P.L., W.Y., Y.C. and C.Q.; software, D.W., B.Z., J.X., C.Q. and R.Y.; formal analysis, P.L., Y.C. and W.Y.; writing—original draft preparation, R.Y.; visualization, D.W., B.Z., J.X. and R.Y.; project administration, R.Y.; funding acquisition, R.Y. and W.Y. All authors have read and agreed to the published version of the manuscript.

Funding: This research was funded by the National Natural Science Foundation of Henan province of China (242300421332), the Science and Technology Program of Henan province of China (232102320111), the Science and Technology Planning Major Project of Fujian province of China (2022N0010), and the Natural Resources Science and Technology Innovation Project of Fujian province of China (KY-090000-04-2022-016).

Institutional Review Board Statement: Not applicable.

Informed Consent Statement: Not applicable.

Data Availability Statement: The original contributions presented in the study are included in the article, further inquiries can be directed to the corresponding author.

Conflicts of Interest: The authors declare no conflicts of interest.

References

1. Petrasch, S.; Knapp, S.J.; van Kan, J.A.; Blanco-Ulate, B. Grey mould of strawberry, a devastating disease caused by the ubiquitous necrotrophic fungal pathogen *Botrytis cinerea*. *Mol. Plant Pathol.* **2019**, *20*, 877–892. [CrossRef] [PubMed]
2. Lazcano, C.; Boyd, E.; Holmes, G.; Hewavitharana, S.; Pasulka, A.; Ivors, K. The rhizosphere microbiome plays a role in the resistance to soil-borne pathogens and nutrient uptake of strawberry cultivars under field conditions. *Sci. Rep.* **2021**, *11*, 3188. [CrossRef] [PubMed]
3. Iqbal, M.; Jamshaid, M.; Zahid, M.A.; Andreasson, E.; Vetukuri, R.A.; Stenberg, J.A. Biological control of strawberry crown rot, root rot and grey mould by the beneficial fungus *Aureobasidium pullulans*. *Bio-Control* **2021**, *66*, 535–545. [CrossRef]
4. Abdel-Sattar, M.A.; El-Marzoky, H.A.; Mohamed, A.I. Occurrence of soilborne diseases and root knot nematodes in strawberry plants grown on compacted rice straw bales compared with naturally infested soils. *J. Plant Prot. Res.* **2008**, *48*, 223–235.
5. Hu, Y.J.; Yang, H.M.; Jin, J.; Yan, H.H.; Wang, J.P.; Zhang, R.Q. Synergistic activity of antagonistic *Trichoderma* spp. and *Rhizoctonia solani* increases disease severity on strawberry petioles. *Eur. J. Plant Pathol.* **2022**, *164*, 375–389.
6. Erper, I.; Ozer, G.; Alkan, M.; Zholdoshbekova, S.; Turkkan, M. First report of *Dactylonectria torresensis* causing black root rot of strawberries in Kyrgyzstan. *J. Plant Pathol.* **2020**, *103*, 379–380. [CrossRef]
7. Fang, X.L.; Phillips, D.; Li, H.; Sivasithamparam, K.; Barbetti, M.J. Severity of crown and root diseases of strawberry and associated fungal and oomycete pathogens in Western Australia. *Australas. Plant Pathol.* **2011**, *40*, 109–119. [CrossRef]
8. Bahar, M.; Shahab, H. Analysis of Iranian isolates of *Fusarium solani* using morphological, pathogenicity and microsatellite DNA marker characterization. *Afr. J. Biotechnol.* **2012**, *11*, 474–482.
9. Chamorro, M.; Aguado, A.; Santos, B.D.L. First report of root and crown rot caused by *Pestalotiopsis clavispora* (*Neopestalotiopsis clavispora*) on strawberry in Spain. *Plant Dis.* **2015**, *100*, 7. [CrossRef]
10. Pastrana, A.M.; Capote, N.; Santos, B.D.L.; Romero, F.; Basallote-Ureba, M.J. First report of *Fusarium solani* causing crown and root rot on strawberry crops in Southwestern Spain. *Plant Dis.* **2014**, *98*, 161. [CrossRef]
11. Zhang, Y.T.; Yu, H.; Hu, M.H.; Wu, J.Y.; Zhang, C.Q. Fungal pathogens associated with strawberry crown rot disease in China. *J. Fungi* **2022**, *8*, 1161. [CrossRef] [PubMed]
12. Zhao, Y.M.; Li, J.T.; Shi, H.; Liang, C.H.; Wang, Z.G.; Wu, X.H. Species identification of *Fusarium* spp. causing root rot on strawberry. *Acta Phytopathol. Sin.* **2024**, *54*, 451–456.
13. Ceja-Torres, L.F.; Mora-Aguilera, G.; Teliz, D.; Mora-Aguilera, A.; Sanchez-Garcia, P.; Munoz-Ruiz, C.; Tlapal-Bolanos, B.; De La Torre-Almaraz, R. Fungi prevalence and etiology of strawberry dry wilt under different crop management systems. *Agrociencia* **2008**, *42*, 451–461.
14. Manici, L.M.; Caputo, F.; Baruzzi, G. Additional experiences toelucidate the microbial component of soil suppressiveness towards strawberr black root rot complex. *Ann. Appl. Biol.* **2005**, *146*, 421–431. [CrossRef]

15. Ayoubi, N.; Soleimani, M.J. Morphological and molecular identification of pathogenic *Fusarium* spp. on strawberry in Iran. *Sydowia* **2016**, *68*, 163–171.
16. De la Lastra, E.; Villarino, M.; Astacio, J.D.; Larena, I.; De Cal, A.; Capote, N. Genetic diversity and vegetative compatibility of *Fusarium solani* species complex of strawberry in Spain. *Phytopathology* **2019**, *109*, 2142–2151. [CrossRef]
17. Shen, T.; Wang, C.; Yang, H.; Deng, Z.L.; Wang, S.M.; Shen, B.; Shen, Q.R. Identification, solid-state fermentation and biocontrol effects of *Streptomyces hygroscopicus* B04 on strawberry root rot. *Appl. Soil Ecol.* **2016**, *103*, 36–43. [CrossRef]
18. Liu, Y.; Tian, Y.; Yue, L.; Constantine, U.; Zhao, X.; Zhou, Q.; Wang, Y.; Zhang, Y.B.; Chen, G.F.; Dun, Z.H.; et al. Effectively controlling *Fusarium* root rot disease of *Angelica sinensis* and enhancing soil fertility with a novel attapulgite-coated biocontrol agent. *Appl. Soil Ecol.* **2021**, *168*, 104121. [CrossRef]
19. Abd-El-Kareem, F.; Elshahawy, I.E.; Abd-Elgawad, M.M. Local *Trichoderma* strains as a control strategy of complex black root rot disease of strawberry in Egypt. *Bull. Natl. Res. Cent.* **2019**, *43*, 160. [CrossRef]
20. Abd-El-Kareem, F.; Elshahawy, I.E.; Abd-Elgawad, M.M. Application of *Bacillus pumilus* isolates for management of black rot disease in strawberry. *Egypt. J. Biol. Pest Control* **2021**, *31*, 25. [CrossRef]
21. Gowtham, H.G.; Murali, M.; Singh, S.B.; Lakshmeesha, T.R.; Murthy, K.N.; Amruthesh, K.N.; Niranjana, S.R. Plant growth promoting rhizobacteria-*Bacillus amyloliquefaciens* improves plant growth and induces resistance in chilli against anthracnose disease. *Biol. Control* **2018**, *126*, 209–217. [CrossRef]
22. Cui, W.Y.; He, P.J.; Munir, S.; He, P.B.; Li, X.Y.; Li, Y.M.; Wu, J.J.; Wu, Y.X.; Yang, L.J.; He, P.F. Efficacy of plant growth promoting bacteria *Bacillus amyloliquefaciens* B9601-Y2 for biocontrol of southern corn leaf blight. *Biol. Control* **2019**, *139*, 104080. [CrossRef]
23. Cui, L.; Yang, C.; Wang, Y.; Ma, T.; Cai, F.; Wei, L.; Jin, M.; Osei, R.; Zhang, J.; Tang, M. Potential of an endophytic bacteria *Bacillus amyloliquefaciens* 3-5 as biocontrol agent against potato scab. *Microb. Pathog.* **2022**, *163*, 105382. [CrossRef] [PubMed]
24. Luo, L.; Zhao, C.; Wang, E.; Raza, A.; Yin, C. *Bacillus amyloliquefaciens* as an excellent agent for biofertilizer and biocontrol in agriculture: An overview for its mechanisms. *Microbiol. Res.* **2022**, *259*, 127016. [CrossRef]
25. Alila-Kolsi, I.; Ben-Mahmoud, A.; Al-Barazie, R. *Bacillus amyloliquefaciens*: Harnessing its potential for industrial, medical, and agricultural applications-a comprehensive review. *Microorganisms* **2023**, *11*, 2215. [CrossRef]
26. Chowdhury, S.P.; Hartmann, A.; Gao, X.; Borriss, R. Biocontrol mechanism by root-associated *Bacillus amyloliquefaciens* FZB42-A Review. *Front. Microbiol.* **2015**, *6*, 780. [CrossRef]
27. Gill, S.S.; Tuteja, N. Reactive oxygen species and antioxidant machinery in abiotic stress tolerance in crop plants. *Plant Physiol. Biochem.* **2010**, *48*, 909–930. [CrossRef] [PubMed]
28. Cun, H.; Munir, S.; He, P.; Wu, Y.; He, P.; Ahmed, A.; Che, H.; Li, J.; He, Y. Diversity of root endophytic bacteria from maize seedling involved in biocontrol and plant growth promotion. *Egypt. J. Biol. Pest Control* **2022**, *32*, 129. [CrossRef]
29. Morocko, I.; Fatehi, J.; Gerhardson, B. *Gnomonia fragariae*, a cause of strawberry root rot and petiole blight. *Eur. J. Plant Pathol.* **2006**, *114*, 235–244. [CrossRef]
30. Ding, Y.; Liu, F.J.; Yang, J.; Fan, Y.Y.; Yu, L.J.; Li, Z.H.; Jiang, N.; An, J.; Jiao, Z.W.; Wang, C. Isolation and identification of *Bacillus mojavensis* YL-RY0310 and its biocontrol potential against *Penicillium expansum* and patulin in apples. *Biol. Control* **2023**, *182*, 105239. [CrossRef]
31. Hong, S.; Kim, T.Y.; Won, S.-J.; Moon, J.-H.; Ajuna, H.B.; Kim, K.Y.; Ahn, Y.S. Control of fungal diseases and fruit yield improvement of strawberry using *Bacillus velezensis* CE100. *Microorganisms* **2022**, *10*, 365. [CrossRef] [PubMed]
32. Weisburg, W.G.; Barns, S.M.; Pelletier, D.A.; Lane, D.J. 16S ribosomal DNA amplification for phylogenetic study. *J. Bacteriol.* **1991**, *173*, 697–703. [CrossRef] [PubMed]
33. Yamamoto, S.; Harayama, S. PCR amplification and direct sequencing of *gyrB* genes with universal primers and their application to the detection and taxonomic analysis of *Pseudomonas putida* strains. *Appl. Environ. Microbiol.* **1995**, *61*, 1104–1109. [CrossRef] [PubMed]
34. Koumoutsi, A.; Chen, X.H.; Henne, A.; Liesegang, H.; Hitzeroth, G.; Franke, P.; Vater, J.; Borriss, R. Structural and functional characterization of gene clusters directing nonribosomal synthesis of bioactive cyclic lipopeptides in *Bacillus amyloliquefaciens* strain FZB42. *J. Bacteriol.* **2004**, *186*, 1084–1096. [PubMed]
35. Hsieh, F.C.; Lin, T.C.; Meng, M.; Kao, S.S. Comparing methods for identifying *Bacillus* strains capable of producing the antifungal lipopeptide iturin A. *Curr. Microbiol.* **2008**, *56*, 1–5. [CrossRef] [PubMed]
36. Kim, P.I.; Pyoung, I.L.; Jaewon, R.; Young, H.K.; Youn, T.C. Production of biosurfactant lipopeptides iturin A, fengycin, and surfactin A form *Bacillus subtilis* CMB32 for control of *Colletotrichum gloeosporioides*. *J. Microbiol. Biotechnol.* **2010**, *20*, 138–145. [CrossRef] [PubMed]
37. Yang, R.; Ye, W.; Liu, P.; Li, J.; Lu, M.M.; Wang, Z.H.; Shao, D.K. Endophytic *Bacillus amyloliquefaciens* Mdgb15 is a potential biocontrol agent against tree peony gray mold caused by *Botrytis cinerea*. *Eur. J. Plant Pathol.* 2024; *in press*. [CrossRef]
38. Pan, H.Y.; Jin, W.Y.; Zhang, X.M.; Song, X.L.; Yan, Y.Z.; Yan, B.; Xue, J. Inhibition of antifungal substances from *Bacillus amyloquefaciens* B15 against *Botrytis Cinerea*-the agent of "gray mold" of grape. *Acta Microbiol. Sin.* **2018**, *58*, 1245–1254.
39. Sritongon, N.; Boonlue, S.; Mongkolthanaruk, W.; Jogloy, S.; Riddech, N. The combination of multiple plant growth promotion and hydrolytic enzyme producing rhizobacteria and their effect on *Jerusalem artichoke* growth improvement. *Sci. Rep.* **2023**, *13*, 5917. [CrossRef]
40. Alexander, D.B.; Zuberer, D.A. Use of chrome azurol S reagents to evaluate siderophore production by rhizosphere bacteria. *Biol. Fertil. Soils* **1991**, *12*, 39–45. [CrossRef]

41. Ahmad, F.; Ahmad, I.; Khan, M.S. Screening of free living rhizospheric bacteria for their multiple plant growth promoting activities. *Microbiol. Res.* **2008**, *163*, 173–181. [CrossRef] [PubMed]
42. Hamdali, H.; Bouizgarne, B.; Hafidi, M.; Lebrihi, A.; Virolle, M.J.; Ouhdouch, Y. Screening for rock phosphate solubilizing Actinomycetes from Moroccan phosphate mines. *Appl. Soil Ecol.* **2008**, *38*, 12–19. [CrossRef]
43. Bhattacharyya, C.; Banerjee, S.; Acharya, U.; Mitra, A.; Mallick, I.; Haldar, A.; Ghosh, A.; Ghosh, A. Evaluation of plant growth promotion properties and induction of antioxidative defense mechanism by tea rhizobacteria of Darjeeling, India. *Sci. Rep.* **2020**, *10*, 15536. [CrossRef] [PubMed]
44. Vestberg, M.; Kukkonen, S.; Saari, K.; Parikka, P.; Huttunen, J.; Tainio, L.; Devos, N.; Weekers, F.; Kevers, C.; Thonart, P.; et al. Microbial inoculation for improving the growth and health of micropropagated strawberry. *Appl. Soil Ecol.* **2004**, *27*, 243–258. [CrossRef]
45. Murage, E.N.; Masuda, M. Response of pepper and eggplant to continuous light in relation to leaf chlorosis and activities of antioxidative enzymes. *Sci. Hortic.* **1997**, *70*, 269–279. [CrossRef]
46. Wang, C.Y. Effect of temperature preconditioning on catalase, peroxidase, and superoxide dismutase in chilled zucchini squash. *Postharvest Biol. Technol.* **1995**, *5*, 67–76. [CrossRef]
47. Zhang, M.; Kong, Z.; Fu, H.; Shu, X.; Xue, Q.; Lai, H.; Guo, Q. Rhizosphere microbial ecological characteristics of strawberry root rot. *Front. Microbiol.* **2023**, *14*, 1286740. [CrossRef] [PubMed]
48. Yadav, D.K.; Devappa, V.; Kashyap, A.S.; Kumar, N.; Rana, V.S.; Sunita, K.; Singh, D. Boosting the biocontrol efficacy of *Bacillus amyloliquefaciens* DSBA-11 through physical and chemical mutagens to control bacterial wilt disease of tomato caused by *Ralstonia solanacearum*. *Microorganisms* **2023**, *11*, 1790. [CrossRef]
49. Wu, Y.M.; Chen, X.; Wang, F.; Hsiao, C.Y.; Yang, C.Y.; Lin, S.T.; Wu, L.H.; Chen, Y.K.; Liang, Y.S.; Lin, Y.H. *Bacillus amyloliquefaciens* strains control strawberry anthracnose through antagonistic activity and plant immune response intensification. *Biol. Control* **2021**, *157*, 104592.
50. Chen, Z.; Huang, J.; Zhao, J.; Hao, Y.S.; Liang, H. Isolation and identification of pathogenic fungi of strawberry root rot and the inhibition of antagonistic bacteria CM3 on these fungi. *Biotechnol. Bull.* **2018**, *34*, 135–141.
51. Van Wees, S.C.; Van der Ent, S.; Pieterse, C.M. Plant immune responses triggered by beneficial microbes. *Curr. Opin. Plant Biol.* **2008**, *11*, 443–448. [CrossRef]
52. Ren, L.; Yuan, Z.; Xie, T.; Wu, D.; Kang, Q.; Li, J.; Li, J. Extraction and characterization of cyclic lipopeptides with antifungal and antioxidant activities from *Bacillus amyloliquefaciens*. *J. Appl. Microbiol.* **2022**, *133*, 3573–3584. [CrossRef] [PubMed]
53. Liu, L.; Medison, R.G.; Zheng, T.W.; Meng, X.J.; Sun, Z.X.; Zhou, Y. Biocontrol potential of *Bacillus amyloliquefaciens* YZU-SG146 from *Fraxinus hupehensis* against *Verticillium* wilt of cotton. *Biol. Control* **2023**, *183*, 105246.
54. Fan, B.; Wang, C.; Song, X.F.; Ding, X.L.; Wu, L.M.; Wu, H.J.; Gao, X.W.; Borriss, R. *Bacillus velezensis* FZB42 in 2018: The Gram-Positive model strain for plant growth promotion and biocontrol. *Front. Microbiol.* **2018**, *9*, 2491. [CrossRef] [PubMed]
55. Al-Mutar, D.M.K.; Alzawar, N.S.A.; Noman, M.; Azizullah; Li, D.; Song, F. Suppression of *Fusarium* wilt in watermelon by *Bacillus amyloliquefaciens* DHA55 through extracellular production of antifungal lipopeptides. *J. Fungi* **2023**, *9*, 336. [CrossRef] [PubMed]
56. Yang, P.; Yuan, P.; Liu, W.; Zhao, Z.; Bernier, M.C.; Zhang, C.; Adhikari, A.; Opiyo, S.O.; Zhao, L.; Banks, F.; et al. Plant growth promotion and plant disease suppression induced by *Bacillus amyloliquefaciens* strain GD4a. *Plants* **2024**, *13*, 672. [CrossRef] [PubMed]
57. Kang, B.R.; Park, J.S.; Jung, W.J. Antifungal evaluation of fengycin isoforms isolated from *Bacillus amyloliquefaciens* PPL against *Fusarium oxysporum* f. sp. *lycopersici*. *Microb. Pathog.* **2020**, *149*, 104509. [PubMed]
58. Jiao, R.; Cai, Y.; He, P.; Munir, S.; Li, X.; Wu, Y.; Wang, J.; Xia, M.; He, P.; Wang, G.; et al. *Bacillus amyloliquefaciens* YN201732 produces lipopeptides with promising biocontrol activity against fungal pathogen *Erysiphe cichoracearum*. *Front. Cell Infect. Microbiol.* **2021**, *11*, 598999. [CrossRef] [PubMed]
59. Chowdhury, S.P.; Uhl, J.; Grosch, R.; Alquéres, S.; Pittroff, S.; Dietel, K.; Schmitt-Kopplin, P.; Borriss, R.; Hartmann, A. Cyclic lipopeptides of *Bacillus amyloliquefaciens* subsp. *plantarum* colonizing the lettuce rhizosphere enhance plant defense responses toward the bottom rot pathogen *Rhizoctonia solani*. *Mol. Plant Microbe. Interact.* **2015**, *28*, 984–995.
60. Pieterse, C.M.; Zamioudis, C.; Berendsen, R.L.; Weller, D.M.; Van Wees, S.C.; Bakker, P.A. Induced systemic resistance by beneficial microbes. *Ann. Rev. Phytopathol.* **2014**, *52*, 347–375. [CrossRef]
61. Niu, D.D.; Wang, X.J.; Wang, Y.R.; Song, X.O.; Wang, J.S.; Guo, J.H.; Zhao, H.W. *Bacillus cereus* AR156 activates PAMP-triggered immunity and induces a systemic acquired resistance through a NPR1-and SA-dependent signaling pathway. *Biochem. Biophys. Res. Commun.* **2016**, *469*, 120–125. [CrossRef] [PubMed]
62. Ahmad, R.; Kim, Y.H.; Kim, M.D.; Kwon, S.Y.; Cho, K.; Lee, H.S.; Kwak, S.S. Simultaneous expression of choline oxidase, superoxide dismutase and ascorbate peroxidase in potato plant chloroplasts provides synergistically enhanced protection against various abiotic stresses. *Physiol. Plant.* **2010**, *138*, 520–533. [CrossRef]
63. Prasannath, K. Plant defense-related enzymes against pathogens: A review. *Agrieast J. Agric. Sci.* **2017**, *11*, 38–48. [CrossRef]
64. Batool, T.; Ali, S.; Seleiman, M.F.; Naveed, N.H.; Ali, A.; Ahmed, K.; Abid, M.; Rizwan, M.; Shahid, M.R.; Alotaibi, M.; et al. Plant growth promoting rhizobacteria alleviates drought stress in potato in response to suppressive oxidative stress and antioxidant enzymes activities. *Sci. Rep.* **2020**, *10*, 16975. [CrossRef] [PubMed]

65. Kazerooni, E.A.; Maharachchikumbura, S.S.; Al-Sadi, A.M.; Kang, S.M.; Yun, B.W.; Lee, I.J. Biocontrol potential of *Bacillus amyloliquefaciens* against *Botrytis pelargonii* and *Alternaria alternata* on *Capsicum annuum*. *J. Fungi* **2021**, *7*, 472. [CrossRef]
66. Pei, D.; Zhang, Q.; Zhu, X.; Zhang, L. Biological control of *Verticillium* wilt and growth promotion in tomato by *Rhizospheric* soil-derived *Bacillus amyloliquefaciens* Oj-2.16. *Pathogens* **2022**, *26*, 37. [CrossRef]

Journal of *Fungi*

MDPI

Article

The Hemibiotrophic Apple Scab Fungus *Venturia inaequalis* Induces a Biotrophic Interface but Lacks a Necrotrophic Stage

Ulrike Steiner and Erich-Christian Oerke *

Institute of Crop Science and Resource Conservation—Plant Pathology, Rheinische Friedrich-Wilhelms-Universitaet Bonn, 53115 Bonn, Germany; u-steiner@uni-bonn.de
* Correspondence: ec-oerke@uni-bonn.de

Abstract: Microscopic evidence demonstrated a strictly biotrophic lifestyle of the scab fungus *Venturia inaequalis* on growing apple leaves and characterised its hemibiotrophy as the combination of biotrophy and saprotrophy not described before. The pathogen–host interface was characterised by the formation of knob-like structures of the fungal stroma appressed to epidermal cells as early as 1 day after host penetration, very thin fan-shaped cells covering large parts of the host cell lumen, and enzymatic cuticle penetration from the subcuticular space limited to the protruding conidiophores. The *V. inaequalis* cell wall had numerous orifices, facilitating intimate contact with the host tissue. Pathogen-induced modifications of host cells included partial degradation of the cell wall, transition of epidermal cells into transfer cells, modification of epidermal pit fields to manipulate the flow of nutrients and other compounds, and formation of globular protuberances of mesophyll cells without contact with the pathogen. The non-haustorial biotrophy was characterised by enlarged areas of intimate contact with host cells, often mediated by a matrix between the pathogen and plant structures. The new microscopic evidence and information on the pathogens' biochemistry and secretome from the literature gave rise to a model of the lifestyle of *V. inaequalis*, lacking a necrotrophic stage that covers and explains its holomorphic development.

Keywords: hemibiotrophy; host response; pit fields; plasmodesmata; subcuticular development; transfer cells

Citation: Steiner, U.; Oerke, E.-C. The Hemibiotrophic Apple Scab Fungus *Venturia inaequalis* Induces a Biotrophic Interface but Lacks a Necrotrophic Stage. *J. Fungi* **2024**, *10*, 831. https://doi.org/10.3390/jof10120831

Academic Editor: Ofir Degani

Received: 14 October 2024
Revised: 22 November 2024
Accepted: 27 November 2024
Published: 29 November 2024

1. Introduction

Venturia inaequalis (Cooke) G. Winter (1875) is an ascomycetous fungus, class Dothideomycetes, that causes apple scab, the most important disease in apple production worldwide [1]. The pathogen described as a hemibiotroph has remarkable features, as (I) it growths subcuticularly without penetrating host cell walls; (II) in its parasitic stage on young leaves and fruits, it neither kills plant cells nor produces haustoria within host cells for nutrient uptake; and (III) it perforates the cuticle for conidia production and uses the increased cuticular transpiration for its nutrition [2]. As germ tubes and subcuticular hyphae are non-melanised and conidia become melanised only in the later stages of their development, host plant cells have contact with non-melanised *V. inaequalis* cell walls only, despite the production of melanised conidia and ascospores [3]. The growth of scab lesions is limited in time (and space) due to ontogenetic resistance, which becomes complete by the time the leaves are fully expanded [4,5]. All apple cultivars exhibit ontogenetic resistance, which appears never to have been overcome by the pathogen [6,7]. In the late stages of the vegetation period of its host plant, *V. inaequalis* resumes hyphal growth, saprophytically colonises dead leaf material by melanised hyphae, and produces pseudothecia on leaf litter for its sexual stage, ascospore dissemination, and new host-plant infections in spring.

In addition to *V. inaequalis*, subcuticular pathogens include other species of the genus *Venturia*, e.g., *V. asperata* on apples [8], *V. effusa* on pecans [9], *V. nashicola* on Asian pears [10], *V. oleaginea* on olives [11], *V. paralias* on *Euphorbia* [12], *V. pyrina* on European pears [13], *Diplocarpon rosae* causing black spots on roses [14,15], and *Rhynchosporium secalis* causing scalding of barley and rye [16,17]. Nevertheless, there are differences between structures (and lifestyles) within the genus *Venturia*, e.g., *V. carpophila* on peaches produces melanised epicuticular hyphae [18], whereas the subcuticular runner hyphae of *V. inaequalis* are non-melanised.

Mechanical pressure is not required, but enzymatic hydrolysis is proposed to allow cuticle penetration [19]. Extracellular cutinase activity has been implicated, as this enzyme is produced by germinating conidia, and a cutinase inhibitor can prevent penetration [20,21]. The enzymatic activity is discussed as being localised and limited by the melanised ring structure at the interface between the appressorium and the plant surface [22]. Subcuticular-growing pathogens like *V. inaequalis* differentiate specialised subcuticular infection structures—primary stroma, runner hyphae, secondary stromata—in close contact with the underlying epidermal host cells [4,23]. These non-melanised hyphae differ from epicuticular germ tubes and hyphae as well as from hyphae produced in artificial cultures. Runner hyphae enable the subcuticular spread from the penetration site and are the initials for secondary stromata. They are reported to be wider and flatter than tubular hyphae and may form 'hyphal superhighways'; stromata are mono- and multi-layered pseudo-parenchymatic structures thought to be involved in nutrient uptake and effector delivery [24]; however, they also give rise to conidia. Secondary stromata produce melanised conidiophores that bulge out of the cuticle before forming conidia outside the plant tissue [3]. Damage to leaf cuticles is strictly limited to the pores necessary for the emerging conidiophores to release their conidia into the ambient air [2].

As the subcuticular hyphal growth of *V. inaequalis* is strictly apoplastic between the cuticle and epidermal cell layers, and without the formation of specific structures entering the plant cells (like haustoria), the sourcing and uptake of nutrients are hardly understood. Like obligate pathogens, *V. inaequalis* influences hormone levels at infection sites, specifically cytokinins, in its biotroph stage, most probably to enable biotrophic nutrition [25,26]. Cytokinin accumulation at infection sites affects the sink-source relations of the plant and results in the translocation of nutrients to the fungus [25].

Apart from cuticle penetration during infection, the host tissue exhibits no obvious damage throughout stroma development. Most damage of the apple tissue has been reported to result from breaching of the cuticle upon sporulation [5,27]. At this time, the epidermal cells underlying the stroma are reported to progressively become depleted of plastids and cytoplasm, accompanied by increasing vacuolation, leading ultimately to necrosis, which is possibly affected by partial cell wall degradation late in the infection cycle [23,28]. However, collapse of epidermal apple cells has rarely been detected in highly compatible *M. domestica*–*V. inaequalis* interactions [2,29]. In contrast, the epicuticular formation and release of masses of conidia requires the re-penetration of the plant cuticle by *V. inaequalis* conidiophores—two to three magnitudes the frequency of penetration sites covered by the tightly attached appressoria—which causes an increased cuticular transpiration, a crucial impact on the water status of the host plant [2].

For in vitro growing hyphae of *V. inaequalis*, the production and secretion of cellulase, β-glucosidase, pectinase, and endo- and exo-polygalacturonase activities have been described [5,30,31]. Provided these enzymes are only expressed late in infection with host cell wall degradation, they are unlikely to play a major role in nutrient acquisition during the establishment of infection [7]. In liquid cultures, *V. inaequalis* also produces melanoproteins [32,33]. A melanoprotein binding iron and copper has been demonstrated also in the host apoplast and has been discussed to be involved in the acquisition of ions and the release of cell-wall-degrading enzymes (CWDEs) [27,34]. They have been speculated to tether CWDEs and to facilitate their slow release [35], and to be involved in diverting the solute flow towards the site of infections, probably by altering the membrane permeability and solute transport system of apples to facilitate the availability of nutrients for pathogen growth and development [5]. However, as subcuticular hyphae of *V. inaequalis* are not melanised, a role of melanin in the acquisition of nutrients is very unlikely.

Plant–pathogenic fungi may modify their cell wall during host colonisation in order to avoid detection [36,37]. One strategy is to deacetylate chitin to chitosan [38–40]. Chitin and other β-glucans are MAMPs—strong elicitors of the plant immune system [41,42]. During subcuticular growth, *V. inaequalis* down-regulates genes for the biosynthesis of chitin and β-1,6-glucan and coats its infection structures with chitosan [43,44], a carbohydrate reported to be less effective in PTI [45,46] and a weak substrate of plant chitinases [47].

The *V. inaequalis* secretome includes carbohydrate active enzymes (CAZymes), proteases, peroxidases, lipases, small secreted proteins (SSPs), and many with an unknown role [48–51]. SSPs account for 40–60% of the secretomes throughout all fungal lifestyles and many of them have been described as effector proteins in diverse diseases. More than 600 SSPs have been reported from the *V. inaequalis* genome [51–53]. Many of these effector candidates are unique to *V. inaequalis*, and a few of them are known to belong to extended gene families. These effectors are expressed in waves at different time intervals during the infectious process [43,48]. When interpreting these secretome data, the non-obligate biotroph lifestyle of *V. inaequalis* has to be respected; the repertoire of enzymes and SSPs may include proteins specific to one of both lifestyles, whereas others may be active in both life stages.

Wang et al. [54] classified *V. inaequalis* as a biotroph member of Dothideomycetes within Ascomycota, with 13,741, 1754, and 486 proteins in the proteome, secretome, and effectome, respectively. The number of genes encoding CAZymes—glycoside hydrolases (GHs), glycosyl transferases (GTs), polysaccharide lyases (PLs), carbohydrate esterases (CEs), auxiliary activities (AAs), and carbohydrate-binding modules (CBMs)—and the number and activity of secondary metabolite biosynthetic gene clusters—nonribosomal peptides, polyketides, and terpenes—were rated below average, except for AAs. The CE, GH, and PL superfamilies are also known as CWDEs due to their role in the disintegration of the plant cell wall by bacterial and fungal pathogens [55].

The chemical composition of the cell walls of apple leaves is unknown; in apple fruit skin the substrate of *V. inaequalis* when causing fruit scab; pectin is the predominant polysaccharide at 65%, compared with 3% for cellulose [56]. *V. inaequalis* CAZymes are optimally adapted to this cell wall composition, with pectin-specific CAZymes predominating; the two most numerous classes of CAZymes in the *Venturia* secretome, after those with cutinase activity (CE5 domains), are GH28 and GH43, with pectin as a substrate [51]. One GH28 enzyme and two putative polysaccharide lyases, that also have pectin as the target substrate [57,58], are up-regulated during infection. This lytic activity may contribute to nutrition via degrading the surface polysaccharides of the epidermal cells beneath stromata.

Proteins similar to the *Aspergillus oryzae* CutL1 polyesterase/cutinase and HsbA are well represented in the *Venturia* infection secretomes [51]. Extracellular cutinase is produced by germinating *V. inaequalis* conidia, and a cutinase inhibitor can prevent penetration [21]. The high number of carbohydrate esterases in the *Venturia* secretomes is in agreement with an enzymatic penetration of the cuticle by *V. inaequalis* [51]. The hypothesis that cuticular degradation during host colonisation by *Venturia* fungi also makes the cuticle or cuticle precursors a source of nutrients for the pathogens [7] has to be revisited in view of the time and spatial limitation of cuticle degradation.

Cutin consists of omega hydroxy acids and their derivatives, which are interlinked via ester bonds, forming a polyester polymer of indeterminate size. Pectin is an acidic heteropolysaccharide in the primary and middle lamella of cell walls of terrestrial plants, with galacturonic acid as the main component. It may be degraded by pectinases; as a consequence, the middle lamellae break down and cells become separated from each other—a process not described for *V. inaequalis* infections so far. Pectinases catalyse either depolymerisation (hydrolases and lyases) or de-esterification (esterases) reactions. An endo-polygalacturonase of *V. inaequalis* has been characterised by Kollar [31]. The restriction of fungal growth to the subcuticular space suggests that pectinases released by infection hyphae or subcuticular hyphae may be important in infection [10]. Also, cellulase activity is induced in *V. inaequalis* in vitro and in vivo [59]. As cutin, pectin, and cellulose, the main components of the plant cell wall, do not contain nitrogen atoms—pathogens that degrade only the cell wall, ignoring embedded or apoplastic proteins— they lack a source of nitrogen essential for fungal growth and development.

Hardham [60] summarised the fungal strategies to gain access to the nutrients they need for growth, development, and reproduction: Necrotrophic fungi obtain nutrients by killing the host cells, forming expanding necrotic lesions. Biotrophic pathogens establish a close and stable relationship with living host cells and redirect the flow of nutrients from the plant cell into the pathogen. They form specialised feeding structures that take the form of haustoria or differentiated intracellular hyphae. Hemibiotrophs initially establish a biotrophic interaction with living host cells before killing the plant cells and turning to a necrotrophic lifestyle.

A key feature of scab pathogens is their ability to differentiate specialised subcuticular infection structures that, to date, remain largely understudied [44]. The subcuticular hyphae of *V. inaequalis* are not in contact with the plasma membranes of epidermal cells of apples but are likely to collect nutrients in the intercellular space. This lifestyle requires (I) *V. inaequalis* structures for nutrient interception; (II) modifications of apple leaf tissue for the release of nutrients for *V. inaequalis*, most probably induced by effectors of the pathogen. Therefore, light microscopy, scanning electron microscopy, and transmission electron microscopy were used to study the host–pathogen interface of apple leaf tissue and subcuticular hyphae of *V. inaequalis* in detail. The objectives were the identification of (I) fungal structures specialised for nutrient uptake that should not impair the flow of water, the medium transporting organic nutrients to infection sites, e.g., differentiation of hyphae to maximise the surface area in contact with the surface of plant cells/the nutrient broth of the apoplastic intercellular fluid; (II) modifications of plant cells—colonised epidermal cells as well as more remote mesophyll cells—able to promote a flow of nutrients to the immediate surroundings of subcuticular hyphae.

2. Materials and Methods

Plant material. Seedlings of apples (*Malus × domestica* Borkh., cvs. Golden Delicious, Cripps Pink) were grown in pots (9 cm × 9 cm × 8 cm) filled with standard potting mixture (Klasmann-Deilmann GmbH, Geeste, Germany) at 20 ± 2 °C and 16 h daylight (300 µmol m^{-2} s^{-1}, Philips SGR 140, Hamburg, Germany) in a glasshouse with 50 to 70% relative humidity (RH). Plants were irrigated and fertilised (liquid fertiliser Flory 2 special, NPK 16 + 9 + 22) as required. Plants with at least four fully expanded leaves were used for the experiments. Inoculated plants were maintained at 20/18 °C (day/night), which is most favourable for rapid scab development, in a climate chamber (RH 50 to 70%).

Pathogen and inoculation. *Venturia inaequalis* (Cooke) G. Wint., isolate HS1, was cultivated on apple leaves by spraying conidia suspensions onto the leaves and incubating the plants at 100% RH for 48 h. Leaves with sporulating scab lesions were stored at −18 °C as inoculum. Conidia were washed off the stored leaves with tap water, passed through two layers of cheese cloth and adjusted to a conidia concentration of 1.0 × 105 conidia mL^{-1} using a Fuchs–Rosenthal hemacytometer. Apple leaves were spray-inoculated using a commercial hand sprayer (0.5 mL per leaf). Depending on the research question, the adaxial and abaxial side, respectively, were sprayed with inoculum before incubation at 100% RH for 48 h. Subsequently, RH was reduced to 50 to 70% until the end of the experiment. Four to eight plants per treatment were inoculated and used for sampling for microscopy.

Microscopy. All microscopic investigations were repeated several times (>3 times). For each experiment, samples for microscopic investigations were taken from similar leaves from at least 4 different apple plants. The images are representative for apple leaf tissue and the fungal structures indicated.

Light microscopy. Leaf discs were excised from the 2nd- to 3rd-youngest leaves of the apple plants by using a cork borer (Ø 1.1 cm). They were cleared in saturated chloral hydrate (Sigma-Aldrich, Darmstadt, Germany) or Farmer's Fixative (95% ethanol/glacial acetic acid, 3:1) for 3 to 5 days. Trypan blue staining was carried out by boiling leaf fragments

(1 cm^2) in a mixture of phenol, lactic acid, glycerol, and distilled water (1:1:1:1) containing 1 mg mL^{-1} trypan blue for 1 min. Then, the tissue was cleared overnight in chloral hydrate. For brightfield microscopy, specimens were stained with acid fuchsin (Merck, Darmstadt, Germany; 0.01% in lactophenol) or aniline blue according to Bruzzesse and Hasan [61] for protein staining. Semi-thin sections were stained with toluidine blue (0.5 g toluidine blue, 0.5 g sodium tetraborate, 50 mL bidistilled water). For fluorescence microscopy of fungal structures and callose, leaf samples were rinsed with distilled water and stained either with Uvitex 2B (0.05%; Polysciences, Warrington, PA, USA) or with aniline blue (0.05% aniline blue in 0.067 M K_2HPO_4) according to Hood and Shew [62]. For imaging of total preparations and conidia, samples were studied under the light microscopes DMRB and DM6000 B (Leica, Wetzlar, Germany), equipped with Nomarski interference contrast and epifluorescence, respectively. The aniline-blue-stained pathogen structures were visualised using filter cube A (excitation 340–380 nm, beam splitter 400 nm, stop filter LP 425 nm). Images were recorded and analysed using the software Diskus (v. 4.60.1611; Technisches Buero Hilgers, Koenigswinter, Germany).

Scanning electron microscopy (SEM). The preparation and examination of apple leaf samples were carried out according to the protocol of Juraschek et al. [63] for grapevine leaves. For studying subcuticular fungal structures, the leaf cuticle was removed by using adhesive tape. In some cases, the removed surface (on the tape) was investigated after sputtering.

Transmission electron microscopy (TEM). The preparation and examination of samples were carried out as described by Schumacher et al. [64].

3. Results

3.1. Hypothesis 1: Formation of Pathogen Structures Suitable for Nutrient Uptake

3.1.1. Primary Stroma and Runner Hyphae

Cuticular penetration from epicuticular appressoria gave rise to a primary subcuticular stroma 2 to 3 days post inoculation (d p.i.) (Figure 1a,b). As early as 72 h p.i., the cells of this monolayer showed knob-like structures at the interface with the epidermal host cells beneath—indicated by the common optical level and central orifice of some of these structures (Figure S1a). At the hyphal apex, the cuticle was separated from the cell wall, which was partially degraded; a maximum of half of the cell wall thickness was affected (Figure 1c) and in many cases no degradation was observed. The interface between the pathogen cell wall and host epidermis exhibited a (rippled) matrix of varying thickness.

For the subcuticular spread of the pathogen, primary stroma produced runner hyphae with a diameter wider than that of the epicuticular germ tubes (7–9 μm vs. 5–6 μm) and also showing knob-like structures at the interface with epidermal cells (Figures 1d,g and S1c). Often, several hyphal strands grew side-by-side (Figure 1e,f). Runner hyphae produced a hyphal network starting on the minor veins and large plated areas of the leaves before the emergence of first conidia (Figures 1h and S1d).

3.1.2. Fan-Shaped Hyphae

As early as 3 to 4 d p.i., the long runner hyphae laterally differentiated into ramified, fan-shaped, ultra-thin hyphae covering large areas of the epidermal cell surfaces beneath (Figure 2a–c). The small and flat cells (height 1–2 μm) were branched several-fold and tapered to the margin that advanced beneath the cuticle (Figure 2f,h,k). Undulation of the fungal cell wall on the epidermal cell increased the surface of the interface (Figure 2j,m) and anticlinal cell walls of fan-shaped hyphae often were thicker than periclinal walls (Figure 2l,n). The fan-shaped structures were produced above the lumen of epidermal cells and rarely exceeded the next anticlinal wall of the epidermal layer (Figure 2e,k). Their lobes/thin constituents were separated from each other as demonstrated by the existence of two cell walls of adjacent hyphae (Figure 2g). The small cells formed a structure with a large surface-to-volume ratio, optimal to capture and take up nutrients from the intercellular fluid like a filter/strainer (Figure 2e,k,m). These fan-shaped hyphal structures on epidermal cells did not act as a barrier but are likely to be porous/permeable for water and substances dissolved within. Similar to primary stromata and runner hyphae, early fan-shaped hyphae had numerous knob-like structures at the interface with epidermal cells (Figure 2o,p).

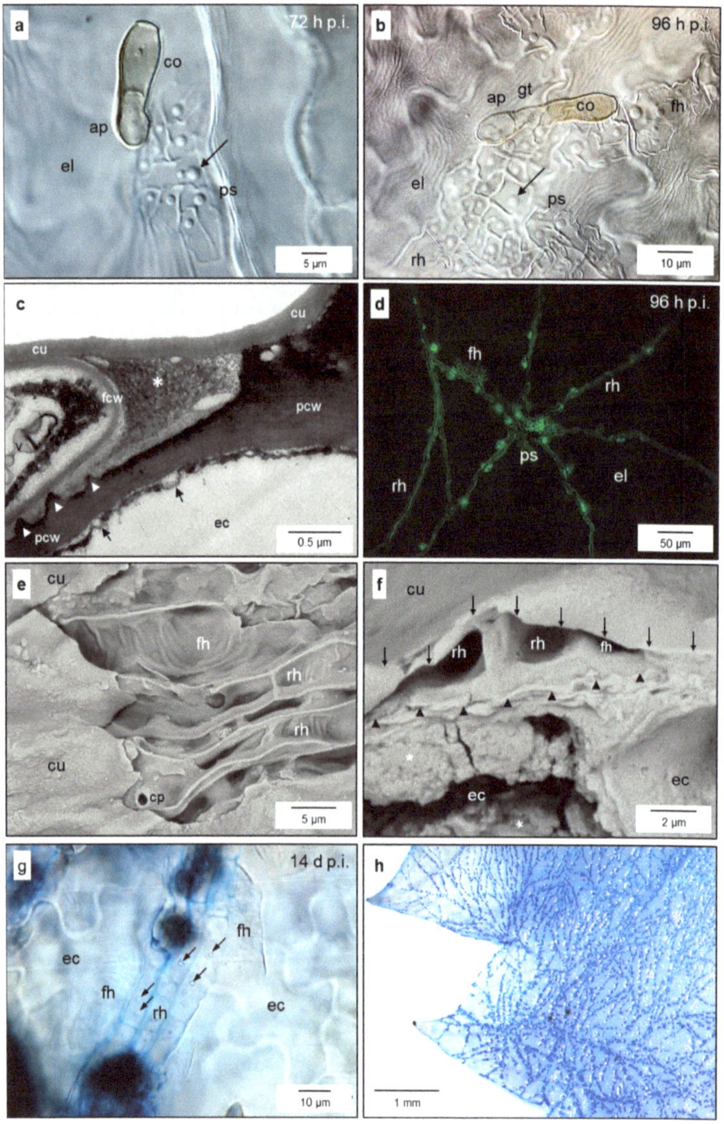

Figure 1. Formation of early subcuticular infection structures by *V. inaequalis*. Primary stroma (ps) 72 h p.i. (**a**) and 96 h p.i. (**b**) with knob-like structures (arrows) at the interface with the epidermal layer; details of the advancing hyphal tip between cuticle (cu) and cell wall (pcw), which is partially degraded, matrix material (asterisk) at the hyphal tip, cavities (arrow heads) at the interface, formation of vesicles at the plant cell membrane (arrows) (**c**); subcuticular spread from the primary stroma by runner hyphae (rh) that start to form lateral fan-shaped hyphae (fh) 96 h p.i. (**d**); runner hyphae, early fan-shaped hyphae, and initial of conidiophore (cp) after partial removal of the plant cuticle (**e**);

cross-section of runner hyphae between plant cuticles (marked by ↓) and upper cell walls (arrow heads) of epidermal cells (ec) showing massive appositions of the cell wall (asterisks) (**f**); runner hyphae with knob-like structures (arrows) and lateral fan-shaped hyphae 14 d p.i. (**g**); spread of runner hyphae on minor leaf veins at the tip of a young apple leaf 7 d p.i. (**h**). Bright-field microscopy (**a**,**b**,**g**) [staining with trypan blue], transmission electron microscopy (**c**), fluorescence microscopy after aniline blue staining (**d**), scanning electron microscopy (**e**,**f**), RGB camera, ink staining (**h**). ap, appressorium; co, conidium; el, epidermal layer; gt, germ tube.

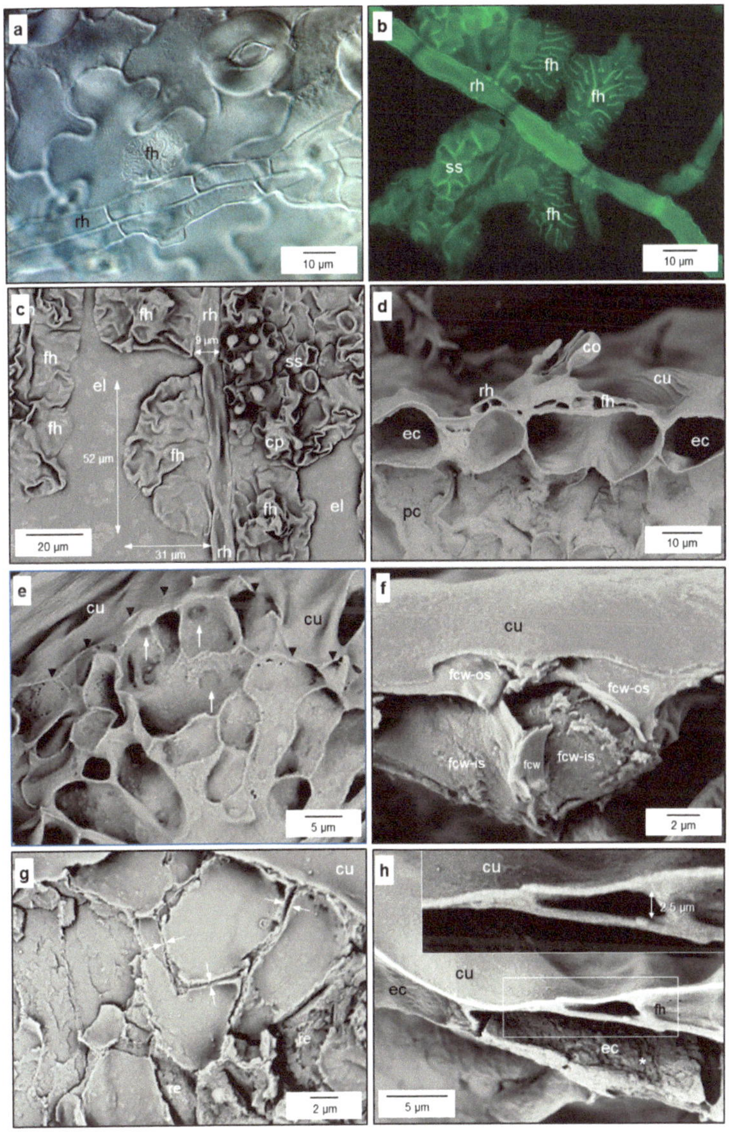

Figure 2. *Cont.*

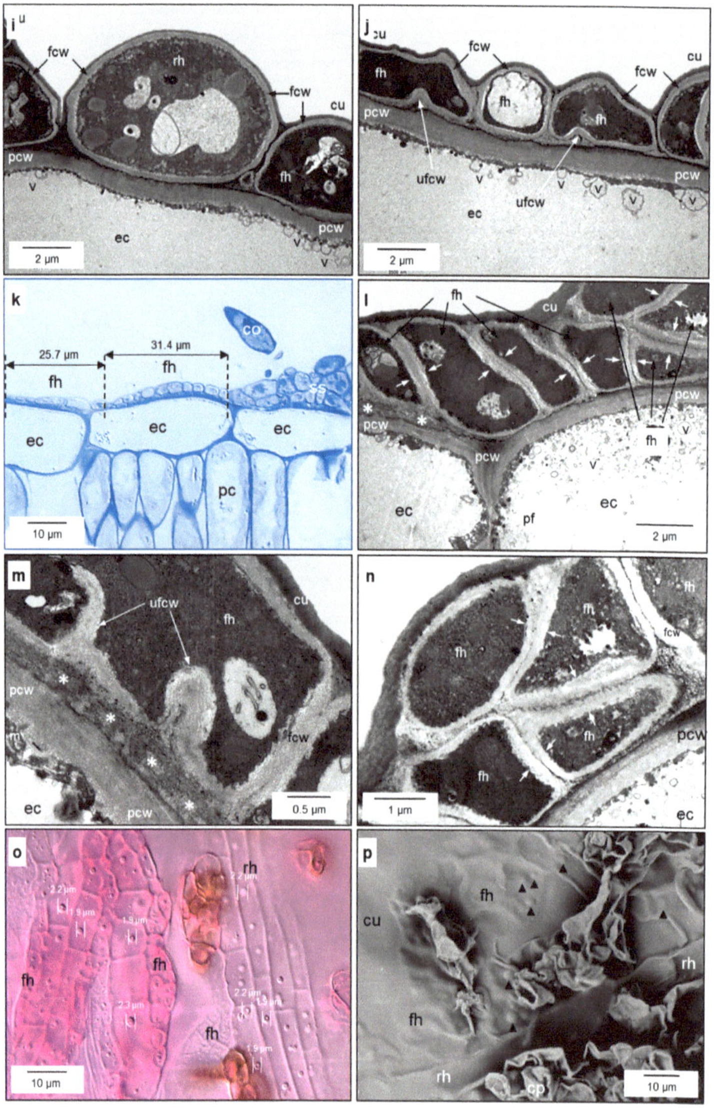

Figure 2. Subcuticular formation of fan-shaped hyphae (fh) by runner hyphae of *V. inaequalis*. Early fan-shaped hyphae laterally expand from a cell of a runner hypha (rh) on the lumen of an epidermal cell (**a**); various sizes and stages of fan-shaped hyphae expanding from a runner hypha; fungal cell walls stained with aniline blue (**b**); dimensions of fan-shaped hyphae without and with conidiophores (cp) formed in later developmental stages (top view; (**c**)); lateral view of infection site with runner hyphae, fan-shaped hyphae, and conidiophores with conidia (co) on top of epidermal cells (ec) and covered by the cuticle (cu) (**d**); top view of fan-shaped hyphae after (partial) removal of plant cuticle (arrow heads mark cuticle margin); arrows indicate vents in the fungal cell wall (**e**); lateral view of hyphal cells advancing underneath the cuticle; the outer surface (fcw-os) and the inner surface (fcw-is) of the fungal cell wall (fcw) demonstrate the low-rise structure (**f**); delineation of fungal cells of fan-shaped hyphae by two cell walls at the contact zone (after partial cuticle removal) (**g**);

tip of fan-shaped hypha advancing above an epidermal cell to build the subcuticular space (**h**); section of runner hypha with neighbouring fan-shaped hyphae (**i**); formation of mem-brane vesicles by the epidermal cell underneath fan-shaped hyphae of. *V. inaequalis* (**j**); expansion of fan-shaped hyphae limited to the lumen of individual epidermal cells (**k**); ultrastructure of fan-shaped hyphae above epidermal cells with a pit field (pf) and increased formation of vesicles (**l**); undulations of the fungal cell wall (ufcw) and matrix material (*) between cell walls of pathogen and plant, respectively (**m**); the cell wall of the contact area of fungal cells was thicker, but less compact than the cell wall in contact with epidermal cell and cuticle, respectively (**n**); knob-like structures at the interface between the subcuticular pathogen and epidermal cell after staining with fuchsin acid (**o**) and in SEM image (**p**). Bright-field (**a**,**k**) [staining with toluidine blue], o [staining with fuchsin acid], and fluorescence (**b**) microscopy, scanning electron microscopy (**c**–**h**,**p**), transmission electron microscopy (**i**,**j**,**l**–**n**). pc, palisade cell; ss, secondary stroma.

Focussing even more closely on the pathogen–host interface, the contact zone between fan-shaped hyphae (FH) and the epidermal cell wall was characterised by the presence of granules fluorescing after staining with aniline blue, which were surrounded by a verge (sealing lip) of FH—thin outer lobes of the FH structure on the epidermal cells (Figures S1e,f and 3). The number of fluorescing granules increased with the size (and age) of the FH (Figure 3b,d). These granules were limited to the FH and could not be detected beneath runner hyphae. Electron microscopy revealed that FH at later stages of pathogenesis were surrounded not only by the nutrient broth of the apoplast but also by residual material (Figures 2m and 4f). The spatial pattern of runner hyphae and fan-shaped hyphae upon the cells of the epidermal layer was spatially oriented in order to maximise the contact area with the pit fields of epidermal cells beneath.

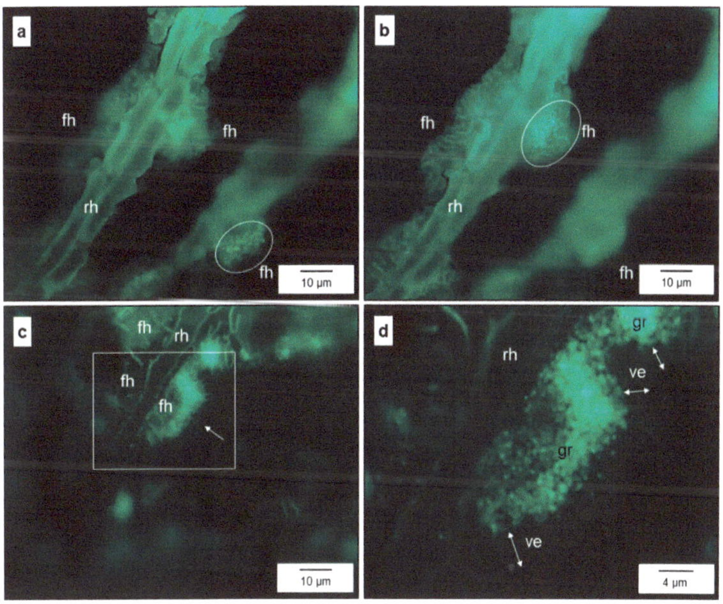

Figure 3. Formation of fan-shaped hyphae at runner hyphae of *V. inaequalis*. Early stage of fan-shaped hyphae (fh) with formation of fluorescent granules at the plant–pathogen interface (**a**,**b**); focus level of b below that of (**a**); later stage with fluorescent granules (gr) at the plant–pathogen interface surrounded by a verge (ve) free from granules (**c**,**d**) [detail of (**c**)]). rh, runner hypha.

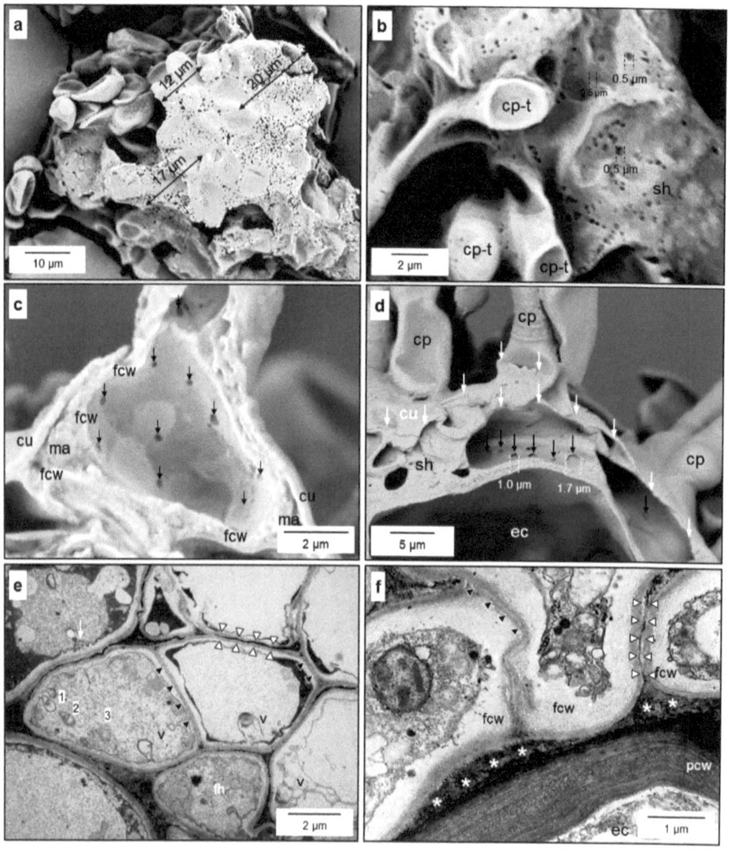

Figure 4. Morphological details of the subcuticular *V. inaequalis* hyphae. Heterogeneous pattern of cell wall orifices at the lower surface of the subcuticular stroma in contact with host epidermal cells [after dislodging of the epidermal layer] (**a**); top view on the surface of fungal secondary stroma with cell wall orifices [after dislodging of the cuticle] (**b**); section through a conidiophore with several small orifices (↓) at the inner surface of the fungal cell wall (fcw; (**c**)); subcuticular hypha (sh) above an epidermal cell (ec) with larger orifices at the interface with the epidermal cell (**d**); vesicle trafficking in the apex of a subcuticular hypha surrounded by other hyphae; 1, 2, 3, stages of vesicles (v) during endocytosis; hyphal septae (black arrowheads) and two cell walls of neighbouring hyphae (white arrowheads) (**e**); deposition of debris material (*) at the interface between the plant cell wall (pcw) and fungal cell wall (fcw; (**f**)). cp-t, top of conidiophore; cu, cuticle; ma, matrix between cuticle and conidiophore; cp, conidiophore. Scanning electron microscopy (**a**–**d**), transmission electron microscopy (**e**,**f**).

3.1.3. Secondary Stromata

Two to three days after their formation, fan-shaped hyphae often developed into multilayered secondary stromata. In the case of the colonisation of intercostal epidermal cells, secondary stromata had three to four layers, whereas infections above leaf veins and of stem tissue could result in the formation of more layers (Figures 4e and S1). Conidiophores were the only melanised fungal cells and were able to penetrate the cuticle from the subcuticular space to the outside. The secondary stroma of *V. inaequalis* was formed by aggregations of hyphae with septae with simple pores (Figure S1h). Experimental removal of the plant cuticle revealed that the fungal cell wall with contact to the epidermal cell wall (lower interface), as well as with the cuticle (upper surface), exhibited numerous notches/orifices visible in SEM images (Figure 4a,b,d). Images of the inner cell wall of *V. inaequalis* confirmed the existence of smaller orifices of varying diameter (Figure 4c,d). Within the subcuticular hyphae, vesicle trafficking crucial for the import of nutrients (endocytose) and the export of effectors (exocytose) could be observed (Figure 4e). TEM images exhibited the unipolar growth of fungal hyphae with a common septum within hyphae and two separate cell walls between adjacent hyphae (Figures 4f and S1h).

During all stages of leaf colonisation, the plant cuticle covered all pathogen structures except mature conidiophores. This was true for infections of the adaxial as well as for the abaxial leaf side (Figure 5a–c). The integrity and functionality of the plant cuticle was preserved and even fixed by the pathogen in case the abaxial epidermal layer was at risk of collapsing and losing its function as a water barrier during fungal colonisation (Figure 5b).

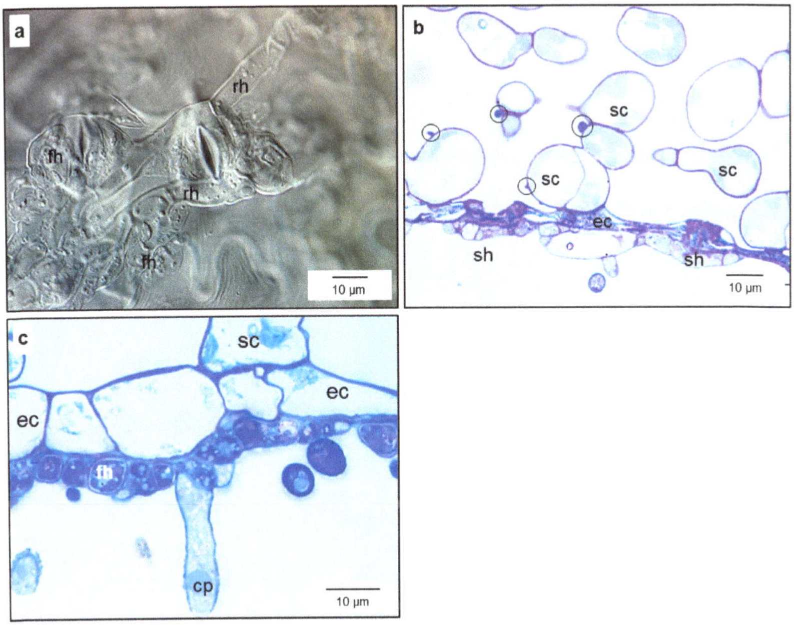

Figure 5. Colonisation of the abaxial epidermal layer by *V. inaequalis*. Formation of runner hyphae (rh) and fan-shaped hyphae (fh) in immediate vicinity of a stoma (**a**); development of subcuticular hyphae (sh) affecting the epidermal cell layer; cells of the spongy parenchyma (sc) respond to infection by the formation of globular protuberances stained blue (circles; (**b**)); subcuticular hyphae with formation of a conidiophore (cp) hardly affecting the epidermal layer (**c**). Scanning electron microscopy (**a**); bright-field microscopy of semi-thin sections stained with toluidine blue (**b**,**c**). ec, epidermal cell.

3.2. *Hypothesis 2: Modifications of Apple Leaf Tissue in Response to Venturia inaequalis Colonisation*

3.2.1. Plasmodesmal Network of Epidermal Cells

Pit fields of anticlinal epidermal cells harbouring primary plasmodesmata were visualised by callose staining of plant cell walls with aniline blue and focussing on the respective level of middle lamellae (Figure 6). Epidermal cells had about 20 (n = 11) pit fields with plasmodesmata in one optical level connecting them to their horizontal neighbours (Figures 6c,d and S2a,b). The average distance between pit fields was 7 µm. The intensity of callose accumulation in pit fields of epidermal cell walls varied and reduced the capacity of plasmodesmata in horizontal transfer (Figure 6b,e,f).

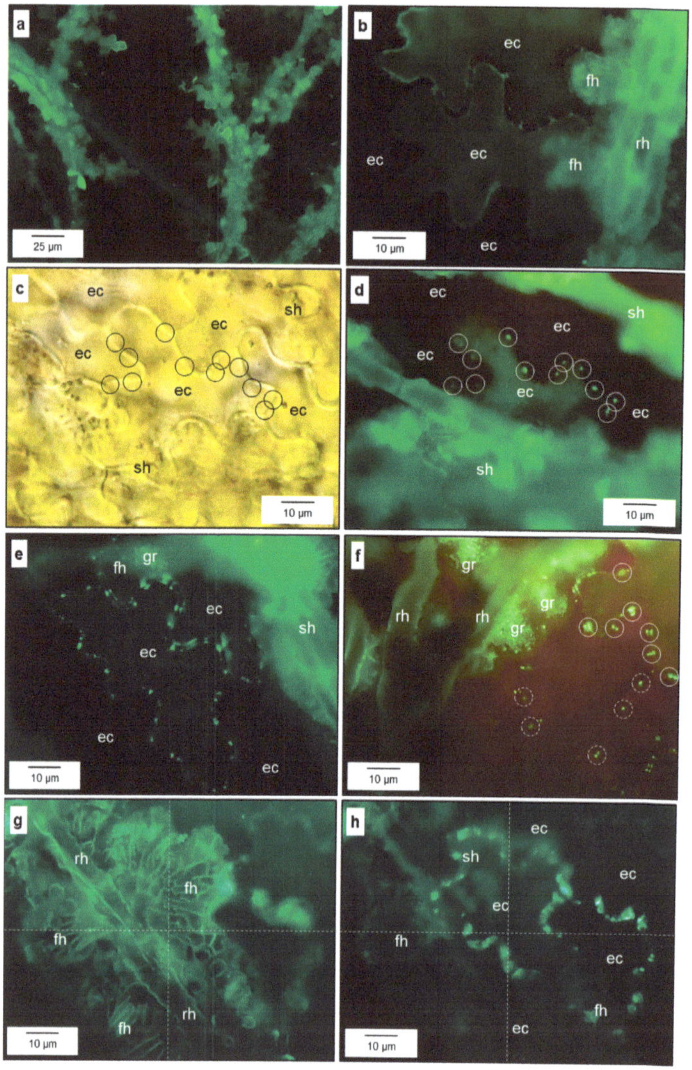

Figure 6. Effect of *V. inaequalis* colonisation on anticlinal pit fields of epidermal cells of apple leaves. Fluorescence of epidermal cells underneath subcuticular hyphae after staining with aniline blue (**a**); formation of callose at pit fields of neighbouring epidermal cells (ec) in immediate vicinity of runner hyphae (rh) and fan-shaped hyphae (fh) above (**b**); spatial pattern of pit fields of an epidermal cell (**c**) and callose deposition at pit fields (**d**); callose accumulation at pit fields of an epidermal cell next to a cell colonised superficially (**e**); variation in the intensity of callose accumulation at pit fields varying from strong (solid line) to light (dotted line) (**f**); spatial pattern of subcuticular colonisation (**g**) and callose accumulation at the pit fields of epidermal cells beneath (**h**). Bright-field microscopy (**c**); fluorescence microscopy (**a**,**b**,**d**–**h**) after staining with aniline blue. gr, fluorescent granules; sh, subcuticular hyphae.

Callose accumulation at pit fields occurred not only beneath subcuticular pathogen structures, but also in adjacent epidermal cells, indicating a plant reaction which was not successful in limiting the pathogenesis of *V. inaequalis* (Figure 6b,g,h). As demonstrated by a varying degree of fluorescence staining, the permeability of the cytoplasm membrane of affected epidermal cells was modified (Figure 6a,b).

3.2.2. Plant–Pathogen Interface

The outer cell wall of colonised epidermal cells revealed disaggregation across the wall, which is probably an indicator of a facilitated transport through the cell wall (Figures 7b,d and S2f). The decreased compactness of the otherwise solid wall was limited to the plant–fungus interface (Figure S2g).

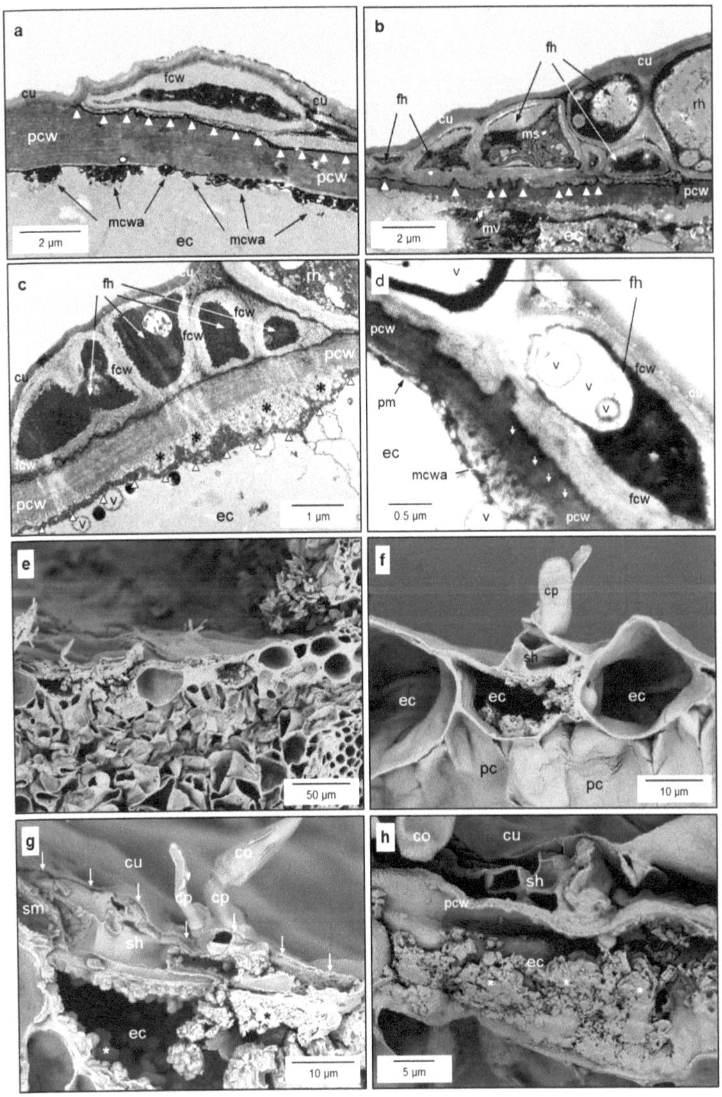

Figure 7. Reaction of epidermal cells of apple leaves to subcuticular development of *V. inaequalis*. Increased formation of cytoplasm membrane vesicles in an epidermal cell (ec) underneath a subcuticular hypha (**a**); formation of membranous vesicles (mv) in an epidermal cell and membrane staples (ms) in the fan-shaped hyphae of *V. inaequalis* above, respectively; rippled host–pathogen interface (**b**);

membranous apposition (*) and increased vesicle (v) formation at the plant cell wall underneath fan-shaped hyphae (fh); anticlinal fungal cell walls (fcw) thicker than periclinal walls (**c**); membranous cell wall appositions (mcwa) and loosening of the plant cell wall in contact with a subcuticular hypha (**d**); spatial pattern of epidermal cells with and without cell wall ingrowth in response to scab development (**e**); close-up of epidermal cells with and without cell wall ingrowth underneath infection structures of *V. inaequalis* (**f**); increase in the deposition of (membranous) material (*) in epidermal cells, superficially colonised by *V. inaequalis* in an early (**g**) and later (**h**) stage of pathogenesis. Images from transmission electron microscopy (**a–d**) and scanning electron microscopy (**e–h**), respectively. co, conidium; cp, conidiophore; cu, cuticle; pc, cell of palisade parenchyma; pcw, plant cell wall; pm, plasma membrane; rh, runner hypha; sh, subcuticular hypha; sm, subcuticular matrix material.

Intricate protuberances/ingrowth of the epidermal cell wall of some epidermal cells beneath subcuticular hyphae of *V. inaequalis* demonstrated a transformation into transfer cells (Figure 7). Modified cells were typically surrounded by epidermal cells without any modification. TEM images revealed membranous cell wall appositions of different intensities at the inner side of the outer epidermal cell wall SEM images (Figure 7a–d). In rare cases, epidermal cells responded to the presence of subcuticular hyphae with the formation of a solid cell wall apposition on the inner side (Figure S2h,i). Adjacent epidermal cells, however, exhibited intensive vesicle trafficking towards the growing pathogen above (Figure S2i). Also, SEM images revealed that only some epidermal cells beneath the subcuticular hyphae responded with the formation of massive modifications of the cell wall (Figure 7e–h). Both periclinal cell walls—the top in contact with the pathogen, the bottom forming the interface with mesophyll cells—were affected by cell wall depositions, whose intensity increased during pathogenesis.

Intensive subcuticular colonisation of leaves by *V. inaequalis* hyphae sometimes resulted in a turgor loss of epidermal cells, and in some cases the epidermal cell layer beneath secondary stromata of *V. inaequalis* was compressed (Figure 8a,b). A loss of cytoplasm integrity was never observed and often epidermal cells beneath hyphae maintained full turgescence (Figure 8c). The vitality of colonised leaf tissue was demonstrated by the formation of green islands under the sites of *V. inaequalis* sporulation, producing the brown, melanised conidia on the surface of apple leaves (Figure 8d). Islands of green leaf tissue beneath and around scab infections were observed on leaves infected for several weeks in apple orchards as well as under controlled conditions (5 weeks after inoculation).

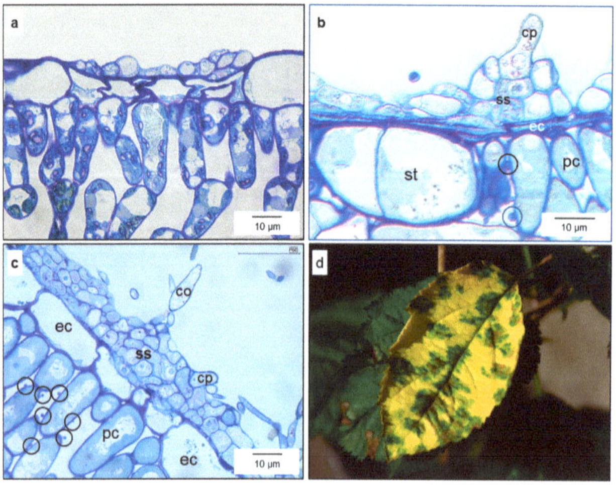

Figure 8. Effects of *V. inaequalis* on the morphology of apple leaf tissue in later stages of pathogenesis. Compression of colonised epidermal cells—fungal stroma with one and two layers, respectively; integrity of the cytoplasm of epidermal cells (**a**); formation of globular protuberances (circles) at the

surface of a palisade cell (pc); epidermal cells (ec) beneath the sporulating stroma (ss) compressed, but not necrotic (**b**); multi-layered stroma with sporulation hardly affecting the morphology of epidermal cells of an apple stem 35 d p.i. (**c**); formation of green islands under scab lesions, which present as brown due to the massive presence of melanised conidia, not because of the formation of plant necroses (**d**). Bright-field microscopy of leaf sections stained with toluidine blue (**a**–**c**); RGB camera (**d**). co, conidium; cp, conidiophore.

3.2.3. Response of Remote Cell Layers of Infected Apple Leaves

Seven to ten days after infection, some mesophyll cells' neighbouring epidermal cells subcuticularly colonised by *V. inaequalis* exhibited globular protuberances with a diameter of up to 2 µm on the cell surface (Figure 9). The formation started in the parenchyma layer adjacent to the infection site and often expanded to the spongy parenchyma and palisade parenchyma, respectively, during further infection. The intensity and size of these protuberances varied considerably. The close contact between palisade cells became loose and the protuberances were formed at both the interface of neighbouring cells as well as at the interface to the intercellular space (Figure 9c). The interior wall of affected palisade cells had small indentions, probably corresponding to the site of outer protuberances (Figure 9b). Some cells of the spongy parenchyma formed a high number of protuberances, with neighbouring cells showing no visible modification (Figure 9a,d).

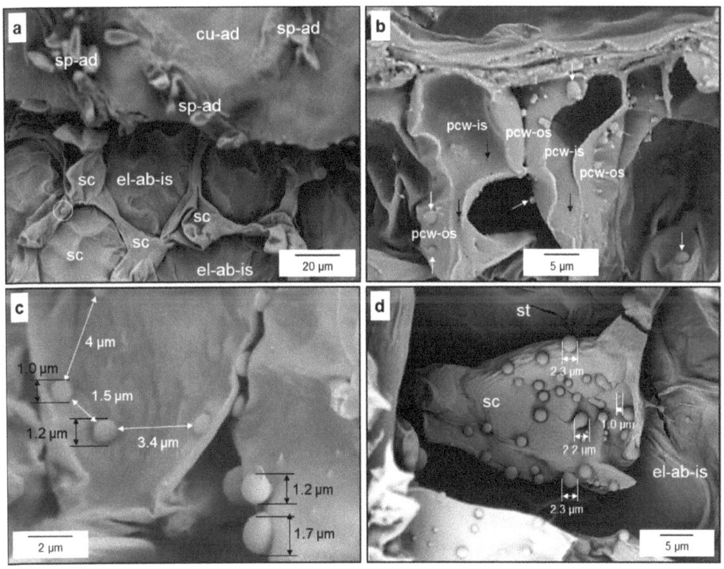

Figure 9. Modifications of mesophyll cells during *V. inaequalis* pathogenesis. Globular protuberance (circle) on the surface of a spongy parenchyma cell (sc); infection of adaxial epidermal cells with sporulation causes modifications of remote cells 16 d p.i. (**a**); mesophyll cells respond to subcuticular hyphae by the formation of globular, knob-like protuberances (white arrows) on the outer surface of cells of the palisade (pcw-os) and spongy parenchyma (white arrow bottom right); cross-section also visualising wall thinning spots (black arrows) on the inner surface of palisade cells (pcw-is), probably corresponding to protuberances at the outer side (**b**); size and number of knob-like structures on mesophyll cells varied; globular protuberances of palisade cells between adjoining cells and oriented to the intercellular space (**c**); individual spongy parenchyma cell intensively covered by protuberances with a diameter up to >2 µm (**d**). Images from scanning electron microscopy. cu-ad, adaxial cuticle; el-ab-is, inner surface of abaxial epidermal layer; sp-ad, adaxial sporulation; st, stoma.

4. Discussion

Subcuticular development of *V. inaequalis* is characterised by strictly apoplastic growth between the cuticle and epidermal layers without producing structures that enter plant cells and without causing obvious damage to apple leaf tissue. Runner hyphae and stromata are thought to be involved in nutrient acquisition and effector delivery [44].

Biotrophic Pathogen Structures

After cuticle penetration from epicuticular appressoria, the subcuticular colonisation of apple leaves by mono-layered primary stroma, thick runner hyphae, and laterally formed, thin, fan-shaped hyphae (FH) is intensive. Runner hyphae preferentially spread above anticlinal host cell walls—probably with a high density of pit fields—and produce lateral fan-shaped hyphae which, in early stages, form a monolayer positioned on top of the lumen of epidermal cells. Rarely, FH exceed the margin of the epidermal cell beneath and expand to adjacent cells. As especially the epidermal layer above leaf veins provides a high nutrient content of the intercellular fluid, the growth and sporulation of *V. inaequalis* is often stronger in veins than on intercostal leaf areas.

The small and very thin FH advance under the cuticle by separating it from the epidermal cell wall beneath, most probably by secretion of hydrolytic enzymes like polygalacturonases and other pectinases. They have a high surface-to-volume ratio, often even increased by undulations of the cell wall in contact with the plant cell, and cover large areas of the surface of epidermal cells. In later stages, FH become the foundation of multi-layered secondary stromata that produce melanised conidiophores and conidia [3]. FH form a layer of cells porous for water (and nutrients) which is considerably larger than the leaf area, with sporulating conidiophores perforating the plant cuticle from inside, and which is likely to function as a filter for nutrients, not as a water barrier. After cuticle perforation by *V. inaequalis* conidiophores, the secondary stroma contributes to the redirection of intercellular water and nutrients to the pathogen and the uptake of organic nutrients facilitated by indentions and orifices of the fungal cell wall. In the case of abaxial scab infections, the dense layer of fan-shaped hyphae may even support the functionality of the fragile abaxial epidermal cells as a layer regulating the loss of water to the ambient air.

The activity of fungal cutinases often described as crucial for pathogenesis [7,21] has to be limited in time and space to the penetration of the cuticle for invasion and for sporulation—essential for both pathogen spread and the redirection of water flow towards the subcuticular pathogen. *V. inaequalis* preserves the functionality of the plant cuticle by minimising its damage to the sites of emerging conidiophores [2]. Other hydrolases (pectinases, etc.) reported for *V. inaequalis* [30,31,51] are essential for the separation of cuticle and epidermal cell walls in order to produce the subcuticular niche.

The Host–Pathogen Interface

The huge interface between subcuticular fan-shaped hyphae and the adaxial side of epidermal cells facilitates the nutrient uptake of the pathogen. High metabolic activity is favoured by small cells. Not only is the abaxial side of the mycelium in contact with the nutrient mixture of the intercellular fluid provided by the plant cells, but all sides of the abutting fungal cells are able to contribute to nutrient uptake. Their straining activity is likely to be highest in anticlinal fungal cell walls, which are thicker than periclinal walls, although disaggregated and less compact.

The first stages of pathogenesis after penetration of the leaf cuticle—primary stroma, runner hyphae, and early fan-shaped hyphae—are likely to have rather low nutrient requirements, which may be matched by the nutrients available in the apoplast; however, knob-like structures of these hyphae appressed to host cells occur already in these early stages. These structures can be detected by using light and scanning electron microscopy and may be interpreted as areas of intimate contact between pathogen and host, probably related to notches in the fungal cell wall, facilitating the exchange of nutrients and effectors. As the subcuticular fungal cells are tightly surrounded by epidermal cell walls and the intact cuticle, respectively, small cavities (vents) in the cell wall do not affect the integrity of the pathogen. In later stages, especially during conidiation with high nutrient requirements, expanded fan-shaped hyphae produce a large interface with epidermal cells, which themselves have to be nourished by mesophyll cells—likely via an increased formation or modifications of secondary plasmodesmata. The realisation of modifications in the metabolism and morphology of plant cells induced by pathogen effectors is a series of events which takes some time.

Partial degradation is only one of the modifications of the cell walls of apple leaves. The areas in direct contact with *V. inaequalis* hyphae often appear loosened and weakened with anticlinal abnormities, and vesicle formation of the cytoplasm membrane is often increased. Some epidermal cells respond by the formation of membranous wall ingrowth, characterising transfer cells [65]. Their enhanced capacity for nutrient transport is conferred by an amplified plasma membrane surface area, enriched in membrane nutrient transporters, supported on an intricate wall labyrinth [66]. They are located at bottlenecks for nutrient transport between apoplastic and symplastic compartments, e.g., at sites of nutrient loading/unloading of vascular pipelines or interfaces between host and biotroph pathogens [66,67]. The formation of transfer cells is induced by external stimuli, e.g., light, or infection by nematodes [68,69] and rust fungi [70]. Adaxial epidermal cells of *Vicia faba* cotyledons undergo trans-differentiation to transfer cell morphology and function within hours [67]. In *Arabidopsis thaliana*, wall ingrowth formation in phloem parenchyma transfer cells occurring in leaf minor veins is induced by their phloem-loading activity, which is regulated by sucrose [71]. In apple leaves, leaf minor veins are the preferred sites of scab development. The fluffy membrane appositions at the inner cell wall of epidermal cells are different from solid cell wall appositions, which rarely occur as a non-successful resistance response of the host.

Fungi may develop structures similar to transfer cells, especially in the case of the Hartig net in ectomycorrhizae, where the penetrating hyphal system is a broad-lobed, fan-like hyphal front with rare cellular septation. The multinucleate coenocytic organisation has wall ingrowths which are actually incomplete septae [72–74]. The structure of the fan-shaped hyphae of *V. inaequalis* produced during subcuticular growth in apple leaves is very similar to the top view of the Hartig net of ectomycorrhizal fungi. Host root rhizodermal and cortical cells in contact with hyphae may also develop wall protuberances [75]. Both structures of the pathogen and changes in plant cells obviously increase the surface of the interface in a corresponding manner.

Pores Regulate the Flow of Nutrients and Information

The heterogeneity in the response of epidermal cells to subcuticular *V. inaequalis* hyphae is in agreement with observations that the intercellular movement of resources and information can be controlled by the density and structure of plasmodesmata, as well as by their opening status. Cell-to-cell connectivity within the epidermal layer is heterogeneous due to variation in the status—open, closed, intermediate—of plasmodesmata and leads to a mosaic of connectivity [76–78]. As plasmodesmata allow the distribution of resources and the passage of signals by hormones, small RNAs, transcription factors, and mRNAs, the restriction of these processes by callose accumulation reduces the interactions between plant cells and reorganises the flux of resources and information on a local scale. The induction of secondary plasmodesmata in the cell walls of epidermal cells in contact with the cuticle and the pathogen, respectively, would optimise the nutrient availability for *V. inaequalis*.

Plasmodesmata connect neighbouring cells to form the plant symplast (primary and secondary plasmodesmata); they are cytoplasmatic strands including the endoplasmatic reticulum. The respective cell wall is thin and forms pit fields with several plasmodesmata. Information on the number of plasmodesmata connecting epidermal cells to the palisade cells of apple leaves is lacking. In other plant species, the number differs between species but is similar to the number of anticlinal connections to other epidermal cells [79]. The number of plasmodesmata connections between palisade and spongy mesophyll cells is three to one-hundred times higher. It is not known whether epidermal cell walls in contact with the cuticle have plasmodesmata primordia (half-plasmodesmata) or pore-like structures facilitating the transport of vesicles or solutes through the cell wall to the exterior/apoplast, or whether these structures may be induced by *V. inaequalis*.

The number and density of plasmodesmata may be actively decreased or increased [80]. The formation of secondary plasmodesmata may be induced by phytohormones [81,82], flower-inducing day length [83], and parasitic plants which use this mechanism to develop new plasmodesmata between themselves and their hosts [84]. *Cuscuta* haustorial cells share plasmodesmata with hosts across chimeric cell walls [85,86], and these have been implicated in the host–parasite mobility of RNA [87]. In Arabidopsis, an increase in cell size increases and decreases the number of complex plasmodesmata of anticlinal and periclinal walls, respectively [81].

Being symplastic pores that facilitate cell-to-cell communication between neighbouring cells, plasmodesmata are a target of pathogen effectors [88]. The *Melampsora larici-populina* effector Mlp37347 accumulate exclusively at the plasmodesmata of *A. thaliana* and increase plasmodesmal flux by reducing callose deposition [89]. Effector Six5 of the wilt pathogen *Fusarium oxysporum* directly targets plasmodesmata [90,91] and the *Phytophthora brassicae* effector RxLR3 increases plasmodesmal flux by blocking the activity of the callose synthases CalS1, CalS2, and CalS3 at plasmodesmata [92]. Hofmann et al. [93] reported that a temporal callose deposition along plasmodesmata impaired the symplastic exchange in young syncytia of Arabidopsis infected by *Heterodera schachtii*. Plasmodesmata closure and transient cell isolation may serve to contain pathogen-derived effectors. Some of the effector proteins introduced by plant pathogens into host cells to counteract defence responses or to modify cellular metabolism to their own advantage can move through plasmodesmata, potentially to prepare host cells for infection [94,95].

Callose deposition/degradation around plasmodesmata necks or/and cell walls often occur during plant defence responses [96–98]. It is typically triggered by conserved PAMPs like fungal chitin and chitosan from the fungal cell wall [99]. Pectin-methyl-esterase and pectinase activities at plasmodesmata may contribute to local loosening of the wall to accommodate plasmodesmata aperture oscillations [95]. In this context, it seems likely that any invading microorganism—parasitic, mutualistic, or beneficial—would benefit from regulating plasmodesmata to maintain connectivity [100].

Host–pathogen interactions of the non-haustorial biotrophic fungi *Taphrina* spp. include an interface consisting of host and fungal cell walls, sometimes separated by a thin intercellular matrix, and are associated with an alteration of host plasma membrane permeability [75]. Electron microscopic images also demonstrate that the fan-shaped hyphae of *V. inaequalis* are embedded in a matrix-separating fungal cell wall and host plant cell wall, similar to an extrahaustorial matrix of biotrophic pathogens [101,102].

During the stroma development in the biotrophic grass endophyte *Epichloe amarillans*, the contacted host cell walls become modified to allow the easier acquisition of host nutrients by the fungus without the need for haustoria [103]. Changes in the permeability of the epidermal cell wall and plasma membrane are very likely, especially in areas of knob-like structures in fungal cells, transfer cells of plants, and fluorescent granules (exclusively underneath FH). The architecture of fan-shaped hyphae—small cells, thickness of anticlinal cell walls, high portion of vesicles, and membrane staples—corresponds to host transfer cells. The fluorescent granules underneath the core of FH are surrounded by a sealing lip from ultra-thin hyphae at the margin, which prevents the horizontal loss of nutrients. The increase in the density of fluorescent granules at the interface between epidermal cells and fan-shaped hyphae emphasises the dynamic nature of the interaction; however, it is not clear whether it is an indicator of increased or decreased transfer of nutrients and other compounds.

V. inaequalis secretes effectors in order to manipulate the metabolism/response of its host. Effectors may be directly released into plant cells/actively taken up by plant cells via endocytosis—symplastic localisation of a resistance protein coded by *Rvi15* has been reported by Schouten et al. [104]—or released into the apoplast, where they may travel longer distances and become active at or in plant cells remote from the infection site. Host cell re-programming by fungal effector molecules is likely to also cause morphological modifications of leaf cells.

Increased Release and Flow of Nutrients from Remote Plant Cells

The formation of globular protuberances of mesophyll cells in response to *V. inaequalis* infection has been already documented in the microscopic classification of scab symptoms [105] and in a textbook [106]; however, these authors had no explanation of their relevance and function or did not mention this phenomenon at all. Globules exuded by hypertrophied mesophyll cells towards intercellular spaces were observed in weakly susceptible host genotypes, whereas susceptible genotypes showed no morphological modifications of the mesophyll at the time of symptom appearance [105]. The globules could be stained with toluidine blue, which is used in plant specimens to detect pectin and lignin [107,108]. Ultrastructural information indicates a link between pit fields/plasmodesmata and the globules on the outside of mesophyll cells—the intimate contact between pathogens and epidermal cell walls hampers the detection of epidermal globules. As an activation of the plasmodesmal transport into the apoplast does not require the formation of cell wall protuberances into the intercellular space, this formation is likely to be associated with another order of magnitude of organic compound release from the symplast.

Pectinaceous beads are protuberances which occur on the outside of cell walls facing intercellular spaces. They are bead-like projections consisting of pectic polysaccharides, proteins, and fatty acids. These beads are a normal feature of epidermal and mesophyll cells of *Xyris* spp. leaves [109] on the surfaces of callus cells during grafting or wound healing, where pectin fragments (via pectinase activity) may function as signalling molecules for the compatibility/incompatibility of grafts, or on the outer surface of mesophyll cells in tobacco and some tree species [75]. The larger size and the lower number per cell distinguish the globules induced by *V. inaequalis* from those described in the literature so far, although a pectinaceous nature is in agreement with the release of pectinases by *V. inaequalis*.

Biotrophy and the Holomorph of *V. inaequalis*

The development of ontogenetic or adult plant resistance of fully expanded apple leaves to scab infection is in agreement with the transition from sink to source. As leaves age, plasmodesmal transport within the epidermal layer decreases gradually and quantitatively, and leaves switch from heterotrophic sinks for carbohydrates, nitrogen, and other resources to photoautotrophic sources of carbohydrates for the rest of the plant [110,111]. Restriction of plasmodesmata transport in source leaves strongly limits the fungal pathogen's capacity to redirect the flow of nutrients. Under experimental conditions, latent infections with subcuticular growth but only rare sporulation may produce slightly chlorotic tissue and no penetration of the cuticle from the inside of the plant. Nevertheless, earlier, successful colonisation events are suitable to result in the formation of green islands on sab-infected leaves in late pathogenesis stages, demonstrating the pathogen effect on the phytohormone balance within leaf tissue typical for biotroph host–pathogen interactions.

Biotrophy depends on the capability of plant pathogens to prevent host defence reactions by modifications of surface properties and to manipulate the host cells to provide the organic nutrients required for growth and reproduction and is rather independent of structures like intracellular, still apoplastic haustoria. Other mechanisms like the redirection of water flow and the induction of plant transport mechanisms into the extracellular space in combination with a large extracellular interface may also be suitable. *V. inaequalis* uses the water vapour pressure deficit between plant tissue and ambient air as the driving force of nutrient flow from plant cells to the pathogen [2]. Nevertheless, this mechanism highly depends on the water status of the host and, thus, also on environmental conditions. Drought conditions result in precocious plant damage and the end of the biotroph stage, but *V. inaequalis* survives under these conditions and resumes activity on fallen leaves in autumn as a saprophyte to produce the sexual stage during hibernation. This type of biotroph lifestyle is well adapted to moderate climates with a sufficient water supply and medium temperatures limiting the overall rate of apple transpiration.

5. Conclusions

From our knowledge on the effect of scabs on the balance and movement of water in apple leaves [2] and the detailed studies on both subcuticular infection structures and morphological modifications of apple tissue in response to pathogen colonisation, a model of the lifestyle of *V. inaequalis* has been conceptualised (Figure 10). The pathogen takes up nutrients from living host cells by (I) intimate contact with a large interface between pathogen and host cells, provided by fan-shaped hyphae with small and very thin cells and facilitated by the presence of local vents in the fungal cell wall (Figure 10a,b); (II) induction of secondary plasmodesmata in the epidermal cell wall and blockage of anticlinal pit fields by callose accumulation, favouring the flow of nutrients (and effectors) in a vertical direction (Figure 10a); (III) fungal effectors modifying the permeability of host cells and causing the transition of some epidermal cells into transfer cells, as well as the formation of globular protuberances (from pit fields?) in mesophyll cells, which support an apoplastic nutrient flow in the direction of the subcuticular hyphae, driven by the water flow through the cuticular pores produced by conidiophores which penetrate the cuticle from the subcuticular space to the outside of the leaf (Figure 10a,c).

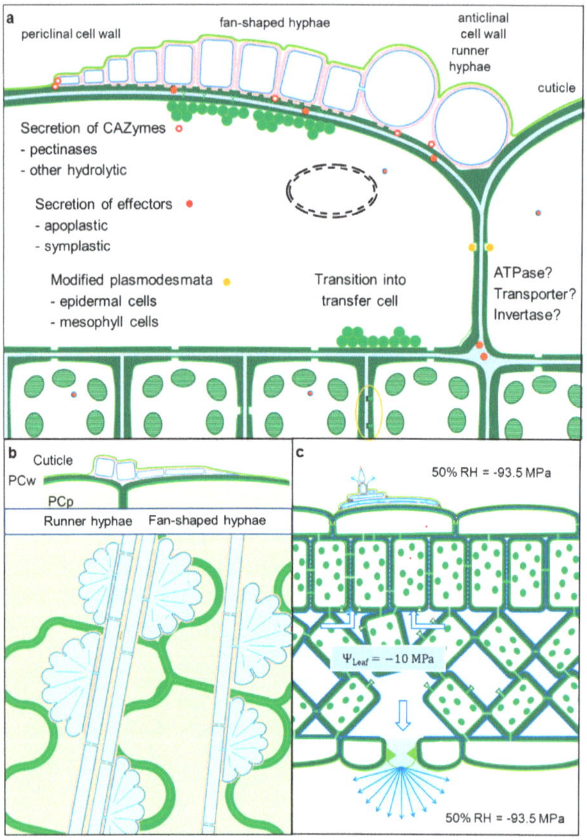

Figure 10. Model of *V. inaequalis* infection structures and apple leaf tissue responses. The primary stroma advances subcuticularly by the secretion of pectinolytic enzymes and forms runner hyphae for spreading; small and very thin fan-shaped hyphae cover large areas of the epidermal cell lumen

for nutrient uptake. The secretion of effectors that may stay apoplastic or become symplastic contributes to modifications of epidermal and mesophyll cells. Intimate contact of fungal hyphae with the plant surface and partial degradation result in cell wall loosening and the formation of membranous cell wall appositions in responsive cells. Modifications of plasmodesmata in anticlinal pit fields favour the vertical flow of water and nutrients, which is strongly increased after the perforation of the cuticle due to conidiophore formation 5 to 6 d p.i. Secondary plasmodesmata at the plant–pathogen interface improve nutrient supply of the biotroph pathogen. Formation of globular protuberances on the surface of mesophyll cells (orange oval) increases the nutrition of the subcuticular fungus in the later stages of pathogenesis (**a**). Growth pattern of runner hyphae and lateral fan-shaped hyphae between cuticle and epidermal cells (**b**). Increased cuticular transpiration due to cuticle perforation by conidiophores redirects the flow of water and nutrient to the subcuticularly growing pathogen (**c**).

Non-obligate biotrophic pathogens may be cultivated on artificial media and live saprotrophically on dead leaf tissue in autumn. *V. inaequalis* produces conidia for polycyclic spread in its biotroph stage as long as young host tissue is available; the pathogen lacks a necrotroph stage with active killing of plant cells by enzymes or toxins but resumes growth with melanised hyphae saprotrophically only after dieback of leaf tissue due to senescence (Figure S3). Hemibiotrophy, therefore, has to be differentiated into the combinations biotrophy–necrotrophy and biotrophy–saprotrophy.

Supplementary Materials: The following supporting information can be downloaded at https://www.mdpi.com/article/10.3390/jof10120831/s1, Figure S1: Subcuticular development of *Venturia inaequalis* hyphae; Figure S2: Reaction of apple tissue to subcuticular leaf colonisation by *Venturia inaequalis*; Figure S3: Life cycle and lifestyles of the holomorph of *Venturia inaequalis*.

Author Contributions: Both authors developed the concept, conducted the experiments, and prepared the manuscript work reported. All authors have read and agreed to the published version of the manuscript.

Funding: This research received no external funding.

Institutional Review Board Statement: Not applicable.

Informed Consent Statement: Not applicable.

Data Availability Statement: The data supporting the findings of this study are available from the first author, Ulrike Steiner, upon reasonable request.

Acknowledgments: The authors thank Maximillian Becker and Jerôme Farhumand-Khunssari for experimental support as well as Carolin Sichtermann and Joachim Hamacher for support in transmission electron microscopy.

Conflicts of Interest: The authors declare no conflicts of interest.

References

1. Carisse, O.; Bernier, J. Effect of environmental factors on growth, pycnidial production and spore germination of Microsphaeropsis isolates with biocontrol potential against apple scab. *Mycol. Res.* **2002**, *106*, 1455–1462. [CrossRef]
2. Oerke, E.C.; Steiner, U. Hyperspectral imaging reveals small-scale water gradients in apple leaves due to minimal cuticle perforation by *Venturia inaequalis* conidiophores. *J. Exp. Bot.* **2024**, *75*, 3125–3140. [CrossRef]
3. Steiner, U.; Oerke, E.C. A melanin-deficient isolate of *Venturia inaequalis* reveals various roles of melanin in pathogen life cycle and fitness. *J. Fungi* **2023**, *9*, 35. [CrossRef] [PubMed]
4. Gessler, C.; Stumm, D. Infection and stroma formation by *Venturia inaequalis* on apple leaves with different degrees of susceptibility to scab. *J. Phytopathol.* **1984**, *110*, 119–126. [CrossRef]
5. MacHardy, W.E. *Apple Scab: Biology, Epidemiology, and Management*; The American Phytopathological Society Press: St. Paul, MN, USA, 1996.
6. MacHardy, W.E.; Gadoury, D.M.; Gessler, C. Parasitic and biological fitness of *Venturia inaequalis*: Relationship to disease management strategies. *Plant Dis.* **2001**, *85*, 1036–1051. [CrossRef] [PubMed]
7. Bowen, J.K.; Mesarich, C.H.; Bus, V.G.M.; Beresford, R.M.; Plummer, K.M.; Templeton, M.D. *Venturia inaequalis*: The causal agent of apple scab. *Mol. Plant Pathol.* **2011**, *12*, 105–122. [CrossRef] [PubMed]
8. Caffier, V.; Le Cam, B.; Expert, P.; Tellier, M.; Devaux, M.; Giraud, M.; Chevalier, M. A new scab-like disease on apple caused by the formerly saprotrophic fungus *Venturia asperata*. *Plant Pathol.* **2012**, *61*, 915–924. [CrossRef]

9. Latham, A.J.; Rushing, A.E. Development of *Cladosporium caryigenum* in pecan leaves. *Phytopathology* **1988**, *78*, 1104–1108. [CrossRef]
10. Park, P.; Ishii, H.; Adachi, Y.; Kanematsu, S.; Ieki, H.; Umemoto, S. Infection behavior of *Venturia nashicola*, the cause of scab on Asian pears. *Phytopathology* **2000**, *90*, 1209–1216. [CrossRef]
11. Lanza, B.; Ragnelli, A.M.; Priore, M.; Aimola, P. Morphological and histochemical investigation of the response of *Olea europaea* leaves to fungal attack by *Spilocaea oleaginea*. *Plant Pathol.* **2017**, *66*, 1239–1247. [CrossRef]
12. Hunter, G.C.; Zeil-Rolfe, I.; Jourdan, M.; Morin, L. Exploring the host range and infection process of *Venturia paralias* isolated from *Euphorbia paralias* in France. *Eur. J. Plant Pathol.* **2021**, *159*, 811–823. [CrossRef]
13. Chevalier, M.; Bernard, C.; Tellier, M.; Lespinasse, Y.; Filmond, R.; Lezec, L.M. Variability in the reaction of several pear (*Pyrus communis*) cultivars to different inocula of *Venturia pirina*. *Acta Hort.* **2004**, *663*, 177–181. [CrossRef]
14. Blechert, O.; Debener, T. Morphological characterization of the interaction between Diplocarpon rosae and various rose species. *Plant Pathol.* **2005**, *54*, 82–90. [CrossRef]
15. Gachomo, E.W.; Dehne, H.W.; Steiner, U. Microscopic evidence for the hemibiotrophic nature of *Diplocarpon rosae*, cause of black spot disease of rose. *Physiol. Mol. Plant Pathol.* **2006**, *69*, 86–92. [CrossRef]
16. Zhan, J.; Fitt, B.D.L.; Pinnschmidt, H.O.; Oxley, S.J.P.; Newton, A.C. Resistance, epidemiology and sustainable management of *Rhynchosporium secalis* populations on barley. *Plant Pathol.* **2008**, *57*, 1–14. [CrossRef]
17. Linsell, K.J.; Keiper, F.J.; Forgan, A.; Oldach, K.H. New insights into the infection process of *Rhynchosporium secalis* in barley using GFP. *Fung. Genet. Biol.* **2011**, *48*, 124–131. [CrossRef] [PubMed]
18. Zhou, Y.; Zhang, L.; Ji, C.Y.; Chaisiri, C.; Yin, L.F.; Yin, W.X.; Luo, C.X. Cytological observation of the infectious process of *Venturia carpophila* on peach leaves. *Plant Dis.* **2022**, *106*, 79–86. [CrossRef] [PubMed]
19. Smereka, K.J.; Kausch, A.P.; MacHardy, W.E. Intracellular junctional structures in germinating ascopores of *Venturia inaequalis*. *Protoplasma* **1988**, *142*, 1–4. [CrossRef]
20. Köller, W.; Parker, D.M. Purification and characterization of cutinase from *Venturia inaequalis*. *Phytopathology* **1989**, *79*, 278–283. [CrossRef]
21. Köller, W.; Parker, D.M.; Becker, C.M. Role of cutinase in the penetration of apple leaves by *Venturia inaequalis*. *Phytopathology* **1991**, *81*, 1375–1379. [CrossRef]
22. Steiner, U.; Oerke, E.C. Localized melanization of appressoria is required for pathogenicity of *Venturia inaequalis*. *Phytopathology* **2007**, *97*, 1222–1230. [CrossRef] [PubMed]
23. Nusbaum, C.J.; Keitt, G.W. A cytological study of host-parasite relations of *Venturia inaequalis* on apple leaves. *J. Agric. Res.* **1938**, *56*, 595–618.
24. Kucheryava, N.; Bowen, J.K.; Sutherland, P.W.; Conolly, J.J.; Mesarich, C.H.; Rikkerink, E.H.; Kemen, E.; Plummer, K.M.; Hahn, M.; Templeton, M.D. Two novel *Venturia inaequalis* genes induced upon morphogenetic differentiation during infection and in vitro growth on cellophane. *Fung. Genet. Biol.* **2008**, *45*, 1329–1339. [CrossRef] [PubMed]
25. Cooper, S.J.; Ashby, A.M. Comparison of cytokinin and cytokinin-O-glucoside cleaving β-glucosidase production in vitro by *Venturia inaequalis* and other phytopathogenic fungi with differing modes of nutrition in planta. *Physiol. Mol. Plant Pathol.* **1998**, *53*, 61–72. [CrossRef]
26. Walters, D.R.; McRoberts, N. Plants and biotrophs: A pivotal role for cytokinins? *Trends Plant Sci.* **2006**, *11*, 581–586. [CrossRef] [PubMed]
27. Jha, G.; Thakur, K.; Thakur, P. The *Venturia*–apple pathosystem: Pathogenicity mechanisms and plant defense responses. *J. Biomed. Biotechnol.* **2009**, *2009*, 680160. [CrossRef] [PubMed]
28. Valsangiacomo, C.; Gessler, C. Role of the cuticular membrane in ontogenic and Vf-resistance of apple leaves against *Venturia inaequalis*. *Phytopathology* **1988**, *78*, 1066–1069. [CrossRef]
29. Corlett, M.; Chong, J.; Kokko, E.G. The ultrastructure of the Spilocea state of *Venturia inaequalis* in vivo. *Can. J. Microbiol.* **1976**, *22*, 1144–1152. [CrossRef]
30. Valsangiacomo, C.; Gessler, C. Purification and characterization of an exo-polygalacturonase produced by *Venturia inaequalis*, the causal agent of apple scab. *Physiol. Mol. Plant Pathol.* **1992**, *40*, 63–77. [CrossRef]
31. Kollar, A. Characterization of an endopolygalacturonase produced by the apple scab fungus, *Venturia inaequalis*. *Mycol. Res.* **1998**, *102*, 313–319. [CrossRef]
32. Hignett, R.C.; Kirkham, D.S. The role of extracellular melanoproteins of *Venturia inaequalis* in host susceptibility. *J. Gen. Microbiol.* **1967**, *48*, 269–275. [CrossRef] [PubMed]
33. Hignett, R.C.; Roberts, A.L.; Carder, J.H. Melanoprotein and virulence determinants of *Venturia inaequalis*. *Physiol. Plant Pathol.* **1984**, *24*, 321–330. [CrossRef]
34. Singh, P.; Piotrowski, M.; Gau, A.E. Purification and partial characterization of an extracellular melanoprotein from the fungus *Venturia inaequalis*. *Z. Naturforsch. C* **2005**, *60*, 109–115. [CrossRef]
35. Jacobson, E.S. Pathogenic roles for fungal melanins. *Clin. Microbiol. Rev.* **2000**, *13*, 708–717. [CrossRef] [PubMed]
36. Rovenich, H.; Zuccaro, A.; Thomma, B.P.H.J. Convergent evolution of filamentous microbes towards evasion of glycan-triggered immunity. *New Phytol.* **2016**, *212*, 896–901. [CrossRef] [PubMed]
37. Oliveira Silva, A.S.L.; Wirsel, S.; Deising, H. Pathogenesis-related cell wall biogenesis with emphasis on the maize anthracnose fungus *Colletotrichum graminicola*. *Plants* **2022**, *11*, 849. [CrossRef] [PubMed]

38. Geoghegan, I.A.; Gurr, S.J. Investigating chitin deacetylation and chitosan hydrolysis during vegetative growth in *Magnaporthe oryzae*. *Cell. Microbiol.* **2017**, *19*, e12743. [CrossRef] [PubMed]
39. Gao, F.; Zhang, B.S.; Zhao, J.H.; Huang, J.F.; Jia, P.S.; Wang, S.; Zhang, J.; Zhou, J.M.; Guo, H.S. Deacetylation of chitin oligomers increases virulence in soil-borne fungal pathogens. *Nat. Plants* **2019**, *5*, 1167–1176. [CrossRef] [PubMed]
40. Noorifar, N.; Savoian, M.S.; Ram, A.; Lukito, Y.; Hassing, B.; Weikert, T.W.; Moerschbacher, B.M.; Scott, B. Chitin deacetylases are required for *Epichloë festucae* endophytic cell wall remodeling during establishment of a mutualistic symbiotic interaction with *Lolium perenne*. *Mol. Plant-Microbe Interact.* **2021**, *34*, 1181–1192. [CrossRef] [PubMed]
41. Zhang, J.; Zhou, J.M. Plant immunity triggered by microbial molecular signatures. *Mol. Plant* **2010**, *3*, 783–793. [CrossRef] [PubMed]
42. Saijo, Y.; Loo, E.P.; Yasuda, S. Pattern recognition receptors and signaling in plant—Microbe interactions. *Plant J.* **2018**, *93*, 592–613. [CrossRef] [PubMed]
43. Rocafort, M.; Bowen, J.K.; Hassing, B.; Cox, M.P.; McGreal, B.; de la Rosa, S.; Plummer, K.M.; Bradshaw, R.E.; Mesarich, C.H. The *Venturia inaequalis* effector repertoire is dominated by expanded families with predicted structural similarity, but unrelated sequence, to avirulence proteins from other plant-pathogenic fungi. *BMC Biol.* **2022**, *20*, 246. [CrossRef] [PubMed]
44. Rocafort, M.; Srivastava, V.; Bowen, J.K.; Díaz-Moreno, S.M.; Guo, Y.A.; Bulone, V.; Plummer, K.M.; Sutherland, P.W.; Anderson, M.A.; Bradshaw, R.E.; et al. Cell wall carbohydrate dynamics during the differentiation of infection structures by the apple scab fungus, *Venturia inaequalis*. *Microbiol. Spectr.* **2023**, *11*, e0421922. [CrossRef] [PubMed]
45. Vander, P.; Varum, K.M.; Domard, A.; Eddine El Gueddari, N.; Moerschbacher, B.M. Comparison of the ability of partially *N*-acetylated chitosans and chitooligosaccharides to elicit resistance reactions in wheat leaves. *Plant Physiol.* **1998**, *118*, 1353–1359. [CrossRef] [PubMed]
46. Gubaeva, E.; Gubaev, A.; Melcher, R.L.J.; Cord-Landwehr, S.; Singh, R.; El Gueddari, N.E.; Moerschbacher, B.M. 'Slipped Sandwich' model for chitin and chitosan perception in Arabidopsis. *Mol. Plant-Microbe Interact.* **2018**, *31*, 1145–1153. [CrossRef]
47. Lopez-Moya, F.; Suarez-Fernandez, M.; Lopez-Llorca, L.V. Molecular mechanisms of chitosan interactions with fungi and plants. *Int. J. Mol. Sci.* **2019**, *20*, 332. [CrossRef] [PubMed]
48. Feldman, D.; Yarden, O.; Hadar, Y. Seeking the roles for fungal small-secreted proteins in affecting saprophytic lifestyles. *Front. Microbiol.* **2020**, *11*, 455. [CrossRef] [PubMed]
49. Akhoon, B.A.; Gupta, S.K.; Dhar, M.K. Dissecting the genome, secretome, and effectome repertoires of *Monilinia* spp. the causal agent of brown rot disease: A comparative analysis. *Postharvest Biol. Technol.* **2023**, *195*, 112120. [CrossRef]
50. Jia, M.; Gong, X.; Fan, M.; Liu, H.; Zhou, H.; Gu, S.; Liu, Y.; Dong, J. Identification and analysis of the secretome of plant pathogenic fungi reveals lifestyle adaptation. *Front. Microbiol.* **2023**, *14*, 1171618. [CrossRef] [PubMed]
51. Deng, C.H.; Plummer, K.M.; Jones, D.A.B.; Mesarich, C.H.; Shiller, J.; Taranto, A.P.; Robinson, A.J.; Kastner, P.; Hall, N.E.; Templeton, M.D.; et al. Comparative analysis of the predicted secretomes of Rosaceae scab pathogens *Venturia inaequalis* and *V. pirina* reveals expanded effector families and putative determinants of host range. *BMC Genom.* **2017**, *18*, 339. [CrossRef]
52. Le Cam, B.; Sargent, D.; Gouzy, J.; Amselem, J.; Bellanger, M.N.; Bouchez, O.; Brown, S.; Caffier, V.; De Gracia, M.; Debuchy, R.; et al. Population genome sequencing of the scab fungal species *Venturia inaequalis*, *Venturia pirina*, *Venturia aucupariae* and *Venturia asperata*. *G3 Genes Genomes Genet.* **2019**, *9*, 2405–2414. [CrossRef]
53. Khajuria, Y.P.; Akhoon, B.A.; Kaul, S.; Dhar, M.K. Secretomic insights into the pathophysiology of *Venturia inaequalis*: The causative agent of scab, a devastating apple tree disease. *Pathogens* **2023**, *12*, 66. [CrossRef] [PubMed]
54. Wang, Y.S.; Wu, J.; Yan, J.C.; Guo, M.; Xu, L.; Hou, L.P.; Zou, Q. Comparative genome analysis of plant ascomycete fungal pathogens with different lifestyles reveals distinctive virulence strategies. *BMC Genom.* **2022**, *23*, 34. [CrossRef]
55. Ospina-Giraldo, M.D.; Griffith, J.G.; Laird, E.W.; Mingora, C. The CAZyome of *Phytophthora* spp.: A comprehensive analysis of the gene complement coding for carbohydrate-active enzymes in species of the genus *Phytophthora*. *BMC Genom.* **2010**, *11*, 525. [CrossRef]
56. Arnous, A.; Meyer, A.S. Quantitative prediction of cell wall polysaccharide composition in grape (*Vitis vinifera* L.) and apple (*Malus domestica*) skins from acid hydrolysis monosaccharide profiles. *J. Agric. Food Chem.* **2009**, *57*, 3611–3619. [CrossRef] [PubMed]
57. Yip, V.; Withers, S. Breakdown of oligosaccharides by the process of elimination. *Curr. Opin. Chem. Biol.* **2006**, *10*, 147–155. [CrossRef]
58. Lombard, V.; Golaconda Ramulu, H.; Drula, E.; Coutinho, P.M.; Henrissat, B. The carbohydrate-active enzymes database (CAZy) in 2013. *Nucleic Acids Res.* **2014**, *42*, D490–D495. [CrossRef] [PubMed]
59. Kollar, A. Characterization of specific induction, activity, and isozyme polymorphism of extracellular cellulases from *Venturia inaequalis* detected in vitro and on the host plant. *Mol. Plant-Microbe Interact.* **1994**, *7*, 603–611. [CrossRef]
60. Hardham, A.R. Cell biology of fungal and oomycete infection of plants. In *Biology of the Fungal Cell*, 2nd ed.; Howard, R.J., Gow, N.A.R., Eds.; The Mycota VIII; Springer: Berlin/Heidelberg, Germany, 2007; pp. 251–289.
61. Bruzzese, E.; Hasan, S.A. A whole leaf clearing and staining technique for host specificity studies of rust fungi. *Plant Pathol.* **1983**, *32*, 335–338. [CrossRef]
62. Hood, M.E.; Shew, H.D. Applications of KOH–aniline blue fluorescence in the study of plant–fungal interactions. *Phytopathology* **1996**, *86*, 704–708. [CrossRef]

63. Juraschek, L.M.; Matera, C.; Steiner, U.; Oerke, E.C. Pathogenesis of *Plasmopara viticola* depending on resistance mediated by *Rpv3_1*, and *Rpv10* and *Rpv3_3*, and by the vitality of leaf tissue. *Phytopathology* **2022**, *112*, 1486–1499. [CrossRef]
64. Schumacher, C.F.A.; Steiner, U.; Dehne, H.W.; Oerke, E.C. Localized adhesion of nongerminated *Venturia inaequalis* conidia to leaves and artificial surfaces. *Phytopathology* **2008**, *98*, 760–768. [CrossRef] [PubMed]
65. Bergareche, D.; Royo, J.; Muñiz, L.M.; Hueros, G. Cell wall invertase activity regulates the expression of the transfer cell-specific transcription factor ZmMRP-1. *Planta* **2018**, *247*, 429–442. [CrossRef] [PubMed]
66. Offler, C.E.; McCurdy, D.W.; Patrick, J.W.; Talbot, M.J. Transfer cells: Cells specialized for a special purpose. *Annu. Rev. Plant Biol.* **2003**, *54*, 431–454. [CrossRef]
67. Andriunas, F.A.; Zhang, H.M.; Xia, X.; Patrick, J.W.; Offler, C.E. Intersection of transfer cells with phloem biology-broad evolutionary trends, function, and induction. *Front. Plant Sci.* **2013**, *4*, 221. [CrossRef] [PubMed]
68. Dorhout, R.; Gommers, F.J.; Kollöffel, C. Phloem transport of carboxyfluorescein through tomato roots infected with *Meloidogyne incognita*. *Physiol. Mol. Plant Pathol.* **1993**, *43*, 1–10. [CrossRef]
69. Rodiuc, N.; Vieira, P.; Banora, M.Y.; de Almeida Engler, J. On the track of transfer cell formation by specialized plant-parasitic nematodes. *Front. Plant Sci.* **2014**, *5*, 160. [CrossRef] [PubMed]
70. Mims, C.W.; Rodriguez-Lother, C.; Richardson, E.A. Ultrastructure of the host-parasite interaction in leaves of *Duchesnea indica* infected by the rust fungus *Frommeëla mexicana* var. indicae as revealed by high pressure freezing. *Can. J. Bot.* **2001**, *79*, 49–57.
71. Wei, X.; Nguyen, S.T.T.; Collings, D.A.; McCurdy, D.W. Sucrose regulates wall ingrowth deposition in phloem parenchyma transfer cells in Arabidopsis via affecting phloem loading activity. *J. Exp. Bot.* **2020**, *71*, 4690–4702. [CrossRef] [PubMed]
72. Blasius, D.; Feil, W.; Kottke, I.; Oberwinkler, F. Hartig net structure and formation in fully ensheathed ectomycorrhizas. *Nord. J. Bot.* **1986**, *6*, 837–842. [CrossRef]
73. Kottke, I.; Oberwinkler, F. The cellular structure of the Hartig net: Coenocytic and transfer cell-like organization. *Nord. J. Bot.* **1987**, *7*, 85–95. [CrossRef]
74. Massicotte, H.B.; Peterson, R.L.; Melville, L.H. Hartig net structure of ectomycorrhizae synthesised between *Laccaria bicolor* (Tricholomaceae) and two hosts: *Betula alleghaniensis* (Betulaceae) and *Pinus resinosa* (Pinaceae). *Am. J. Bot.* **1989**, *76*, 1654–1667. [CrossRef]
75. Fink, S. *Pathological and Regenerative Plant Anatomy*; Encyclopedia of Plant Anatomy XIV (6); Gebrüder Borntraeger: Berlin, Germany, 1999.
76. Cheval, C.; Samwald, S.; Johnston, M.G.; de Keijzer, J.; Breakspear, A.; Liu, X.; Bellandi, A.; Kadota, Y.; Zipfel, C.; Faulkner, C. Chitin perception in plasmodesmata characterizes submembrane immune-signaling specificity in plants. *Proc. Natl. Acad. Sci. USA* **2020**, *117*, 9621–9629. [CrossRef] [PubMed]
77. Huang, C.; Sede, A.R.; Elvira-Gonzalez, L.; Yan, Y.; Rodriguez, M.; Mutterer, J.; Boutant, E.; Shan, L.; Heinlein, M. DsRNA-induced immunity targets plasmodesmata and is suppressed by viral movement proteins. *Plant Cell* **2023**, *35*, 3845–3869. [CrossRef] [PubMed]
78. Tee, E.E.; Johnston, M.G.; Papp, D.; Faulkner, C. A PDLP-NHL3 complex integrates plasmodesmal immune signaling cascades. *Proc. Natl. Acad. Sci. USA* **2023**, *120*, e2216397120. [CrossRef] [PubMed]
79. Voitsekhovskaja, O.V.; Melnikova, A.N.; Demchenko, K.N.; Ivanova, A.N.; Dmitrieva, V.A.; Maksimova, A.I.; Lohaus, G.; Tomos, A.D.; Tyutereva, E.V.; Koroleva, O.A. Leaf epidermis: The ambiguous symplastic domain. *Front. Plant Sci.* **2021**, *12*, 695415. [CrossRef] [PubMed]
80. Sager, R.E.; Lee, J.Y. Plasmodesmata at a glance. *J. Cell Sci.* **2018**, *131*, jcs20934. [CrossRef] [PubMed]
81. Fitzgibbon, J.; Beck, M.; Zhou, J.; Faulkner, C.; Robatzek, S.; Oparka, K. A developmental framework for complex plasmodesmata formation revealed by large-scale imaging of the Arabidopsis leaf epidermis. *Plant Cell* **2013**, *25*, 57–70. [CrossRef] [PubMed]
82. Ormenese, S.; Bernier, G.; Perilleux, C. Cytokinin application to the shoot apical meristem of *Sinapis alba* enhances secondary plasmodesmata formation. *Planta* **2006**, *224*, 1481–1484. [CrossRef] [PubMed]
83. Ormenese, S.; Havelange, A.; Deltour, R.; Bernier, G. The frequency of plasmodesmata increases early in the whole shoot apical meristem of *Sinapis alba* L. during floral transition. *Planta* **2000**, *211*, 370–375. [CrossRef]
84. Kim, G.J.; LeBlanc, M.L.; Wafula, E.K.; de Pamphilis, C.W.; Westwood, J.H. Genomic-scale exchange of mRNA between a parasitic plant and its hosts. *Science* **2014**, *354*, 808–811. [CrossRef] [PubMed]
85. Vaughn, K.C. Dodder hyphae invade the host: A structural and immunocytochemical characterization. *Protoplasma* **2003**, *220*, 189–200. [CrossRef] [PubMed]
86. Birschwilks, M.; Haupt, S.; Hofius, D.; Neumann, S. Transfer of phloem-mobile substances from the host plants to the holoparasite *Cuscuta* sp. *J. Exp. Bot.* **2006**, *57*, 911–921. [CrossRef]
87. David-Schwartz, R.; Runo, S.; Townsley, B.; Machuka, J.; Sinha, N. Long-distance transport of mRNA via parenchyma cells and phloem across the host–parasite junction in *Cuscuta*. *New Phytol.* **2008**, *179*, 1133–1141. [CrossRef]
88. Iswanto, A.B.B.; Vu, M.H.; Pike, S.; Lee, J.H.; Kang, H.; Son, G.H.; Kim, J.Y.; Kim, S.H. Pathogen effectors: What do they do at plasmodesmata? *Mol. Plant Pathol.* **2022**, *23*, 795–804. [CrossRef]
89. Rahman, M.S.; Madina, M.H.; Plourde, M.B.; dos Santos, K.C.G.; Huang, X.; Zhang, Y.; Laliberté, J.F.; Germain, H. The fungal effector Mlp37347 alters plasmodesmata fluxes and enhances susceptibility to pathogen. *Microorganisms* **2021**, *9*, 1232. [CrossRef] [PubMed]

90. Cao, L.; Blekemolen, M.C.; Tintor, N.; Cornelissen, B.J.C.; Takken, F.L.W. The Fusarium oxysporum Avr2-Six5 effector pair alters plasmodesmatal exclusion selectivity to facilitate cell-to-cell movement of Avr2. *Mol. Plant* **2018**, *11*, 691–705. [CrossRef] [PubMed]
91. Blekemolen, M.C.; Cao, L.; Tintor, N.; de Groot, T.; Papp, D.; Faulkner, C.; Takken, F.L.W. The primary function of Six5 of *Fusarium oxysporum* is to facilitate Avr2 activity by together manipulating the size exclusion limit of plasmodesmata. *Front. Plant Sci.* **2022**, *13*, 910594. [CrossRef]
92. Tomczynska, I.; Stumpe, M.; Doan, T.G.; Mauch, F. A *Phytophthora* effector protein promotes symplastic cell-to-cell trafficking by physical interaction with plasmodesmata-localised callose synthases. *New Phytol.* **2020**, *227*, 1467–1478. [CrossRef] [PubMed]
93. Hofmann, J.; Youssef-Banora, M.; de Almeida-Engler, J.; Grundler, F.M.W. The role of callose deposition along plasmodesmata in nematode feeding sites. *Mol. Plant Microbe Interact.* **2010**, *23*, 549–557. [CrossRef] [PubMed]
94. Khang, C.H.; Berruyer, R.; Giraldo, M.C.; Kankanala, P.; Park, S.Y.; Czymmek, K.; Kang, S.; Valent, B. Translocation of *Magnaporthe oryzae* effectors into rice cells and their subsequent cell-to-cell movement. *Plant Cell* **2010**, *22*, 1388–1403. [CrossRef]
95. Tilsner, J.; Nicolas, W.; Rosado, A.; Bayer, E.M. Staying tight: Plasmodesmal membrane contact sites and the control of cell-to-cell connectivity in plants. *Annu. Rev. Plant Biol.* **2016**, *67*, 337–364. [CrossRef] [PubMed]
96. Kumar, D.; Kumar, R.; Hyun, T.K.; Kim, J.Y. Cell-to-cell movement of viruses via plasmodesmata. *J. Plant Res.* **2015**, *128*, 37–47. [CrossRef]
97. Ngou, B.P.M.; Ahn, H.K.; Ding, P.; Jones, J.D.G. Mutual potentiation of plant immunity by cell-surface and intracellular receptors. *Nature* **2021**, *592*, 110–115. [CrossRef]
98. Reagan, B.C.; Burch-Smith, T.M. Viruses reveal the secrets of plasmodesmal cell biology. *Mol. Plant–Microbe Interact.* **2020**, *33*, 26–39. [CrossRef] [PubMed]
99. Gong, B.Q.; Wang, F.Z.; Li, J.F. Hide-and-seek: Chitin-triggered plant immunity and fungal counterstrategies. *Trends Plant Sci.* **2020**, *25*, 805–816. [CrossRef] [PubMed]
100. Tee, E.E.; Faulkner, C. Plasmodesmata and intercellular molecular traffic control. *New Phytol.* **2024**, *243*, 32–47. [CrossRef]
101. Yi, M.; Valent, B. Communication between filamentous pathogens and plants at the biotrophic interface. *Annu. Rev. Phytopathol.* **2013**, *51*, 587–611. [CrossRef] [PubMed]
102. Martinez-Cruz, J.; Romero, D.; Davila, J.C.; Perez-Garcia, A. The *Podosphaera xanthii* haustorium, the fungal Trojan horse of cucurbit-powdery mildew interactions. *Fung. Genet. Biol.* **2014**, *71*, 21–31. [CrossRef]
103. White, J.F., Jr.; Bacon, C.W.; Hinton, D.M. Modifications of host cells and tissues by the biotrophic endophyte *Epichloe amarillans* (Clavicipitaceae; Ascomycotina). *Can. J. Bot.* **1997**, *75*, 1061–1069. [CrossRef]
104. Schouten, H.J.; Brinkhuis, J.; van der Burgh, A.; Schaart, J.G.; Groenwold, R.; Broggini, G.A.L.; Gessler, C. Cloning and functional characterization of the *Rvi15* (*Vr2*) gene for apple scab resistance. *Tree Genet. Genomes* **2014**, *10*, 251–260. [CrossRef]
105. Chevalier, M.; Lespinasse, Y.; Renaudin, S. A microscopic study of the different classes of symptoms coded by the Vf gene in apple for resistance to scab (*Venturia inaequalis*). *Plant Pathol.* **1991**, *40*, 249–256. [CrossRef]
106. Lucas, J.A. *Plant Pathology and Plant Pathogens*, 4th ed.; Wiley Blackwell: Hoboken, NJ, USA, 2020.
107. O'Brien, T.P.; Feder, N.; McCully, M.E. Polychromatic staining of plant cell walls by toluidine blue O. *Protoplasma* **1964**, *59*, 368–373. [CrossRef]
108. Ribeiro, V.C.; Leitao, C.A.E. Utilisation of Toluidine blue O pH 4.0 and histochemical inferences in plant sections obtained by free-hand. *Protoplasma* **2020**, *257*, 993–1008. [CrossRef] [PubMed]
109. Machado, S.R.; Sajo, M.G. Intercellular pectic protuberances in leaves of some *Xyris* species (Xyridaceae). *Can. J. Bot.* **1996**, *74*, 1539 1541. [CrossRef]
110. Masclaux, C.; Valadier, M.H.; Brugière, N.; Morot-Gaudry, J.F.; Hirel, B. Characterization of the sink/source transition in tobacco (*Nicotiana tabacum* L.) shoots in relation to nitrogen management and leaf senescence. *Planta* **2000**, *211*, 510–518. [CrossRef]
111. Brunkard, J.O.; Zambryski, P. Plant cell-cell transport via plasmodesmata is regulated by light and the circadian clock. *Plant Physiol.* **2019**, *181*, 1459–1467. [CrossRef]

Journal of *Fungi*

MDPI

Article

Contrasting Performance of Two Winter Wheat Varieties Susceptible to Leaf Rust under Diverse Pathogen Pressure, Fungicide Application, and Cultivation Practices

Radivoje Jevtić [1,2,*], Vesna Župunski [1], Dragan Živančev [1], Emilija Arsov [2], Sasa Mitrev [2], Ljupco Mihajlov [2] and Branka Orbović [1]

[1] Institute of Field and Vegetable Crops, 21000 Novi Sad, Serbia; vesna.zupunski@ifvcns.ns.ac.rs (V.Ž.); dragan.zivancev@ifvcns.ns.ac.rs (D.Ž.); branka.orbovic@ifvcns.ns.ac.rs (B.O.)
[2] Faculty of Agriculture, Goce Delcev University, 2000 Stip, North Macedonia; emilija.arsov@ugd.edu.mk (E.A.); sasa.mitrev@ugd.edu.mk (S.M.); ljupco.mihajlov@ugd.edu.mk (L.M.)
* Correspondence: radivoje.jevtic@ifvcns.ns.ac.rs; Tel.: +381-21-4898207

Abstract: This study investigated the relationship between yield, thousand kernel weight (TKW), and crude protein of soft white winter wheat–club variety (Barbee) and soft white winter wheat common variety (Zvezdana) susceptible to leaf rust and powdery mildew under different cultivation practices. Results revealed divergence in associations between yield, TKW, and crude protein loss of winter wheat varieties susceptible to obligate pathogens. Under the same level of leaf rust infection, N-input limited yield loss of the two varieties but not to the same extent. TKW loss was affected only by variety×cultivation practice and was significantly correlated with yield loss (r = −0.727, p = 0.011) and crude protein loss (r = −0.600, p = 0.05) only in club winter wheat. We suspected that Ninput affects the difference in the relationship between yield and TKW loss among varieties. Crude protein and yield loss had a low association (R^2 = 18%, p = 0.05). Finally, this study indicated that more attention should be paid to the determination of pathogen pressure that triggers yield loss. It also pointed out that yield, TKW, and crude protein response to fungicides could differ in susceptible varieties. The contribution of fungicide to yield enhancement was highly associated with the specific reaction of the variety to pathogen infection rather than solely the disease level itself.

Keywords: yield; TKW; crude protein; leaf rust; cultivation practice; susceptibility; fungicide

Citation: Jevtić, R.; Župunski, V.; Živančev, D.; Arsov, E.; Mitrev, S.; Mihajlov, L.; Orbović, B. Contrasting Performance of Two Winter Wheat Varieties Susceptible to Leaf Rust under Diverse Pathogen Pressure, Fungicide Application, and Cultivation Practices. *J. Fungi* **2024**, *10*, 401. https://doi.org/10.3390/jof10060401

Academic Editor: Ofir Degani

Received: 21 May 2024
Accepted: 27 May 2024
Published: 2 June 2024

1. Introduction

Wheat is considered an important field crop since it provides more than 20% of the calorific intake for almost two-thirds of the human population and contributes to nearly 30% of the world's grain production and 50% of the world grain trade [1]. The factors influencing yield, yield components, and quality parameters have been studied for decades, but the most influencing factors were analysed separately without consideration of their interactions [2,3]. Breeding high-yielding wheat varieties with good quality is not an easy task since these traits can be negatively correlated or not correlated at all [4,5].

One of the yield components associated with grain quality is thousand kernel weight (TKW). Higher TKW indicates better germination of wheat grain and better milling quality [6]. Although the significant effect of varieties, agro-ecological conditions, and diseases on TKW has been reported in previous studies [7,8], the difference in response of wheat varieties on the combined effect of interactions between disease indices and cultivation practices has rarely been taken into consideration [9]. The crude protein content is usually used as a minimum quality assurance check and is also proven to be highly influenced by climatic factors and cultivation practices [10,11]. However, the combined effect of wheat pathogens and cultivation practices on quality parameters is rarely reported [9].

In addition to powdery mildew (*Blumeria graminis* f. sp. *tritici*), rust diseases (*Puccinia graminis* f. sp. *tritici, Puccinia triticina, Puccinia striiformis* f. sp. *tritici*) are the most important obligate pathogens of wheat [12]. The reduction in winter wheat yields caused by powdery mildew could reach 45% [13]. Leaf rust (*P. triticina*) was the prevalent rust species in Serbia until 2014 when yellow rust (*P. striiformis* f. sp. *tritici*) predominated [14]. The predominance of yellow rust in Serbia was also recorded in 2018, while in the rest of the years, co-occurrence patterns of these two diseases were observed [14]. Yield losses caused by leaf rust were as high as 50%, but they also reached 60% under yellow rust infection [14].

According to U.S. statute, there are three subclasses of soft white wheat: "Soft White Wheat", "White Club Wheat", and "Western White Wheat" [15]. Head (spike) morphology defines features of the first two subclasses. Club wheats are characterised by the reduced overall length of the spike since they possess the "club" or C locus that reduces the internode length of the rachis. Soft white wheat is referred to as "common soft white" since it does not possess the C locus, which results in a "common" spike. The C locus influences kernel size and shape and does not have any direct bearing on the quality of club wheat. In contrast to that, the quality of club wheat varieties is defined by breeding and selection [15]. Although "common" and "club" winter wheat share a number of similarities, they can be distinguished by grain and flour parameters and distinctly different end-use qualities [15]. Consequently, it could be expected that these two classes of wheat react differently to the same cultivation practice, climatic conditions, and pathogen attack.

Zwer et al. [16] investigated differences in spike characteristics, yield, and yield components of common and club wheat varieties and indicated that club wheat had greater adaptive potential than common wheat in areas with marginal moisture. Fewer moisture requirements for seed germination, faster germination, strong culms, and firm spikes made club wheat more adaptive to arid agronomic zones when compared to common wheat. Although resistant to drought with outstanding quality and resistance to common and some races of dwarf bunt, the production of early club wheat varieties such as Omar decreased after the major stripe rust epidemic of 1960 in the Pacific Northwest. Afterward, a multiline approach utilising a mixture of lines with different resistance genes to stripe rust provided durable resistance to stripe rust in newer varieties of club wheat [17]. The advantage of club wheat over common wheat with respect to resistance to *F. culmorum* and *F. graminearum* causing root rot is also associated with the better resistance of club wheat to water stress [18].

Nowadays, pest control is challenging since wheat pathogens continuously change under selection pressures of agro-ecological conditions, resistant varieties, and applied pesticides. Rust fungi are highly adaptive to new resistance genes and are also highly genetically diverse for races [12,19]. In addition, Jevtić et al. [14] indicated that genotypes and their reaction to climatic elements in certain phenological stages could have a strong impact on interactions among obligate pathogens and their predominance. Although the disease development of obligate pathogens and their impacts on yield and quality performance are greater in very susceptible and susceptible varieties, it has also been reported that the yield loss and disease index of obligate pathogens are not always as closely correlated as expected in susceptible varieties. The effects of cultivation practices on these differences are still understated in studies [20]. Current varieties are generally adapted to high-input systems, but there are conflicting reports on whether elite varieties adapted for optimal cropping conditions have good performance under suboptimal environments [21].

In addition to diverse susceptibility, the variability in yield responses of varieties to obligate pathogens is also associated with seasonal weather fluctuations, yield potential, and timing of infection [22]. Early-season infections by powdery mildew can significantly diminish yield in susceptible varieties, lowering it by up to 25% [22]. During the early stages of crop growth, infection with leaf rust could lead to more than a 50% loss of production [23]. Powdery mildew affects a decrease in photosynthetic leaf area, the depletion of crop-available nutrients, and, critically, hampers yield potential via excessive tiller production that fails to yield heads. Severe powdery mildew infections during initial growth phases can also stunt plant growth and delay maturity, amplifying the risk of reinfection. Early infection by leaf rust results in weak plants and poor root and tiller development [22]. Leaf rust causes the most damage when severe rusting covers the upper leaves before flowering. Early defoliation can also occur, reducing the time for grain fill and leading to the creation of smaller kernels. Later-season infections with powdery mildew, occurring between stem elongation and flowering, diminish photosynthetic leaf area, leading to diminished grain size, decreased yields, and quality deterioration. The earlier the onset of infection, the more prolonged its persistence and the greater its spread up the plant, intensifying potential yield losses [22].

Ploughing was considered to be a promising cultivation practice in reducing the level of disease inoculum [24–26], even for foliar diseases that can overwinter in the stubble from the previous crop, such as powdery mildew (*Blumeria graminis* f. sp. *tritici)* and stem rust (*P. graminis* f. sp. *tritici)* [27]. Nowadays, the choice of cultivation practice that will be applied in wheat growing areas depends on the complexity of agro-ecological conditions and socioeconomic aspects, but there is a still lack of knowledge on factors that could jeopardise the expected effects of tillage practices [28]. The effectiveness of fungicide application in controlling powdery mildew and leaf rust is also highly associated with factors that impact their life cycles, infection potentials, and interactions with the host plant. Understanding these differences makes preconditions for more effective disease management strategies [22].

Knowing that breeding of varieties with advantageous genetic traits for low input is often overlooked and that the regulatory network for plant responses to abiotic and biotic stress consists of many mechanisms and components that may function antagonistically [29–31], it was hypothesised in this study that susceptible wheat varieties could react differently on infection of obligate pathogens when cultivated under different cultivation practices (monoculture with reduced N input and conventional N input with crop rotation). Consequently, the main objectives of this study were to measure (1) the effect cultivation practices of susceptible varieties belonging to different wheat classes and their association with yield, thousand kernel weight (TKW), crude protein content, and disease indexes (leaf rust and powdery mildew); (2) the difference in the relationship between yield, TKW, and crude protein response to fungicide application of "club" (Barbee) and "common" (Zvezdana) winter wheat varieties susceptible to powdery mildew and leaf rust. The Barbee is a bearded, semi-dwarf soft white winter wheat—club variety (*Triticum aestivum compactum*), and Zvezdana is soft white winter wheat—common variety (*Triticum aestivum aestivum*).

2. Materials and Methods

Data were obtained from research and development trials (R&D) on fungicide efficacy against obligate pathogens (*Blumeria* and *Puccinia*). Fungicide efficacy trials were conducted in 2021 in the municipality of Novi Sad (Vojvodina, northern province of Serbia) under the direction of the Institute of Field and Vegetable Crops, Novi Sad, Serbia. All permissions for trial conductance were defined by contracts with chemical companies. Research also complied with relevant institutional, national, and international regulations. The trials were carried out at two locations differing in cultivation practices: (1) monoculture + reduced N input and (2) conventional N input with crop rotation. The trials included fungicide-treated and fungicide-untreated plots at both locations, and fungicides were applied in the same manner. More details on model varieties, conditions of trial setup, and assessments of disease levels, yield, TKW, crude protein, and oil content of soft white winter wheat—club variety (Barbee) and soft white winter wheat—common variety (Zvezdana) under naturally occurring inoculum of obligate pathogens are given below.

2.1. Model Varieties

In this study, we used two varieties susceptible to powdery mildew [32] and leaf rust [33] but belonging to different wheat classes. The variety Barbee is a bearded, semi-dwarf soft white winter wheat–club (*Triticum aestivum* ssp. *compactum*). It was originated by USDA-ARS and Washington State University and was released in 1976. Its flour quality is suited for pastries, cookies, and other white wheat products but not for bread. It was developed for production in Idaho and Washington state (USA) and was expected to be cultivated instead of variety Paha in areas where stripe races jeopardised wheat production. Barbee is mid-season in maturity. On a scale of 1 (low or poor) to 10 (high or good), Barbee is indexed at 8 for straw strength, 5 for emergence, 5 for winter hardiness, and 5 for test weight. Concerning disease resistance, Barbee is resistant to common bunt, moderately susceptible to dwarf bunt, resistant to stripe rust, and moderately resistant to flag smut. It was also used as a susceptible check in field trials in many countries because of its high susceptibility to leaf rust [33] and powdery mildew [32,34,35]. In the study of Jevtić et al. [20] the DIs of leaf rust in Barbee ranged from 22% to 68% in untreated plots during the period from 2006 to 2013 at the locality Rimski šančevi (Serbia). DIs of powdery mildew in the same study ranged from 21% to 30%.

Zvezdana is a soft white winter wheat–common (*Triticum aestivum* ssp. *aestivum*). It was released by the Institute of Field and Vegetable Crops, Serbia, in 2005, and it is still commercially available. Zvezdana is high-yielding and resistant to lodging. Zvezdana is characterised as a medium–early cultivar with very good winter hardiness. Stem height has a range from 75 to 83 cm, the test weight is 80–84 kg, and it belongs to the A1-A2 Farinograph quality class (according to the Hungarian Standard method 6369/653 (MSZ 1988) [36]). In Germany, wheat varieties are classified into four classes based on quality parameters: E-class, i.e., elite quality; A-class, i.e., A-quality; B-class, i.e., bread making, and C-class, i.e., used for stock-feed purposes https://www.bundessortenamt.de (accessed on 13 March 2023). Parameters that are usually tested for quality criteria are protein content, grain hardness, sedimentation values, falling number, wet gluten content, dough rheology, etc. Zvezdana is also reported as being highly susceptible to variation in climatic factors during the growing season [37].

Zwer et al. [16] examined variances in spike characteristics, yield, and yield components between common and club wheat varieties. They suggested that club wheat exhibited superior adaptive capabilities over common wheat in regions with limited moisture. Club wheat required less moisture for seed germination, displayed quicker germination, and possessed robust culms and firm spikes, rendering it more suitable for arid agricultural zones compared to common wheat. These findings are also supported by Bafus et al. [38], who reported that club wheat can be the highest yielding variety in Central Oregon during years with temperature extremes, especially when fertilised equally to soft wheat with 350 lb/acre of 30-10-0-7.

2.2. Field Trial

Fungicide efficacy trials were conducted under naturally occurring inoculum of obligate pathogens and under the same climatic conditions but using different cultivation practices: (1) 20-year monoculture + reduced N input; and (2) long-term crop rotation + conventional N input. Monoculture and crop rotation were applied in order to provide different levels of pathogen pressure. Reduced N input included the usage of 68 kg of nitrogen per ha, while conventional N input cultivation included the usage of 100 kg of nitrogen per ha. Long-term crop rotation included a 4-year crop rotation design that is applied for decades.

The soil type in both localities was a slightly carbonated loamy chernozem. The soil properties include a predominantly loamy and crumbly structure with stable aggregates and a neutral to slightly alkaline reaction. Field trials were arranged in a randomised block design comprising four replicates. The plot size of each replicate was 10 m^2. The trials included 10 fungicide-sprayed and non-sprayed check treatments.

All fungicides were applied independently in BBCH 51–59 (inflorescence emergence, heading) and also in combination in three growth stages: BBCH 32–33 (node 2 at least 2 cm above node 1; node 3 at least 2 cm above node 2); BBCH 39 (flag leaf stage: flag leaf fully unrolled, ligule just visible) and BBCH 61–65 (beginning of flowering: first anthers visible; full flowering: 50% of anthers mature). Fungicides were applied in the same manner at both locations, which differed in cultivation practices and the amount of nitrogen applied. Since contracts with chemical companies do not allow preliminary publication of the data, the trade names of tested fungicides are not provided in this manuscript. Active ingredients used in the study were prothioconazole, spiroxamine, trifloxystrobin, bixafen, fluopyram, mefentrifluconazole, fluksapiroksad, pyraclostrobin, and benzovindiflupyr. Fungicides were applied using calibrated field crop sprayers with fan nozzles at 300 kPa pressure and 200 L of water per hectare. The sowing date for winter wheat was 20 October (optimal time of sowing), and the harvest date was 30 June.

2.3. Climatic Conditions

The majority of the territory of the Republic of Serbia is under the Cfb climate type according to the Köppen–Geiger Climate Classification. It is characterised as a warm temperate–fully humid climate type with warm summers, but from 1961 until the present, a significant increase in temperature change and in precipitation patterns has been observed [39]. In the year 2021, when the trial was conducted, lower total precipitation than the eight-year average was recorded in the time of the flowering and grain-filling period (Table 1).

Table 1. Climatic conditions at Novi Sad (Vojvodina, north province of Serbia).

	2021	Aver. *	2021	Aver.	2021	Aver.	2021	Aver.	2021	Aver.	2021	Aver.
	January		February		March		April		May		June	
Temperature (°C)	3.3	1.4	5.1	4.6	6.2	8.0	9.6	13.0	16.0	17.0	23.3	21.8
Rel. hum.	84.0	85.8	78.0	80.5	68.0	71.4	71.0	65.3	66.0	70.3	58.0	69.0
Total rainfall (mm)	44.7	39.1	59.5	43.5	42.8	46.7	55.1	46.0	62.9	110.5	23.9	85.8

* Aver. indicates eight-year averages of temperature, relative humidity, and total precipitation.

2.4. Disease Assessments

Assessments of leaf disease severity were made using a modified Cobb's scale [40] at the growth stage 71–73 BBCH (kernel watery; early milk), known to be highly related to yield [41]. The disease indices (DI %) of leaf rust and powdery mildew were calculated by taking into consideration disease incidence and average disease severity according to [7] as follows:

$$\text{DI}(\%) = [\text{sum}(\text{class frequency} \times \text{score of rating class})]/[(\text{total number of plants}) \times (\text{maximal disease index})] \times 100$$

2.5. Yield

The yield was measured for each plot at 15% water content. Yield loss (%) was determined as yield reduction in untreated plots compared with yield response to fungicide treatment, which provided the best control of wheat diseases (Equation (1)). Yield losses are determined for untreated plots at each locality differing in cultivation practices.

$$Y(\%) = ((Y_1 - Y_2)/Y_1) \times 100 \quad (1)$$

Y_1—the highest grain yield after fungicide treatment
Y_2—grain yield of the non-sprayed check treatment
Yield response to fungicide application (%) using the following formula (Equation (2))

$$Y(\%) = ((Y_1 - Y_2)/Y_2) \times 100 \quad (2)$$

Y_1—grain yield of fungicide treatment for the best wheat disease control
Y_2—grain yield of the non-sprayed check treatment

2.6. Crude Protein and Oil Content Assessments

Crude protein and oil content were determined by handheld GrainSense NIR Analyzer (Oulu, Finland) and expressed as a percentage of dry matter. According to the study of

Živančev et al. [42], the correlation between the standard Kjeldahl digestion method for protein determination and the handheld GrainSense NIR Analyzer was positive (r = 0.951). This device is the first of its kind in the world since it weighs only 820 g without batteries. The time of measuring is only three seconds, and the required sample size is between 60 and 80 kernels. It is based on near-infrared (NIR) technology with a 360° light penetration method (integrating sphere).

2.7. Statistical Methods

The effects of variety, locality (differing in cultivation practices), variety × locality and the disease index of leaf rust × variety on yield, TKW, crude protein and oil content were examined using multiple regression modelling. Regression modelling was also applied to test the most influencing factors on yield, TKW, and crude protein loss in untreated plots as well as yield, TKW, and crude protein responses to fungicide application. In all regression models, the disease index of leaf rust was used as a continuous predictor, while variety (Barbee, Zvezdana) and locality (20-year monoculture + reduced N input and long-term crop rotation + conventional N input) were used as categorical predictors. Stepwise regression was applied since abiotic and biotic factors could be correlated not just to the yield, TKW, and crude protein loss/gain but also to each other (multicollinearity). Multicollinearity is addressed in this study because it can increase the variance of the regression coefficients. We also used VIFs in regression models to detect multicollinearity. The VIFs measure how much the variance of an estimated regression coefficient increases if predictors are correlated and will be 1 if there is no correlation between factors. In our study, VIFs did not exceed 1.2. The alpha level to enter and the alpha level to remove the influencing factors in the stepwise multiple regression were set by default to 0.15. Bursac et al. [43] reported that in stepwise multiple regression, an alpha level of 0.05 to enter and to remove the influencing factors could fail to identify important variables.

Knowing that PCA requires quantitative and MCA qualitative variables only, associations between qualitative variables (variety, cultivation practice) and quantitative variables (yield loss, TKW loss, crude protein loss, disease index of leaf rust) were analysed using principal correspondence analysis with mixed data (PCAmix). PCAmix includes ordinary principal component analysis (PCA) and multiple correspondence analysis (MCA) as special cases.

To provide more information on which means are significantly different and to estimate how much they differ, we used a general linear model and Tukey's pairwise comparisons with 95% confidence. The correlation between variables was analysed using Spearman's coefficient of correlation.

The analysis was performed using Minitab 17 Statistical Software (trial version) and XLSTAT 2022.5.1 (1392) in Microsoft Excel 16.0.17425 [44]. Visualisation of PCAmix analysis was performed using Package 'ggplot2' in RStudio software (RStudio 2021.09.2+382 "Ghost Orchid" Release) [45].

3. Results

On average, when both fungicide-treated and fungicide-untreated plots were taken into consideration under both types of cultivation, yield (8.6 t/ha), TKW (31.8 g), and oil content (1.7%) of Barbee were significantly lower ($p < 0.001$) than yield (10.3 t/ha), TKW (43.0 g), and oil content (2.1%) of Zvezdana. However, the average crude protein of Barbee (11.7%) was significantly higher ($p < 0.001$) than that of Zvezdana (10.1%).

The average yield and crude protein content of Zvezdana were significantly higher when cultivated under conventional nitrogen input with crop rotation than under monoculture with reduced nitrogen input. In contrast, the yield and crude protein content of Barbee did not differ when grown under monoculture with reduced nitrogen input and conventional nitrogen input with crop rotation (Figure 1). The oil content of Zvezdana was significantly higher than that of Barbee and did not differ between the two cultivation practices. The oil content of Barbee was significantly affected by different cultivation systems. Thousand kernel weights (TKWs) were significantly different between the varieties. Cultivation practices did not affect the TKW of a single variety, but fungicide treatment did, only in the variety Zvezdana. Consequently, yield, crude protein, and oil content were significantly influenced by variety, locality (differing in cultivation practices), and the interaction between variety and locality. TKW was significantly influenced only by variety ($p < 0.001$) without interaction with cultivation practices (Table 2).

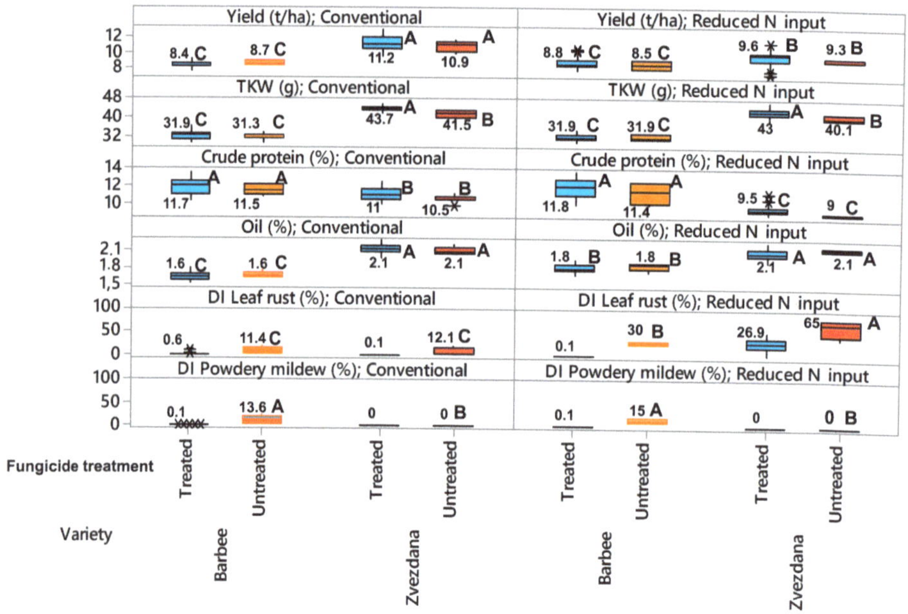

Figure 1. Yield, TKW, crude protein, oil content, disease index of leaf rust, and disease index of powdery mildew in "club" winter wheat Barbee and "common" winter wheat Zvezdana in fungicide-treated (blue color) and fungicide-untreated plots (orange color for Barbee and red color for Zvezdana), cultivated under monoculture with reduced N input and crop rotation with conventional production. Tukey's method was used to create confidence intervals for all pairwise differences between factor level means. The traits designated with different letters are significantly different. Sign "*" is outlier data label.

Barbee and Zvezdana are both varieties susceptible to leaf rust and powdery mildew, but their reactions to obligate pathogens differed in two cultivation systems. Since we intended to measure how cultivation practices, differing in N input and previous crop rotation, affected infection levels of naturally occurring leaf rust and powdery mildew, we evaluated the effects of two cultivation systems on leaf rust and powdery mildew occurrence in untreated plots. In untreated plots, the disease index (DI) of leaf rust was greater under monoculture with reduced N input for both Zvezdana (65%) and Barbee (30%) than under conventional N input with crop rotation, where the DI of leaf rust for Zvezdana and Barbee were 12% and 11%, respectively. Regression modelling indicated that variety ($p < 0.001$), locality ($p < 0.001$), and the interaction between variety and locality ($p < 0.001$) influenced the DI of leaf rust (Table 2). Powdery mildew infected only Barbee, and the average disease index did not exceed 15% (monoculture with reduced N input) and 13.6% (conventional N input with crop rotation). Since the DI of powdery mildew in Barbee under two types of cultivation was not significantly different and did not significantly affect the yield and quality properties of Barbee in untreated plots (Table 2), the association of the DI of powdery mildew with cultivation practices, yield, TKW, and crude protein was not further analysed.

Table 2. Regression analysis of the most influential factors on yield, TKW, crude protein, and oil content, along with the yield, disease index of leaf rust, and powdery mildew in untreated plots of winter wheat varieties Barbee and Zvezdana.

Dependent variable: yield					
Source	DF	Adj SS	Adj MS	F-Value	*p*-Value
Regression	3	156.89	52.27	110.11	0.000
Variety	1	115.90	115.902	244.14	0.000
Locality (cultivation system)	1	14.18	14.18	29.87	0.000
Variety × locality (cultivation system)	1	32.42	32.42	68.29	
Error	148	70.26	0.475		
Total	151	227.08			
R^2 = 69.06%					
Dependent variable: TKW					
Source	DF	Adj SS	Adj MS	F-Value	*p*-Value
Regression	4	4838.96	1209.74	545.13	0.000
Leaf rust	1	7.14	7.14	3.22	0.075
Variety	1	1949.39	1949.39	878.44	0.000
Fungicide treatment	1	12.81	12.81	5.77	0.018
Variety × fungicide treatment	1	13.38	13.38	6.03	0.015
Error	148	328.44	2.22		
Total	152	5167.40			
R^2 = 93.64%					
Dependent variable: Crude protein					
Source	DF	Adj SS	Adj MS	F-Value	*p*-Value
Regression	4	140.86	35.21	52.67	0.000
Variety	1	91.98	91.98	137.56	0.000
Locality (cultivation system)	1	20.82	20.82	31.13	0.000
Fungicide treatment	1	3.09	3.09	4.63	0.033
Variety × locality (cultivation system)	1	22.81	22.81	34.11	0.000
Error	148	98.963	0.6687		
Total	152	239.82			
R^2 = 58.74%					
Dependent variable: Oil content					
Source	DF	Adj SS	Adj MS	F-Value	*p*-Value
Regression	7	6.15	0.87	151.17	0.000
Variety	1	2.56	2.56	440.85	0.000
Locality (cultivation system)	1	0.14	0.14	24.94	0.000
Variety × locality (cultivation system)	1	0.11	0.11	19.04	0.000
Error	145	0.84	0.01		
Total	152	6.99			
R^2 = 87.95%					
Dependent variable: Disease index of leaf rust in untreated plots					
Source	DF	Adj SS	Adj MS	F-Value	*p*-Value
Regression	3	8601	2867.01	30.12	0.000
Variety	1	1229.9	1229.9	12.92	0.003
Locality (cultivation system)	1	5948.6	5948.6	62.49	0.000
Variety × locality (cultivation system)	1	1643.9	1643.9	17.27	0.001
Error	15	1428	95.20		
Total	18	10,029			
R^2 = 85.76%					
Dependent variable: Disease index of powdery mildew in untreated plots					
Source	DF	Adj SS	Adj MS	F-Value	*p*-Value
Regression	1	919.62	919.62	39.99	0.000
Variety	1	919.62	919.62	39.99	0.000
Error	17	390.91	22.995		
Total	18	1310.53			
R^2 = 70.17%					

Table 2. *Cont.*

Dependent variable: Yield in untreated plots					
Source	DF	Adj SS	Adj MS	F-Value	*p*-Value
Regression	2	12,933,148	6,466,574	31.78	0.000
DI leaf rust	1	3,922,916	3,922,916	19.28	0.000
Variety	1	12,586,749	12,586,749	61.85	0.000
Error	16	3,255,852	203,491		
Total	18	16,189,000			
R^2 = 79.89%					

It has to be pointed out that the average yields of fungicide-treated and fungicide-untreated plots of variety Barbee grown in the same location and under conventional N input with crop rotation were constantly lower than average yields of variety Zvezdana over the six-year period (from 2017 to 2022) (Figure 2). Consequently, variability in climatic conditions among the years was not expected to affect the yield relationship between Zvezdana and Barbee. Since the oil content is a parameter that is associated with germ size [46], and in the study of Moore et al. [46] did not correlate with grain yield, its association with yield, TKW, and crude protein was not further analysed, too.

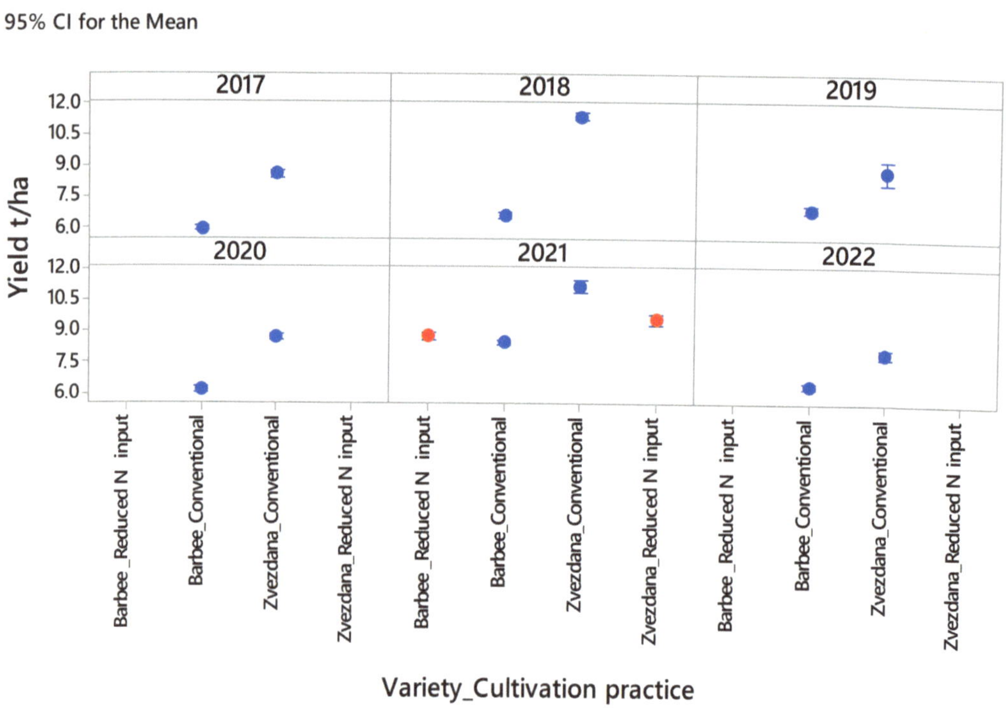

Figure 2. The yield relationship between common winter wheat Zvezdana and club winter wheat Barbee over a six-year period (2017–2022) in the locality of Rimski šančevi under conventional cultivation practice (blue marks). Red marks indicate yield reactions of varieties Barbee and Zvezdana when grown under monoculture with reduced N input.

3.1. Association between Losses in Yield, Thousand Kernel Weight (TKW), and Crude Protein Content, along with Leaf Rust Disease Index and Cultivation Practices in Barbee and Zvezdana Belonging to Different Wheat Classes

In order to compare the reaction of susceptible varieties to leaf rust while grown under different cultivation practices (monoculture + reduced-N-input) and (crop rotation + conventional N-input), the yield, TKW, and crude protein losses were estimated in plots where fungicide treatments were not applied. It is important to notice that comparing only average absolute values of yield, TKW, and crude protein of two varieties in fungicide-treated and fungicide-untreated plots under the same type of cultivation will not produce exact information on how yield/TKW/crude protein in untreated plots changed with respect to the highest values of yield/TKW/crude protein obtained after fungicide treatment, providing the best control of wheat diseases. Thus, relative values in the form of losses of yield, TKW, and crude protein are used in further analysis instead of absolute values.

PCAmix was performed to visualise a global pattern within the data and their associations. The first two dimensions contributed 65.99% to the overall variability (Figure 3). Overall, different production practices (reduced N input with monoculture and conventional N input with crop rotation) grew on opposite sides of the plot origin and contributed differently to the first two dimensions. Zvezdana and Barbee were also positioned on opposite sides of the plot origin. Yield and crude protein loss were not grouped together with TKW loss on a factor map, indicating that they had been affected by different factors. Yield and crude protein loss were more associated with the first dimension, while TKW loss was more associated with the second one. The disease index of leaf rust was grouped with monoculture growing with reduced N input, but its effect on yield, crude protein, and TKW loss of susceptible varieties was not straightforward and was positioned on the opposite side from both varieties. The quality of association between variable categories and a particular axis was presented with squared cosine (cos^2). Squared cosine showed that yield loss, disease index of leaf rust, and production systems had the highest degree of association with the first dimension (red colour), while the variety and TKW had the lowest association with the first two dimensions (blue colour) (Figure 3). Crude protein loss was moderately associated with the first two dimensions. Regression analysis indicated that differences in crude protein loss between varieties and localities were non-significant in statistical terms, although variety Zvezdana suffered higher crude protein loss under monoculture with reduced N-input cultivation when compared with variety Barbee (Table S1).

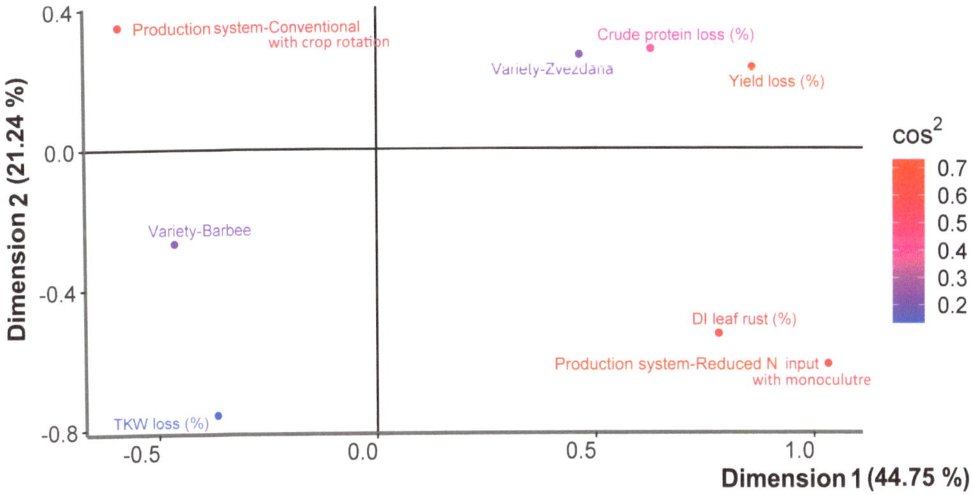

Figure 3. Graphical representation of PCAmix analysis between yield loss, TKW loss, crude protein loss, disease index of leaf rust, production system, and varieties differing in wheat classes.

Since yield loss, disease index of leaf rust, and cultivation practices had the highest contribution to the first dimension of PCAmix analysis and were also differently associated with Zvezdana and Barbee, their relationship is further presented in more detail.

3.1.1. The Effect of Cultivation Practices on Difference in Yield Loss of Barbee and Zvezdana Varieties under Diverse Levels of Leaf Rust Infection

Regression analysis of influencing factors on yield loss of two susceptible varieties was performed with the data originating from untreated plots and indicated that the most influencing factors on yield loss were variety ($p = 0.013$), locality ($p < 0.001$) and interaction variety × locality ($p = 0.007$) (Table S1). The coefficient of determination for the regression model was 69.9%.

Although Barbee and Zvezdana are both varieties susceptible to leaf rust, their reaction to the pathogen pressure in a single locality under the same cultivation system was not the same. When cultivated under a conventional production system (crop rotation + conventional N input), the average yield loss of variety Barbee (4.7%) was significantly lower than that of variety Zvezdana (16.1%), although the average disease index of leaf rust on both Barbee (11.9%) and Zvezdana (11.3%) was not significantly different (Figure 4a).

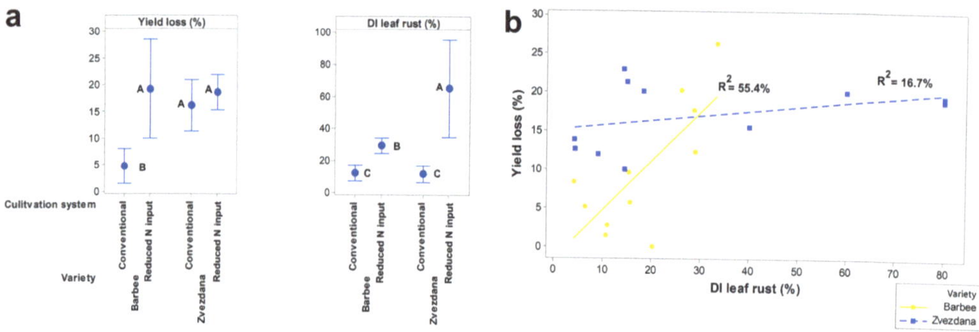

Figure 4. Yield loss of susceptible varieties under diverse levels of leaf rust infection and cultivation practices. (**a**) Interval plots show difference in yield loss and disease index of leaf rust in "club" winter wheat Barbee and "common" winter wheat Zvezdana cultivated under monoculture with reduced N input and crop rotation with conventional N input. (**b**) Linear regression of disease index of leaf rust on yield loss in "club" winter wheat Barbee and "common" winter wheat Zvezdana cultivated under monoculture with reduced N input and conventional production. The data originated from the plots where fungicides were not applied. Tukey's method is used to create confidence intervals for all pairwise differences between factor level means. The traits designated with different letters are significantly different.

Under monoculture with reduced N-input farming without crop rotation, the average disease index (DI) of leaf rust for variety Barbee (29.3%) and variety Zvezdana (65%) was higher than that under the conventional production system. However, the higher DIs of leaf rust did not result in higher yield losses in both varieties compared to yield losses under lower pathogen pressure and conventional N input. The yield loss of variety Barbee under monoculture with reduced N input (19.2%) was significantly higher than the yield loss (4.7%) under conventional N input. However, variety Zvezdana did not exhibit significantly different yield losses when grown under lower pathogen pressure and reduced N input (18.6%) compared to higher pathogen pressure with conventional N input (16.1%) (Figure 4a). Spearman's coefficient of correlation also showed a moderately significant positive correlation between yield loss and the disease index of leaf rust only in variety Barbee ($r = 0.636$, $p = 0.035$). After regression modelling, a significantly higher association between the disease index of leaf rust and yield loss was observed in variety

Barbee ($p = 0.009$), contrary to variety Zvezdana (Figure 4b). Moreover, there was no significant correlation between these two traits in variety Zvezdana ($p = 0.212$).

Since diverse levels of leaf rust infection and different levels of N input did not cause differences in yield losses of variety Zvezdana, the climatic conditions during flowering and grain filling were also considered as additional factors highly influencing yield loss of variety Zvezdana. The total rainfall during flowering and grain filling in 2021 (62.9 mm and 23.9 mm) was less than the eight-year averages (110.5 mm and 85.8 mm) (Table 1). Consequently, it is plausible to suppose that dry conditions during flowering and grain filling affected the yield performance of variety Zvezdana more than that of variety Barbee. Significantly higher yield losses of variety Barbee when grown under higher leaf rust infection and reduced nitrogen input in monoculture, compared to conventional nitrogen input with lower levels of pathogen pressure, indicated that the yield performance of variety Barbee was more affected by these factors rather than extreme climatic conditions.

3.1.2. The Difference in the Relationship between Losses in Yield, Thousand Kernel Weight (TKW), and Crude Protein Content of 'Club' and 'Common' Winter Wheat Varieties Susceptible to Leaf Rust

Since PCA is mostly used as a tool in exploratory data analysis, the relationship between losses of yield, TKW, and crude protein of two varieties belonging to different wheat classes was also analysed with regression modelling (Table S1).

Regression modelling showed that factors influencing yield and TKW loss are not the same. In contrast to yield loss, which was significantly affected by variety ($p = 0.013$), locality ($p < 0.001$), and interaction variety × locality ($p = 0.007$), TKW loss was significantly affected only by interaction variety × locality ($p = 0.021$) (Table S1). In general, when both susceptible varieties were taken into consideration, Spearman's coefficient of correlation between losses of TKW and yield was moderately significantly negative ($r = -0.455$, $p = 0.034$). However, when individual varieties were tested, the correlation between TKW loss and yield loss was moderately significantly negative only in the Barbee variety, while it was not significant in Zvezdana (Table 3). TKW loss of variety Barbee under cultivation with conventional N input (13.3%) was higher than that under reduced N input (8.6%), but according to the Tukey test, these differences were not statistically significantly different from those of variety Zvezdana under conventional N input (9.4%) and reduced N input (13.1%). Interestingly, Barbee and Zvezdana suffered higher TKW losses under different types of cultivation: Barbee under conventional N input with crop rotation and Zvezdana under reduced N input with monoculture.

Table 3. Correlation of yield loss and crude protein loss with TKW loss of "club" winter wheat Barbee and "common" winter wheat Zvezdana.

	Barbee	**Zvezdana**
	TKW loss	TKW loss
Yield loss	$r = -0.727, p = 0.011$	$r = -0.391, p = 0.235$
Crude protein loss	$r = -0.600, p = 0.05$	$r = 0.041, p = 0.905$

The association between crude protein loss and yield loss was low ($R^2 = 18\%$, $p = 0.05$) (Figure S1) but showed a similar trend in both Barbee and Zvezdana. The low association between yield and crude protein losses indicated that different factors influenced these two traits. Regression analysis showed that the effect of the disease index of leaf rust on crude protein loss ($p = 0.148$) should not be understated, although it was not significant at an alpha level of 0.05.

The association between crude protein loss and TKW loss was also different in the two varieties. Spearman's coefficient of correlation indicated a significant moderate negative correlation between TKW loss and crude protein loss only in the variety Barbee (Table 3).

3.2. The Difference in the Relationship between Yield, TKW, and Crude Protein Response to Fungicide Application of "Club" and "Common" Winter Wheat Varieties Susceptible to Leaf Rust

In our study, fungicides were applied independently at BBCH 51–59 (inflorescence emergence, heading) and also in combination at three growth stages: BBCH 32–33, BBCH 39, and BBCH 61–65. The disease index (DI) of leaf rust on the Barbee variety, when all combinations of fungicide treatments were considered, averaged 2.5% under conventional nitrogen input with crop rotation and 3% under reduced nitrogen input with monoculture at BBCH 73–75. For the Zvezdana variety, when all fungicide treatments were considered, the average DI of leaf rust was 0.4% under conventional nitrogen input with crop rotation and 25% under reduced nitrogen input with monoculture at BBCH 73–75.

When grown under conventional N input with crop rotation, the yield gain of Zvezdana after fungicide application (4.7%) was significantly higher than the average yield response to fungicide application of the Barbee variety (−1.3%) (Figure 5). At the same time, in untreated plots, leaf rust infection of Zvezdana and Barbee did not exceed 12%, and DIs were not significantly different. The greater contribution of fungicide treatment to yield gain in variety Zvezdana than in variety Barbee under the same level of pathogen pressure and cultivation practice suggests that the effect of fungicide treatment on yield response is variety-specific and not associated only with pathogen pressure and the ability of a fungicide to prevent or stop the development of pathogen infection but also with the reaction of varieties to environmental factors. Contrary to variety Barbee, Zvezdana is highly susceptible to unfavourable climatic conditions during flowering and grain filling, which, together with pathogen infection, affected yield loss in untreated plots and yield gain in treated ones. This result indicates that fungicide treatments can have a beneficial effect on yield response even under a low level of leaf rust infection, especially for varieties that are more susceptible to environmental stressors, such as variety Zvezdana. Variety Barbee suffered a low level of yield loss in untreated plots under conventional N input with crop rotation, and fungicide treatment had no effects on its yield response in treated ones.

When varieties were grown under reduced nitrogen input and monoculture, the disease indices (DIs) of leaf rust in fungicide-untreated plots reached 29.3% in Barbee and 65% in Zvezdana, significantly higher than those in untreated plots under conventional nitrogen input with crop rotation. In fungicide-treated plots, the average DI of leaf rust was 3% in Barbee and 25% in Zvezdana. Nevertheless, yield gain after fungicide application was not significantly different between the varieties, equalling 3.3% in Barbee and 3.6% in Zvezdana. Interestingly, the yield gain in Barbee was significantly higher under conventional cultivation, while it remained at the same level of significance under two types of cultivation in Zvezdana (Figure 5). This suggests that the higher pathogen pressure of leaf rust and reduced nitrogen input with monoculture cultivation affected yield performance in Barbee more directly, making their relationship more correlated than in Zvezdana.

The same level of yield gain obtained under diverse levels of leaf rust pressure and also under different types of cultivation in a single variety (Zvezdana) indicated that there is an infection level that triggers yield loss for each variety and is associated with its susceptibility not only to the pathogen itself but also to other environmental factors. Barbee exhibited a positive response to fungicide application when disease pressure in untreated plots led to DIs of leaf rust of 29.3%, while in Zvezdana, yield gain was expressed when DIs of leaf rust reached 11.3%.

Contrary to the yield response of Zvezdana to fungicide treatment, the TKW achievements of Zvezdana in fungicide-treated plots were more influenced by the different types of cultivation. TKW gain in Zvezdana was more prominent under monoculture with reduced N-input conditions than under conventional N input (Figure 5). However, there was no difference in the TKW response of the Barbee variety to fungicide treatment under the two types of cultivation. It indicated that fungicide treatments contributed differently to yield and TKW responses in two susceptible varieties, which was also proven with regression modelling (Table S2). Regression analysis showed that interaction between fungicide treatment × disease index of leaf rust ($p = 0.009$) and variety × locality ($p < 0.001$) influenced yield response in treated plots. However, the effect of treatment × disease index of leaf rust ($p = 0.104$) and variety × locality ($p = 0.140$) on TKW response had less significance than the effect of variety ($p < 0.001$) and locality ($p = 0.001$) (Table S2).

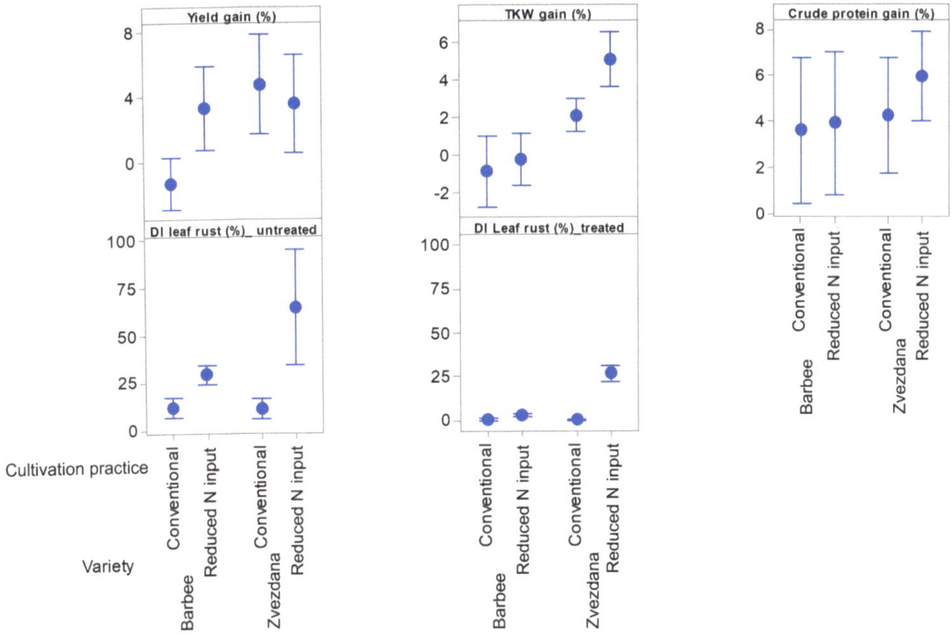

Figure 5. Averages of yield, TKW, crude protein, and disease index of leaf rust responses to fungicide treatment in "club" winter wheat Barbee and "common" winter wheat Zvezdana cultivated under monoculture with reduced N-input farming and crop rotation with conventional N input. Tukey's method is used to create confidence intervals for all pairwise differences between factor level means.

Crude protein response to fungicide treatments was recorded in both varieties in the range from 3.8% (Barbee) to 5.2% (Zvezdana), but their differences were not statistically significant due to the broad range of confidence intervals (Figure 5). The broad range of confidence intervals for yield and crude protein responses in both varieties resulted from different contributions of 10 fungicide treatments and their combinations on these two traits. In this study, crude protein response to a fungicide application was significantly influenced only by leaf rust ($p = 0.067$) (Table S2).

Interestingly, although within the same cultivar (Barbee), the association between thousand kernel weight (TKW) loss and crude protein loss was more pronounced than the association of TKW and crude protein gains. In Barbee, TKW loss and crude protein loss exhibited a high correlation ($r = -0.727$, $p = 0.011$), whereas the correlation between TKW and crude protein responses to fungicide application was lower ($r = -0.367$, $p = 0.002$) (Tables 3 and 4). In Zvezdana, these two traits showed no correlation at all.

Table 4. Correlation of crude protein response with TKW response and yield response of "club" winter wheat Barbee and "common" winter wheat Zvezdana after fungicide application.

	Barbee	Zvezdana
	Crude protein response	Crude protein response
Yield response	r = 0.088. p = 0.481	r = 0.695. p < 0.001
TKW response	r = −0.367. p = 0.002	r = −0.075. p = 0.551

4. Discussion

This study investigated the relationship between yield, thousand kernel weight (TKW), and crude protein of winter wheat varieties susceptible to obligate pathogens under different cultivation practices and the same climatic conditions. It was found that each variety exhibited a specific threshold of pathogen pressure that triggered yield loss and that the reaction of susceptible varieties to the same level of pathogen infection, in the same locality and under the same cultivation system, was not the same. Additionally, it was observed that susceptible varieties highly susceptible to unfavourable climatic conditions could exhibit the same level of yield loss under a broad range of pathogen pressures and a variable nitrogen supply. Differences in the associations among losses in yield, TKW, and crude protein of two winter wheat varieties susceptible to obligate pathogens were observed. Finally, it highlighted that the response of yield, TKW, and crude protein to fungicides could vary among susceptible varieties, emphasising that the contribution of fungicides to yield achievements is closely linked to the specific reaction of each variety to pathogen infection and environmental conditions rather than solely the disease level.

The integrated pest management (IPM) regime urges the reduction of fungicide application, and, consequently, newly developed varieties should express satisfactory disease resistance in addition to good yield and quality performances. However, there is still a lack of knowledge on factors that could affect crop performances when grown under reduced cultivation inputs. In our study, yield loss of two susceptible varieties under the same level of leaf rust infection, under conventional levels of N input, and without the interference of fungicide action was not the same. In addition, we showed that a single susceptible variety could suffer the same level of yield loss regardless of the cultivation practice and under a broad range of leaf rust infection. The results of our study support previous publications reporting that the predictions of disease effects on yield will be more accurate if the effects of pathogens on the main physiological variables (i.e., radiation interception and radiation-use efficiency) are considered [9] as well as the susceptibility of varieties to the combined effect of abiotic and biotic stressors [20,30].

Photosynthesis rate is one of the physiological variables related to radiation-use efficiency. Foliar nitrogen content is reported to be one of the major determinants of photosynthesis rate, and the net photosynthesis rate at light saturation (Pmax) of healthy leaves was shown to be significantly higher in high with respect to low nitrogen treatment [47,48]. However, information on the effects of foliar diseases on the photosynthesis of wheat leaves with different nitrogen content is extremely limited. Carretero et al. [9] reported that changes in leaf nitrogen concentration do not modify the effects of leaf rust on net photosynthesis since leaf rust affects net photosynthesis mainly through non-stomatal events, such as chlorophyll reduction. Carretero et al. [9] also suspected that leaf rust affects light interception rather than radiation-use efficiency at the crop level. Independence in metabolic actions of leaf rust and foliar nitrogen content on net photosynthesis could be one of the reason why susceptible varieties could act differently under same level of pathogen pressure and N availability.

However, we also indicated that the effects of pathogens and N availability on net photosynthesis should not be analysed individually without considering the overall yield potential and genotype stability under unfavourable environmental conditions. The total rainfall during flowering and grain filling in 2021 was less than the eight-year averages. Varieties Barbee and Zvezdana exhibited significantly different yield losses when grown under similar levels of leaf rust infection, the same climatic conditions, without a reduction of N availability, and without the impact of fungicide action. Knowing that club wheat varieties have a higher tolerance to dry weather conditions than common wheat varieties [16], it could be expected that under the same level of leaf rust infection, a low level of total rainfall during the flowering and grain-filling time did not affect the yield loss of club winter wheat Barbee likewise it did in common wheat Zvezdana. Variety Barbee experienced higher yield losses only when the DI of leaf rust reached 29.3% and when nitrogen availability was restricted.

Additionally, Zvezdana exhibited consistent percentage yield losses across a wide range of pathogen pressures and diverse cultivation practices. It is plausible that the high-yielding potential of variety Zvezdana contributed to the stability of yield losses, which ranged from 16% to 18.6% under various levels of nitrogen availability and pathogen pressure ranging from 11.3% to 65%. However, it should be also expected that the lower tolerance of common wheat to dry climatic conditions than that of club wheat Barbee affected the yield loss of variety Zvezdana in addition to pathogen attack, making it less correlated than in variety Barbee. In previous studies, Zvezdana is reported as highly susceptible to extreme variation in climatic factors during growing season [37]. It could be expected that the susceptibility of Zvezdana to abiotic stressors resulted in higher yield loss under a much lower disease index when compared with the response of variety Barbee. Additionally, a leaf rust infection of 11% was identified as the threshold that triggers yield reduction in this variety under unfavourable growing conditions. These results complemented the findings of Jevtić et al. [49], who suspected that there might be a threshold of pathogen pressure below which a specific cultivar will not respond with significant yield loss if other requirements necessary to exhibit its yield potential are met.

Our study also revealed that the relationship between yield losses and thousand kernel weight (TKW) losses, were not the same for the two varieties. These results are in accordance with those of Lopes et al. [50], who indicated that variety and the environment affect TKW. However, in contrast to the study of Lopes et al. [50], where TKW and yield were correlated significantly with a positive direction, the relationship between yield and TKW loss in our study had a negative direction and was significant only for variety Barbee. There are also studies where TKW, yield, and yield components had a significant negative association [5,51,52]; thus, it could be suspected that the difference in the direction of the relationship between yield/TKW and yield/TKW loss could be attributed to differences in cultivar properties, cultivation practices, and environmental factors among studies.

Sugár et al. [52] tested the relationship between yield and TKW under different levels of N fertilisation, and a negative association between TKW and yield was observed in all trials where N fertilisation was applied. In contrast to that, check trials without N fertilisation showed no significant correlation between TKW and yield. The tendency of decrement of TKW with rising levels of N input is also reported by Protić et al. [6]. In our study, yield and TKW loss were estimated under both monoculture with reduced-N low-input and conventional N input with crop rotation production systems. Under the same level of leaf rust infection and conventional increasing levels of N input, Barbee suffered a lower yield loss than Zvezdana. TKW loss (13.3%) and yield loss (4.7%) were significantly negatively correlated only in variety Barbee. Supply in conventional production systems significantly lowered yield loss only in the Barbee variety. Consequently, it could be suspected that conventional N-input fertilisation contributed to the lowering of yield loss in variety Barbee at the expense of TKW loss. The correlation between crude protein loss and TKW loss was also different in the two varieties and was significantly moderately negative only in variety Barbee. It could be suspected that higher levels of N input in conventional cultivation affected limited crude protein loss in variety Barbee, but it should be investigated in more detail in future studies.

In our study, leaf rust significantly affected crude protein loss. This is in accordance with previous investigations indicating that at the beginning of grain development, N accumulation is source-and-sink regulated, but during the grain filling, N accumulation is always limited by the source supply from vegetative tissues [53]. Consequently, N accumulation during grain filling could be affected by pathogens colonising vegetative organs. Although it has been reported that foliar fungal diseases in general affect nitrogen dynamics and carbohydrate accumulation in grain [53], our study found that fungicide application had differing effects on the relationship between yield and crude protein gains. This variation could be partly explained by the different effects of leaf rust on yield and crude protein achievements in the two varieties. Regression modelling in this study indicated that localities with different cultivation practices did not affect crude protein gain, thereby excluding the possibility that nitrogen inputs influenced fungicide treatments and, consequently, the differences in fungicide effects on nitrogen dynamics and crude protein response. Barbottin et al. [54] reported that nitrogen yield depends not only on nitrogen remobilisation but also on biotic and abiotic stresses during the grain-filling period. Consequently, it is plausible that during the grain-filling stage, Barbee and Zvezdana reacted differently to combined biotic and abiotic factors, resulting in different contributions of fungicide treatment to the association between yield and crude protein responses to fungicide application.

Knowing that the determination of genetic variability may not always explain the complexity of phenotype markers, phenotyping in plant breeding will always be an important practice. In addition, there is still a lack of knowledge on divergence in winter wheat responses to combined abiotic and biotic stressors, although it would facilitate decision-making in the breeding of varieties suitable for reduced-input production. Consequently, this study intended to investigate if there is a difference in the relationship between yield, TKW, and crude protein of club and common winter wheat varieties susceptible to obligate pathogens when cultivated under different cultivation practices. To our knowledge, the results of the study revealed, for the first time, the specificity in the response of susceptible varieties to leaf rust infection when cultivated under both monoculture with reduced N input and conventional cultivation practices under the same climatic conditions. This finding provides direction for further investigations focusing on (1) determining the threshold level of pathogen pressure that induces yield loss in susceptible varieties; (2) assessing the contribution of nitrogen fertilisation in reducing yield loss at the expense of thousand kernel weight loss; and (3) exploring the effects of combined biotic and abiotic factors during the grain-filling stage on the crude protein response to fungicide application.

5. Conclusions

The main conclusions that can contribute to the decision-making during the evaluation of winter wheat varieties' performance are given below:

- Susceptible varieties, showing similar levels of leaf rust infection and without the impact of fungicide action, could exhibit significantly different yield losses when grown with the same cultivation practices and under the same climatic conditions;
- Contrary to Barbee, variety Zvezdana, which is more susceptible to unfavourable climatic conditions, suffered the same level of yield loss regardless of cultivation practice and under a broad range of leaf rust infection. This suggests that the level of pathogen pressure that triggers the susceptibility reaction is cultivar-specific and highly associated with the susceptibility of varieties to abiotic stressors, which should be investigated in future studies in more detail;
- Conventional cultivation could affect the reduction of yield loss in wheat varieties susceptible to leaf rust. In this study, conventional cultivation with 100 kg N per ha affected the reduction of yield loss in club winter wheat Barbee but not in Zvezdana;
- The correlation between TKW loss and yield loss was significantly moderately negative only for club winter wheat Barbee. It was suspected that N fertilisation affected the lowering of yield loss at the expense of TKW loss;
- The contribution of fungicide to yield enhancement is highly associated with the specific reaction of the variety to pathogen infection rather than solely the disease level itself.

Supplementary Materials: The following supporting information can be downloaded at https://www.mdpi.com/article/10.3390/jof10060401/s1, Figure S1: Regression of yield loss on crude protein loss of winter wheat varieties Barbee and Zvezdana; Table S1: Regression analysis of the most influencing factors on yield loss, TKW loss, and crude protein loss in winter wheat varieties Barbee and Zvezdana; Table S2: Regression analysis of the most influencing factors on yield, TKW, and crude protein responses to fungicide application in winter wheat varieties Barbee and Zvezdana.

Author Contributions: Conceptualisation, R.J.; methodology, R.J. and V.Ž.; software, V.Ž.; validation, R.J.; formal analysis, V.Ž.; investigation, R.J., V.Ž. and D.Ž.; resources, R.J.; data curation, B.O. and D.Ž.; writing—original draft preparation, V.Ž.; writing—review and editing, R.J., E.A., S.M. and L.M.; visualisation, V.Ž.; supervision, R.J.; project administration, R.J.; funding acquisition, R.J. All authors have read and agreed to the published version of the manuscript.

Funding: This research was funded by the Ministry of Science, Technological Development and Innovation of the Republic of Serbia, institutional funding—200032 (Institute of Field and Vegetable Crops, Novi Sad) (RS-MESTD-inst-2020-200032).

Institutional Review Board Statement: Not applicable.

Informed Consent Statement: Not applicable.

Data Availability Statement: The original contributions presented in the study are included in the article/Supplementary Materials, further inquiries can be directed to the corresponding author.

Acknowledgments: We acknowledge all colleagues from the Institute of Field and Vegetable Crops, the National Institute of the Republic of Serbia, who provided technical support for this research.

Conflicts of Interest: The authors declare no conflicts of interest. The funders had no role in the design of the study; in the collection, analyses, or interpretation of data; in the writing of the manuscript; or in the decision to publish the results.

References

1. Akter, N.; Rafiqul Islam, M. Heat Stress Effects and Management in Wheat. A Review. *Agron. Sustain. Dev.* **2017**, *37*, 37. [CrossRef]
2. Juroszek, P.; von Tiedemann, A. Climate Change and Potential Future Risks through Wheat Diseases: A Review. *Eur. J. Plant Pathol.* **2013**, *136*, 21–33. [CrossRef]
3. White, J.W.; Hoogenboom, G.; Kimball, B.A.; Wall, G.W. Methodologies for Simulating Impacts of Climate Change on Crop Production. *Field Crops Res.* **2011**, *124*, 357–368. [CrossRef]
4. Laidig, F.; Piepho, H.-P.; Rentel, D.; Drobek, T.; Meyer, U.; Huesken, A. Breeding Progress, Environmental Variation and Correlation of Winter Wheat Yield and Quality Traits in German Official Variety Trials and on-Farm during 1983–2014. *Theor. Appl. Genet.* **2017**, *130*, 223–245. [CrossRef]
5. Mladenov, N.; Hristov, N.; Kondic-Spika, A.; Djuric, V.; Jevtic, R.; Mladenov, V. Breeding Progress in Grain Yield of Winter Wheat Cultivars Grown at Different Nitrogen Levels in Semiarid Conditions. *Breed. Sci.* **2011**, *61*, 260–268. [CrossRef]
6. Protić, R.; Jovin, P.; Protić, N.; Jovanovic, Ž. Mass of 1,000 Grains in Several Winter Wheat Genotypes, at Different Dates of Sowing and Rates of Nitrogen Fertilizer. *Rom. Agric. Res.* **2007**, *24*, 39–42.
7. Cao, X.; Yao, D.; Duan, X.; Liu, W.; Fan, J.; Ding, K.; Zhou, Y. Effects of Powdery Mildew on 1000-Kernel Weight, Crude Protein Content and Yield of Winter Wheat in Three Consecutive Growing Seasons. *J. Integr. Agric.* **2014**, *13*, 1530–1537. [CrossRef]
8. Draz, I.S.; Abou-Elseoud, M.S.; Kamara, A.-E.M.; Alaa-Eldein, O.A.-E.; El-Bebany, A.F. Screening of Wheat Genotypes for Leaf Rust Resistance along with Grain Yield. *Ann. Agric. Sci.* **2015**, *60*, 29–39. [CrossRef]
9. Carretero, R.; Bancal, M.O.; Miralles, D.J. Effect of Leaf Rust (Puccinia Triticina) on Photosynthesis and Related Processes of Leaves in Wheat Crops Grown at Two Contrasting Sites and with Different Nitrogen Levels. *Eur. J. Agron.* **2011**, *35*, 237–246. [CrossRef]
10. Taub, D.R.; Miller, B.; Allen, H. Effects of Elevated CO2 on the Protein Concentration of Food Crops: A Meta-Analysis. *Glob. Chang. Biol.* **2008**, *14*, 565–575. [CrossRef]
11. Yang, F.; Jørgensen, A.D.; Li, H.; Søndergaard, I.; Finnie, C.; Svensson, B.; Jiang, D.; Wollenweber, B.; Jacobsen, S. Implications of High-Temperature Events and Water Deficits on Protein Profiles in Wheat (*Triticum Aestivum* L. Cv. Vinjett) Grain. *Proteomics* **2011**, *11*, 1684–1695. [CrossRef] [PubMed]
12. Kolmer, J. Leaf Rust of Wheat: Pathogen Biology, Variation and Host Resistance. *Forests* **2013**, *4*, 70. [CrossRef]
13. Conner, R.L.; Kuzyk, A.D.; Su, H. Impact of Powdery Mildew on the Yield of Soft White Spring Wheat Cultivars. *Can. J. Plant Sci.* **2003**, *83*, 725–728. [CrossRef]
14. Jevtić, R.; Župunski, V.; Lalošević, M.; Jocković, B.; Orbović, B.; Ilin, S. Diversity in Susceptibility Reactions of Winter Wheat Genotypes to Obligate Pathogens under Fluctuating Climatic Conditions. *Sci. Rep.* **2020**, *10*, 19608. [CrossRef] [PubMed]

15. Morris, C.F.; Engle, D.A.; Kiszonas, A.M. Breeding, Selection, and Quality Characteristics of Soft White Wheat. *Cereal Food World* **2020**, *65*, 53. [CrossRef]
16. Zwer, P.K.; Sombrero, A.; Rickman, R.W.; Klepper, B. Club and Common Wheat Yield Component and Spike Development in the Pacific Northwest. *Crop Sci.* **1995**, *35*, 1590–1597. [CrossRef]
17. Chen, X.M. Epidemiology and Control of Stripe Rust [Puccinia Striiformis f. Sp. Tritici] on Wheat. *Can. J. Plant Pathol.* **2005**, *27*, 314–337. [CrossRef]
18. Cook, R.J. Fusarium Foot Rot of Wheat and Its Control in the Pacific Northwest. *Plant Dis.* **1980**, *64*, 1061–1066. [CrossRef]
19. Kolmer, J.A. Virulence of Puccinia Triticina, the Wheat Leaf Rust Fungus, in the United States in 2017. *Plant Dis.* **2019**, *103*, 2113–2120. [CrossRef]
20. Jevtić, R.; Župunski, V.; Lalošević, M.; Župunski, L. Predicting Potential Winter Wheat Yield Losses Caused by Multiple Disease Systems and Climatic Conditions. *Crop Prot.* **2017**, *99*, 17–25. [CrossRef]
21. Voss-Fels, K.P.; Stahl, A.; Wittkop, B.; Lichthardt, C.; Nagler, S.; Rose, T.; Chen, T.-W.; Zetzsche, H.; Seddig, S.; Majid Baig, M.; et al. Breeding Improves Wheat Productivity under Contrasting Agrochemical Input Levels. *Nat. Plants* **2019**, *5*, 706–714. [CrossRef] [PubMed]
22. Ciara, B.; Jason, B. Managing Powdery Mildew in Wheat. Available online: https://www.agric.wa.gov.au/spring/managing-powdery-mildew-wheat (accessed on 14 May 2024).
23. Delfan, S.; Bihamta, M.R.; Dadrezaei, S.T.; Abbasi, A.; Alipour, H. Exploring Genomic Regions Involved in Bread Wheat Resistance to Leaf Rust at Seedling/Adult Stages by Using GWAS Analysis. *BMC Genom.* **2023**, *24*, 83. [CrossRef] [PubMed]
24. Maiorano, A.; Blandino, M.; Reyneri, A.; Vanara, F. Effects of Maize Residues on the Fusarium Spp. Infection and Deoxynivalenol (DON) Contamination of Wheat Grain. *Crop Prot.* **2008**, *27*, 182–188. [CrossRef]
25. Vogelgsang, S.; Beyer, M.; Pasquali, M.; Jenny, E.; Musa, T.; Bucheli, T.D.; Wettstein, F.E.; Forrer, H.-R. An Eight-Year Survey of Wheat Shows Distinctive Effects of Cropping Factors on Different Fusarium Species and Associated Mycotoxins. *Eur. J. Agron.* **2019**, *105*, 62–77. [CrossRef]
26. Wegulo, S.N. Factors Influencing Deoxynivalenol Accumulation in Small Grain Cereals. *Toxins* **2012**, *4*, 1157. [CrossRef] [PubMed]
27. Schumann, G.L.; Leonard, K.J. Stem Rust of Wheat (Black Rust). In *Plant Health Instructor*; The American Phytopathological Society: St. Paul, MN, USA, 2000; Updated 2011. [CrossRef]
28. Župunski, V.; Jevtić, R.; Lalošević, M.; Mikić, S.; Orbović, B. The Applicability of Species- and Trichothecene-Specific Primers in Monitoring the Fusarium Graminearum Species Complex and Its Impact on the Surveillance of Fusarium Head Blight in Winter Wheat in Serbia. *Agronomy* **2021**, *11*, 778. [CrossRef]
29. Glazebrook, J. Contrasting Mechanisms of Defense Against Biotrophic and Necrotrophic Pathogens. *Annu. Rev. Phytopathol.* **2005**, *43*, 205–227. [CrossRef] [PubMed]
30. Kissoudis, C.; van de Wiel, C.; Visser, R.G.F.; van der Linden, G. Enhancing Crop Resilience to Combined Abiotic and Biotic Stress through the Dissection of Physiological and Molecular Crosstalk. *Front. Plant Sci.* **2014**, *5*, 207. [CrossRef] [PubMed]
31. Yasuda, M.; Ishikawa, A.; Jikumaru, Y.; Seki, M.; Umezawa, T.; Asami, T.; Maruyama-Nakashita, A.; Kudo, T.; Shinozaki, K.; Yoshida, S.; et al. Antagonistic Interaction between Systemic Acquired Resistance and the Abscisic Acid–Mediated Abiotic Stress Response in Arabidopsis. *Plant Cell* **2008**, *20*, 1678–1692. [CrossRef]
32. Chantret, N.; Mingeot, D.; Sourdille, P.; Bernard, M.; Jacquemin, J.M.; Doussinault, G. A Major QTL for Powdery Mildew Resistance Is Stable over Time and at Two Development Stages in Winter Wheat. *Theor. Appl. Genet.* **2001**, *103*, 962–971. [CrossRef]
33. Peterson, C.J.; Vogel, O.A., Jr.; George, D.W.; Morrison, K.J.; Rubenthaler, G.L.; Allan, R.E. Barbee Wheat. Washington State University Cooperative Extension Service. Young, Director, and the U.S. Department of Agriculture in Furtherance of the Acts of May 8 and June 30, 1914. Extension Programs are Available to All Persons Without Regard to Race, Color, or National Origin. Published September 1976. Available online: https://content.libraries.wsu.edu/digital/collection/ext/id/20550/ (accessed on 14 May 2024).
34. Robe, P.; Pavoine, M.T.; Doussinault, G. Early Assessment of Adult Plant Reaction of Wheat (*Triticum Aestivum* L.) to Powdery Mildew (*Erysiphe Graminis* f Sp *Tritici*) at the Five-Leaf Seedling Stage. *Agronomie* **1996**, *16*, 441–451. [CrossRef]
35. Villaréal, L.M.M.A.; Lannou, C. Selection for Increased Spore Efficacy by Host Genetic Background in a Wheat Powdery Mildew Population. *Phytopathology* **2000**, *90*, 1300–1306. [CrossRef] [PubMed]
36. *MSZ 6369/6-1988*; Flour Testing Methods. Determination of Water Absorption Capacity and Baking Quality. Hungarian Standards Institution (Magyar Szabványügyi Testület (MSZT): Budapest, Hungary, 1988.
37. Kondić-Špika, A.; Mladenov, N.; Grahovac, N.; Zorić, M.; Mikić, S.; Trkulja, D.; Marjanović-Jeromela, A.; Miladinović, D.; Hristov, N. Biometric Analyses of Yield, Oil and Protein Contents of Wheat (*Triticum Aestivum* L.) Genotypes in Different Environments. *Agronomy* **2019**, *9*, 270. [CrossRef]
38. Bafus, R.; Bassinette, J.; Bohle, M.G.; Karow, R.S. State-Wide Cereal Variety Testing Program Trials in Central Oregon. In *Central Oregon Agricultural Research Center 2001 Annual Report*; Oregon State University: Corvallis, OR, USA, 2002.
39. Vukovic, A.; Vujadinovic, M.; Rendulic, S.; Djurdjevic, V.; Ruml, M.; Babic, V.; Popovic, D. Global Warming Impact on Climate Change in Serbia for the Period 1961–2100. *Therm. Sci.* **2018**, *22*, 2267–2280. [CrossRef]
40. Peterson, R.F.; Campbell, A.B.; Hannah, A.E. A Diagrammatic Scale for Estimating Rust Intensity on Leaves and Stems of Cereals. *Can. J. Res.* **1948**, *26*, 496–500. [CrossRef]

41. Wegulo, S.N.; Breathnach, J.A.; Baenziger, P.S. Effect of Growth Stage on the Relationship between Tan Spot and Spot Blotch Severity and Yield in Winter Wheat. *Crop Prot.* **2009**, *28*, 696–702. [CrossRef]
42. Živančev, D.; Živanović, M.; Aćin, V.; Mikić, S.; Mirosavljević, M.; Jocković, B.; Dražić, T. The Use of the Handheld Grain Quality Analyzer for Determination Protein Content in Wheat Cultivars. In Proceedings of the Biochemical Insight into Molecular Mechanisms, Kragujevac, Serbia, 24 September 2021; p. 189.
43. Bursac, Z.; Gauss, C.H.; Williams, D.K.; Hosmer, D.W. Purposeful Selection of Variables in Logistic Regression. *Source Code Biol. Med.* **2008**, *3*, 17. [CrossRef] [PubMed]
44. *XLSTAT 2022.5.1 (1392) Statistical and Data Analysis Solution*; XLSTAT: Paris, France, 2022.
45. RStudio Team. *RStudio: Integrated Development for R*; RStudio 2021.09.2+382 "Ghost Orchid" Release (fc9e217980ee9320126e33cdf-334d4f4e105dc4f, 2022-01-04) for Windows; RStudio Team: Boston, MA, USA, 2022.
46. Moore, C.M.; Richards, R.A.; Rebetzke, G.J. Phenotypic Variation and QTL Analysis for Oil Content and Protein Concentration in Bread Wheat (*Triticum Aestivum* L.). *Euphytica* **2015**, *204*, 371–382. [CrossRef]
47. Connor, D.; Hall, A.; Sadras, V. Effect of Nitrogen Content on the Photosynthetic Characteristics of Sunflower Leaves. *Funct. Plant Biol.* **1993**, *20*, 251–263. [CrossRef]
48. Evans, J.R. Nitrogen and Photosynthesis in the Flag Leaf of Wheat (*Triticum Aestivum* L.). *Plant Physiol.* **1983**, *72*, 297–302. [CrossRef]
49. Jevtić, R.; Skenderović, N.; Župunski, V.; Lalošević, M.; Orbović, B.; Maširević, S.; Bagi, F. Association between Yield Loss and Fusarium Head Blight Traits in Resistant and Susceptible Winter Wheat Cultivars. *J. Plant Dis. Prot.* **2021**, *128*, 1013–1022. [CrossRef]
50. Lopes, M.; Reynolds, M.; Manes, Y.; Singh, R.; Crossa, J.; Braun, H.-J. Genetic Yield Gains and Changes in Associated Traits of CIMMYT Spring Bread Wheat in a "Historic" Set Representing 30 Years of Breeding. *Crop Sci.* **2012**, *52*, 1123. [CrossRef]
51. Slafer, G.A. Genetic Basis of Yield as Viewed from a Crop Physiologist's Perspective. *Ann. Appl. Biol.* **2003**, *142*, 117–128. [CrossRef]
52. Sugár, E.; Berzsenyi, Z.; Árendás, T.; Bónis, P. Effect of Nitrogen Fertilization and Genotype on the Yield and Yield Components of Winter Wheat. *Bodenkult. J. Land Manag. Food Environ.* **2016**, *67*, 25–34. [CrossRef]
53. Simón, M.R.; Fleitas, M.C.; Castro, A.C.; Schierenbeck, M. How Foliar Fungal Diseases Affect Nitrogen Dynamics, Milling, and End-Use Quality of Wheat. *Front. Plant Sci.* **2020**, *11*, 569401. [CrossRef]
54. Barbottin, A.; Lecomte, C.; Bouchard, C.; Jeuffroy, M.-H. Nitrogen Remobilization during Grain Filling in Wheat: Genotypic and Environmental Effects. *Crop Sci.* **2005**, *45*, 1141–1150. [CrossRef]

Journal of
Fungi

MDPI

Review

Isothermal Detection Methods for Fungal Pathogens in Closed Environment Agriculture

Aylwen Cotter [1], Peter Dracatos [2], Travis Beddoe [1,2] and Kim Johnson [1,2,*]

[1] Australian Research Council Industrial Transformation Research Hub for Medicinal Agriculture, Bundoora 3083, Australia; w.cotter@latrobe.edu.au (A.C.); t.beddoe@latrobe.edu.au (T.B.)
[2] La Trobe Institute for Sustainable Agriculture and Food, Department of Ecological, Plant and Animal Sciences, La Trobe University, Bundoora 3083, Australia; p.dracatos@latrobe.edu.au
* Correspondence: k.johnson@latrobe.edu.au

Abstract: Closed environment agriculture (CEA) is rapidly gaining traction as a sustainable option to meet global food demands while mitigating the impacts of climate change. Fungal pathogens represent a significant threat to crop productivity in CEA, where the controlled conditions can inadvertently foster their growth. Historically, the detection of pathogens has largely relied on the manual observation of signs and symptoms of disease in the crops. These approaches are challenging at large scale, time consuming, and often too late to limit crop loss. The emergence of fungicide resistance further complicates management strategies, necessitating the development of more effective diagnostic tools. Recent advancements in technology, particularly in molecular and isothermal diagnostics, offer promising tools for the early detection and management of fungal pathogens. Innovative detection methods have the potential to provide real-time results and enhance pathogen management in CEA systems. This review explores isothermal amplification and other new technologies in detection of fungal pathogens that occur in CEA.

Keywords: pathogenic fungus; closed environment agriculture; molecular methods; detection; isothermal; LAMP

Citation: Cotter, A.; Dracatos, P.; Beddoe, T.; Johnson, K. Isothermal Detection Methods for Fungal Pathogens in Closed Environment Agriculture. *J. Fungi* **2024**, *10*, 851. https://doi.org/10.3390/jof10120851

Academic Editor: Ofir Degani

Received: 4 November 2024
Revised: 5 December 2024
Accepted: 7 December 2024
Published: 10 December 2024

1. Introduction

Fungi are a large group of eukaryotic organisms, with a diverse range of taxa that play essential roles in the environment and many human activities [1]. These include, but are not limited to, aiding with the decomposition of organic matter and the production of antibiotics and sources of food [2]. Other fungal species can contaminate or cause toxicity in ingested plant-based foods or medicines [3]. Fungi can also be pathogenic and cause disease in both animals and plants. Some of the most destructive diseases of crops are caused by fungal pathogens. It has been estimated that around 15,000 fungal species cause between 70–80% of all plant diseases [4]. For example, wheat is the most widely produced food crop globally but is the primary host for three different rust pathogen species from the genus *Puccinia: P. striiformis* f. sp. *tritici* (stripe rust), *P. graminis* f. sp. *tritici* (stem rust) and *P. triticina* (leaf rust). Yield losses can vary, up to 100% for stripe rust, 90% for stem rust and 70% for leaf rust, leading to major yield losses, food insecurity and the disruption of global markets [5–8]. Pest and diseases can cost food producers billions of dollars each year, with countries like the Unites States of America estimating losses of USD 21 billion through prevention techniques such as insecticides, fungicides, and/or crop losses [9].

Increased urbanization, land degradation and limited resources are reducing the amount and quality of arable land for agriculture, along with the additional challenges of climate change [10]. Indoor crop growing systems, termed closed or controlled environment agriculture (CEA) and protected cropping environments (PCEs), are proposed as one of the solutions to maintain sustainable food production [10]. Indoor agriculture designs have the potential to reduce the overall environmental footprint with approaches such as recycled water, renewable energy, and urban locations to reduce transport [10].

CEA control growth conditions such as temperature, watering and nutrient supply, leading to increased productivity and yields of greater than 50% when compared to field yield [11]. In CEA systems such as greenhouses and vertical farms, approaches to mitigate the introduction of pests and disease include manual or automated sanitisation systems. To reduce the presence of fungi in these environments, methods such as increasing air flow using fans, the use of humidifiers to encourage plant health, removing infected plants and the use of beneficial insects such as fungal gnats that feed on fungi in soils aim to keep the plants healthier and reduce the likelihood of infection. PCEs are similar to closed environments, however, they allow air to flow freely from one end of the structure to the other, which can lead to the infestation of pests and diseases [12]. It is challenging to exclude all pathogens entirely from PCEs as they are often introduced through watering systems, soils, fertilisers, footwear, clothing and seeds. For example, the oomycete pathogen *Phytophthora cryptogea* that causes a root disease in hydroponic lettuces was shown to infect seedlings grown repeatedly on the same site through the watering systems [13]. Disease outbreaks are shown to occur more frequently in CEAs when plants become stressed, reducing the effectiveness of the plant immune response leading to increased fungal colonisation [14]. There is currently limited information about the amount of crop loss and associated costs to indoor cropping industries as a result of fungal disease-causing pathogens. CEA losses from the Australian hydroponic lettuce industry were recorded to be between 20 and 30% for root diseases alone, with complete crop losses occurring during the warmer months [13]. Diseases such as the tomato brown rugose virus can result in 100% crop loss in CEA, emphasizing how damaging exotic pathogens can be to growers [15].

There are several crop types that are currently being grown in CEA including: vegetables, cereals, flowers, herbs, fruits and medicinal crops [16]. Cereal crops such as rice grown indoors have been shown to be infected by *Magnaporthe grisea* that causes rice blast disease, which can infect young seedlings, killing the plant by infecting nodes, and stems causing grain loss [17,18]. Fruits such as strawberries, melons and blueberries are prone to *Botrytis cinerea* infection in CEA, and herbs such as oregano, basil and mint are prone to *Fusarium* spp. infection [16]. When grown indoors, medicinal crops such as cannabis and aloe vera have been reported to show diseases such as *Alternaria alternata* and *Fusarium* spp. [19,20]. The indoor medicinal cannabis industry is frequently affected by the necrotrophic fungus *Botrytis cinerea* through its ability to infect susceptible cultivars both at post-harvest or whilst in storage [21]. The ornamental flower industry that includes roses and carnations are prone to powdery mildews and *Fusarium* spp. [22].

In order to prevent crop damage and cost to the indoor agriculture industry, rapid and early detection methods of pathogens can assist with strategies for control and prevention of further disease spread. Traditional methods of fungal identification in agriculture have been through the visualization of signs or symptoms on plants based on morphological and/or taxonomic features. Molecular approaches are increasingly being adopted to detect fungal presence prior to disease development [23]. In a controlled environment setting, the use of environmental (eDNA) sampling can be adopted to aid in the obtaining of DNA samples for molecular processing. eDNA sampling targets spores moving through the air or run off from watering systems. Unlike broadacre agriculture settings where fungal contaminants from birds, insects, faecal matter and other flora can complicate eDNA sampling, indoor areas are suitable for targeted disease detection. Furthermore, combining more than one method of detection can lead to point of care diagnostics. Point of care diagnostics is the carrying out of tests using specific devices or kits next to or near the patient in health settings, or in field for the agriculture industry in a quick and efficient manner without the need to send the samples off to a laboratory [24,25]. In the following sections we outline fungal pathogens common to the indoor agriculture setting and molecular approaches used for detection.

2. Pathogenic Fungi Affecting CEA

Botrytis and powdery mildew causing pathogens (*Golovinomyces* spp. and *Blumeria* spp.), *Pythium*, *Sclerotinia*, downy mildew (*Peronospora* spp.) and *Fusarium* spp. are the most prevalent fungi affecting CEA industries (Figure 1) [21,26]. Some are host specific, such as the powdery mildew, and others such as *Botrytis cinerea* have a broad host range [19,27]. Diseases caused by fungal pathogens can infect plants at every stage of growth and development however infection depends on many factors including conditions, age, stress, genetics and the fungal species itself [28]. Some species of *Botrytis* and *Fusarium* can cause health issues in humans if consumed through exposure to spores or mycotoxins [19,29].

Figure 1. Microscopic and macroscopic images of fungal pathogens detected in protected cropping environments. (**A**) Laboratory-grown *Botrytis cinerea* on a potato dextrose agar plate and viewed under

a Leica M205 FA Stereomicroscope. (**B**) Strawberry fruit infected with *Botrytis cinerea* in field (photo taken by Marlo Molinaro La Trobe University). (**C**) *Sclerotinia sclerotiorum* grown in the laboratory on a potato dextrose agar plate and viewed under a Leica M205 FA Stereomicroscope. (**D**) Bok choy leaf infected with *Sclerotinia sclerotiorum*, 8 days after inoculation. (**E**) Laboratory grown putative *Fusarium oxysporum* collected from CEA, grown on a potato dextrose agar plate and viewed under a Leica M205 FA Stereomicroscope. (**F**) *Fusarium* growing on a tomato, reprinted with permission from [30]. Copyright 2021 Zahir Shah Safari from Leibniz Universität Hannover. (**G**) Laboratory grown *Pythium irregulare* now *Globisporangium irregulare* on a potato dextrose agar plate viewed under a Leica M205 FA Stereomicroscope, obtained from Dr. Niloofar Vaghefi from University of Melbourne. (**H**) *Pythium irregularre* now known as *Globisporangium irregulare* growing on geraniums taken and unchanged from CABI Plantwise Plus website (https://plantwiseplusknowledgebank.org/doi/10.1079/PWKB.Species.46152, accessed on 16 October 2024) under a Attribution-NonCommercial-NoDerivatives 4.0 International (CC BY-NC-ND 4.0) license taken by Michael Evans, University of Arkansas. (**I**) Suspected *Golovinomyces cichoracearum* grown on onion viewed under a microscope, taken from the Global Biodiversity Information Facility website (https://www.gbif.org/occurrence/4891810393, accessed on 16 October 2024) by Schmidt Dávid (licensed under http://creativecommons.org/licenses/by-nc/4.0/, accessed on 17 October 2024). *Golovinomyces* is an obligate fungus and therefore unable to be cultured in the laboratory on a growing media. (**J**) Suspected *Golovinomyces cichoracearum* grown on *Cannabis sativa* L. courtesy of industry partner. Scale = 200 µm on (**A**,**C**,**E**,**G**), 1 cm on (**B**,**D**,**F**,**H**,**J**) and ~100 µm for (**I**).

2.1. *Botrytis* sp.

There are approximately 35 species of Botrytis [31]. *Botrytis cinerea* is one of the most prominent and destructive pathogens with a broad host range comprising approximately 200 different plant species [32]. Hosts include fruits such as strawberries and tomatoes, and cut flowers such as roses [33,34]. *Botrytis* was first described in 1794 by Christiaan Hendrik Persoon, and can sometimes be confused with *Sclerotinia* spp. [32,35]. *B. cinerea* is distributed all over the world and infection can be present without the appearance of symptoms [36]. This fungal species can lay dormant for days to months and when conditions are of high humidity with cool to moderate temperatures, a grey mould starts to form on infected plants (Figure 1B) [37]. *Botrytis* releases spores, known as conidia, and a distinguishing feature of *Botrytis* is the oval shaped conidia with colonies of dark grey and aerial mycelium [38]. During infection, *Botrytis* conidia released from the conidiophore into the air and land on the hosts surface. The initiation and formation of appressorium subsequently leads to penetration and infection of plant tissue at points of damage or natural openings such as the stomata [32]. Depending on the plant host, the type of plant tissue that is infected varies. For example, on fruiting plants the site of infection is generally the floral organs [32]. Studies have determined that fungal pathogens such as *B. cinerea* can grow asymptomatically within the host subverting the plant defence response prior to switching to killing the host cells to feed on dead cells through acquisition of nutrients as part of its necrotrophic lifestyle [39]. Multiple virulence mechanisms, such as the use of effector proteins, toxins and lytic enzymes and the secretion of oxalic acid to reduce pH in host tissue are involved in infection [34]. *Botrytis* spores can also induce an immune response in humans that can affect health and are known as allergens [19]. Although air is a major form of transmission, the conidia can also travel through watering systems. *B. cinerea* is considered a disease of agricultural importance due to its broad host range leading to widespread crop losses especially to the horticultural industry and tendency to develop fungicide resistance [32].

2.2. *Powdery Mildew*

Over 900 species of powdery mildew and even more sub-species have been described [40,41]. In contrast to *Botrytis* spp., mildews are an obligate biotrophic fungus meaning they obtain their nutrients from intact living cells and require living plant host tissue to complete their lifecycle [41]. Closely related powdery mildew species, such as

Golovinomyces and *Blumeria* spp. are difficult to distinguish morphologically, with the main differences being host range and conidial and hyphal structures, which can only be seen under a microscope [41,42]. As the name suggests the symptoms mimic small white powder patches, commonly found on leaves but can be found on stems and buds and can cover large amounts of plant tissue (Figure 1J) [43]. Powdery mildew affects a diverse array of crops such as grapes, wheat, barley, oat, cannabis and some fruit and vegetable species. The pathogen acts as a sink for carbon and micronutrients causing the depletion of nutrients and carbon that in turn causes yield and grain quality to decrease [44]. Powdery mildew species reproduce mainly via asexual production of conidiospores. The spore lands on the plant host and the appressorium produces sufficient pressure to lyse the cuticle of the epidermal cell layer [45]. Successful colonisation of the epidermal cell by the powdery mildew fungus is characterised by the production of haustoria which facilitates the acquisition of nutrients from the host and in parallel the secretion of effectors by the pathogen [46].

2.3. *Pythium*

Pythium spp. were first discovered in 1858 by Pringsteim and can affect animals and humans depending on species [47,48]. *Pythium* spp. cause root rot or dampening off resulting in yield losses to many varieties of vegetables and ornamental crops including corn, potato, wheat, oats, tobacco and peanuts to name a few [49,50]. On plants, the infection develops into necrotic lesions on the roots (Figure 1H), that ends up affecting the stem and leaves, by producing hyphae that extract nutrients from the host [51,52]. *Pythium* spp. use carbohydrate active enzymes (CAZymes) that aid in the penetration of the plants cell walls and assists in further disease establishment [52]. Other symptoms can include reduced growth, black or brown discolouration visible on seeds and young roots as well as wilting of young or juvenile tissue such as in seedlings [52]. *Pythium* can contaminate irrigation water systems as well as soilless systems and can also be spread through gnats [53]. There are many species of *Pythium* that have been identified that are saprophytic and are not plant pathogens [50]. *Pythium* prefers environmental conditions that are moist with high temperatures, with some species having the ability to remain dormant for long periods of time [50]. Effects on the plants depend on the species infecting them. Some species such as *P. spinosum*, *P. arrhenomanes*, *P. myriotylum*, and *P. dissotocum* affect the seeds, while those causing issues such as dampening off for seedlings and root rot in more mature plants include *P. afertile*, *P. arrhenomanes*, *P. dissotocum*, *P. elongatum*, and *P. spinosum* [54–56].

2.4. *Sclerotinia*

Sclerotinia is a necrotrophic fungus with a wide host range, infecting >450 plant species including but not limited to: soybean, rapeseed, onion, garlic, canola, sunflower and, more recently, cannabis [57–59]. The diseases caused by either *S. sclerotium* or *S. minor* infection are deemed of high economic importance due to the vastness of the hosts infected, and the damage caused by the diseases [57]. Infection causes stem rot and cankers, which can then turn to shoot necrosis. *Sclerotinia* is often referred to as white mould, though has many other names such as crown rot and cottony rot [60]. The infection begins with pale or dark brown lesions and develops into white mycelium that look like cotton wool (Figure 1D). The hyphae then develop small black balls called sclerotia [57]. The sclerotia are renowned for staying dormant in the soil ecosystems for >10 years providing additional challenges for controlling the disease during favourable environmental conditions [61]. The disease cycle is characterised by ascospores landing on host plant tissues and germination begins causing infection preferentially in temperate conditions [62].

2.5. *Other Notable Fungal Pathogens*

Other pathogens that can occur in CEA include downy mildew, caused by pathogens such as *Plasmopara viticola* and *Hyaloperonospora arabidopsidis*, as well as *Fusarium* species. Downy mildew affects a variety of agriculture crops including brassicas, cucurbits, lettuce, peas, grapevine and tobacco [63]. The disease creates mosaic-like patterns on the leaves of a susceptible host and can cause 40–90% yield loss in field given favourable temperature and humidity. Infection is caused by oomycetes that germinate via oospores [64]. The genus and species of this disease depends on the host plant it infects.

Diseases caused by *Fusarium* spp. are estimated to be around 1500 species and therefore signs and symptoms can vary depending on the infected host [65]. Not all species of *Fusarium* are pathogenic, however, some can produce mycotoxins that can contaminate food and harm humans and animals [29]. *Fusarium* spp. can effect a variety of agricultural crops including cabbage, tomato, cotton, spinach, peas and cereal grain crops [66]. For example *Fusarium oxysporum* (Figure 1F) have further evolved *formae speciales ff.* spp. based on the ability to infect a specific range of closely related host plant species and cause wilt disease [66]. Specific examples include *F. oxysporum f.* sp. *lypersici* (tomato) and *F.o f.* sp. *vasinfectum* that causes wilt in cotton [67]. For some *F. oxysporum f.* spp. such as *F. oxysporum f.* sp. *conglutinans* that effect brassica and Arabidopsis plants, are influenced by genetic diversity that leads to phenotypic variations. These variations can be shaped by environmental conditions significantly impacting the interactions with host plants; this, in turn, causes natural accessions within a species and genus and only certain *formae speciales* will affect these accessions [68,69].

3. Fungal Identification and Treatment in Agricultural Settings

Fungal disease identification in agriculture settings has largely been through visualization of disease signs or symptoms. Regional networks can then be involved in alerting other growers to the occurrence of disease in the area. In addition, weather warning systems make growers aware of climatic conditions that favour disease occurrence, such as increased rainfall and humidity. To date, there is limited information and capability to quantify with any accuracy the prevalence and occurrence of specific fungal pathogens over a growing period. Closed environments may limit some of the complexity by maintaining a relatively stable environment so that the use of diagnostics can be used to identify when pathogens are most likely to occur during plant growth cycle and take preventative action.

Fungal identification and detection methods encompass a range of techniques, including traditional microscopy, culture-based approaches, air and water sampling, chemical treatments and molecular tools such as PCR and isothermal assays, in order to enable accurate identification and monitoring of fungal species in various environments. Historically the identification and diagnosis of fungal pathogens was determined by Kochs postulates, a process of identifying the causal relationship between microbe and disease [70,71]. More recently, identification has been achieved through the observation of different pathogen morphology, culturing to determine lifecycle characteristics, microscopic examination and, then, through the development of DNA sequencing and molecular identification [72–74]. The first fungus-like organism to be investigated using a light microscope was the oomycete pathogen *Phytophthora infestans*, the causal agent of potato blight. Detected on potatoes in the 1840s, the strain of *P. infestans* that caused the great potato famine has been traced using DNA analysis [72,75,76].

The disease triangle outlines the relationship between the environment, the plant host and the pathogen [77]. This approach has been used as a method of predicting epidemiological outcomes since the 1960s [77]. More recently there is a fourth aspect to consider, the human effect or society, as well as a fifth element, time. The human effect is a concept based on human potential for transporting pathogens across borders, excessive use of fungicides and plant monocultures that lead to the breakdown in the genetic resistance [77]. Time refers to the event of disease development and the length of time it takes for the disease to grow or latency period (LP). LP, or the time taken for the infection cycle to develop, is affected by numerous factors including seasonal environmental conditions, stage of growth and genetic resistance status in the host plant [77].

Once disease is detected, growers must decide what action to take, and this can include fungicide application, integrated pest management and/or plant destruction [78].

Information through government websites and reputable agriculture associations provide farmers with the latest information and guides to aid in pest and disease management. To validate the identity of the pathogen growers, agronomists and plant pathologists can send samples to a laboratory for accurate identification if needed. In some cases, it is important to notify the appropriate authorities as some pests and pathogens breach biosecurity protocols and are considered exotic incursions, that could potentially have large consequences, this is different depending on the location. The use of point of care (POC) diagnostics for pest and pathogen testing is limited in field settings due to initial development costs and commercialisation, determining which device is best for which pest or pathogen, the ability to sufficiently validate the results and the fact tests are developed to target a few specific pathogens [79–82]. However, once developed, POC diagnostics can rapidly and cheaply be performed in field with the ability to obtain results in a shorter timeframe than current pathology laboratories [83].

More recent non-destructive approaches include the use of biosensors and imaging for diagnostic detection of fungal disease prior to the appearance of symptoms [84]. Such methods can complement integrated pest management plans. These approaches have been covered in reviews such as [85–87]. This review focusses on the molecular approaches such as, but not limited to, Loop-Mediated Isothermal Amplification (LAMP), Helicase-Dependent Amplification (HDA), Whole Genome Amplication (WGA), Nucleic Acid Sequence-Based Amplification (NASBA), Strand Displacement Amplification (SDA), Recombinase Polymerase Amplification (RPA) and Rolling Circle Amplification (RCA) (Table 1). Fungal sporulation is the most efficient detection stage/target due to the period of time it takes for the spores to be released which is defined as the latency period of infection. The development of early detection methods for the sporulation stage of pathogenic fungal growth can reduce the necessity for fungicides and other potentially harmful methods of fungal eradication, as well as potentially reducing the likelihood of human allergic responses and contamination of food and medicinal products. Air detection methods utilize the sampling of spores from air using equipment such as the SKC Biosampler, a liquid cyclonic impinging device that captures spores in a buffer which can then be used to extract nucleic acids for further diagnostic applications [88]. These methods are being adapted for the detection of fungal spores and for DNA testing to determine pathogen genera and species [89]. For outdoor sampling, changes to air sampler design and type, such as the impaction spore trap (BioScout, Marrickville, New South Wales, Australia), have been effective for broadacre agricultural crops [90]. eDNA is increasingly being used to determine microbial biodiversity in a range of settings, such as soils and air [91]. The principles of eDNA sampling could be transferred to the agriculture industry [92]. There is limited published information on fungal in-field sampling in agricultural settings. Most of the current research is aimed at the medical industry focussing on fungal pathogens causing human infection [93,94].

Table 1. Current list of isothermal based assays used in closed environment agriculture for fungal detection.

Molecular Assay	Summary	Fungal Pathogen/s Detected	Genes Targets	References
Loop-mediated isothermal amplification (LAMP)	Uses 4–6 primers for quick amplification. Performed at 60–68 °C	*Sclerotinia sclerotiorum* *Fusarium oxysporum* *Botrytis cinerea*	Suppressor Of Cytokine Signaling 5 (Ssos5) Intergenic spacer spacer (IGS)	[4,73,95,96]
Nucleic acid sequence-based amplification (NASBA):	Amplifies mRNA by using RNA polymerase, performed at 41 °C	*Candida* sp. *Cryptococcosis*	T7 promoter Capsular-associated protein (CAP10)	[4,23,97,98]
Rolling circle ampification (RCA)	Enzymatic assay that generates ssDNA. Performed between 25 °C and 37 °C	*Candida* sp. *Fusarium* sp.	Translation elongation factor 1-alpha (TEF-1α) Internal transcribed spacer (ITS)	[74,99–101]
Whole genome amplification (WGA)	Amplifies entire genome. Has no temperature modulated denaturation	*Penicillum paxilli* *Epichloe festucae* *Pythium insidiosum*	No gene target Requires DNA of 1–10 copies for initial input	[97,102–104]

Table 1. *Cont.*

Molecular Assay	Summary	Fungal Pathogen/s Detected	Genes Targets	References
Strand displacement amplification (SDA)	Performs enzymatic site specific nicks at 37–50 °C	*Erysiphales* spp.	β-tubulin1 (Tub1) Internal transcribed spacer (ITS)	[101,105–107]
Helicase dependent amplification (HDA)	Uses a thermostable helicase enzyme. Performed at 65 °C	*Pythium insidiosum*	Cytochrome c oxidase subunit 2 (COX2) BNI1 product (Bni1p)	[104,108,109]
Recombinase Polymerase Amplification (RPA)	Invades the double stranded DNA using enzymes and is performed at 30–42 °C	*Bipolaris sorokiniana* *Candida albicans*	Internal transcribed spacer (ITS) Calmodulin (cal)	[107,110,111]

Water detection methods for closed environment agriculture have been used for the detection of root diseases such as those caused by *Fusarium* and *Pythium* spp. in hydroponic lettuces [13]. Current methods test the growing mediums used for the propagation of cuttings, such as coco fibre [21]. Coconut fibre and Rockwool are commonly used growing mediums for indoor agriculture as they can be sterilised at high temperatures [112]. Growing mediums absorb nutrients and water to help the plants grow. The potential to test for fungal pathogens after watering using a filtration system is yet to be determined.

Chemical treatments to control fungal diseases in crops depends on the plant species being grown and disease in question [113]. Many producers would have to wait until signs of fungal disease damage appeared on crops to be able to identify the causal pathogen species then apply specific fungicide chemistry accordingly. Some growers spray fungicides intermittently or weekly to prevent an outbreak. These sprays are generally synthetic and can be toxic to natural habitats. Fungicide applications can result in long-term toxic residues in water systems, mammals, including humans, and affect other microorganisms such as bacteria and algae [113–115]. One study analysed multiple fungicide groups, including but not limited to Benzimidazole, organophosphate and chlorophenyl in aquatic systems and found that substances from these groups had a detection frequency up to 96% in water catchments dominated by agriculture [114]. Another study found that, when feed was contaminated with a fungicide from the ethylene(bis) dithiocarbamate family, the chance of various types of tumour significantly increased [116]. Fungicides are divided into two broad classes, contact and systematic, based on their chemical components and mode of action [115,117]. The contact group focusses on preventing the fungi or fungal spores from developing or growing on plant tissue [115]. The systematic class is absorbed and translocated around the plant to infection sites [115]. Some major classes of fungicide include the methyl benzimidazole carbamates (MBCs), the demethylation inhibitors (DMIs), the quinone outside inhibitors (QoIs) and the succinate dehydrogenase inhibitors (SDHIs) [118]. America, Europe and Australia are major agricultural producers and significant users of fungicides, employing these chemicals to protect crops from fungal diseases that can severely impact yields and quality [119]. However, the extensive use of fungicides in these regions has led to the banning of certain chemicals. Banned fungicides are chemicals that have been prohibited due to their harmful effects on human health, wildlife and the environment. In Europe, substances like Chlorothalonil and Mancozeb have been banned because of their potential to cause cancer and environmental damage, particularly to aquatic ecosystems [120,121]. Similar action has been taken in regions like the U.S. and Australia with the prohibited use of the fungicide Carbendazim, with regulatory bodies continuing to monitor and restrict fungicides with harmful side effects [114,122]. These bans reflect an increasing global emphasis on safer agricultural practices and environmental protection. Fungicides can be further broken down to single or multi-site specific due to their modes of action. Single site fungicides such as DMIs and QoIs are used to disrupt a single metabolic process or structure and are more prone to the development of resistance in rapidly evolving pathogen populations. Multi-site specific fungicides, as the name suggests, target multiple sites within the fungal pathogen [118]. MBCs target the β-tubulin gene and affect mitosis resulting in toxicity in fungal cells.

DMIs target the cytochrome P450-dependent sterol 14α-demethylase (Cyp51) gene and affects the cell membrane of fungi [123]. SDHIs and QoIs effect the respiration resulting in decreased energy efficiency in the cytochrome bc1 enzyme complex [123]. Though the chemical treatments are effective in recent years there has been an increase in fungicide resistance [124,125]. Resistance can occur when fungal pathogens have been overexposed to a chemical [126]. This overexposure is predicted to increase due to changing climate conditions and seasonal variation leading to higher disease incidence, in turn increasing fungicide resistance [114]. Climate change and seasonal variations are exacerbating plant disease dynamics by creating favourable conditions such as warmer temperatures, altered humidity, and increased CO_2 levels that enhance fungal growth and sporulation [114,124]. These changes result in longer infection periods and more severe outbreaks, leading to heightened reliance on fungicides. This overuse accelerates the development of fungicide resistance through mechanisms like genetic mutations in target-site genes or increased efflux pump activity, often resulting in cross-resistance across fungicide classes [124]. As resistance grows, crop yields decline, production costs rise, and environmental risks from overapplication intensify [114,124]. Effective mitigation requires integrated disease management strategies, including cultural practices, biological controls and judicious fungicide use, alongside advancements in predictive modelling [118].

The added challenge to resistance is that a single pathogen can become resistant to multiple fungicides. For example, it was reported that *Botrytis cinerea* is resistant to 15 different fungicide classes with resistant mutations appearing in genes such as *Cyp51*, and cytochrome B (*Cyt B*) [105,118,124]. *Sclerotinia* sp. has shown resistance to the fungicides from the SDHI class with mutations in the *SdhB* and *SdhC* target genes that effect components of the succinate dehydrogenase complex affecting respiration [127–129].

4. Use of Molecular Diagnostics in Fungal Detection and Identification

Molecular diagnostics utilize techniques to analyse genetic material and include PCR, qPCR, sequencing, LAMP and other isothermal assays, providing precise and rapid identification of pathogens. Advantages of molecular diagnostics include increased sensitivity and high specificity to specific pathogens, important in biosecurity and treatment contexts and overcoming some of the limitations of visible-based approaches whereby different fungi cause similar symptoms and/or are difficult to distinguish at the microscopic level.

Molecular diagnostics are techniques used to identify biological markers in the genome, transcriptome or proteome of viruses, bacteria, plants, humans and fungi [130]. DNA sequencing is commonly used to identify regions of the genome that can be used as a unique identifier for a particular species. Mycologists regularly use the *internal transcribed spacer* (*ITS*) operon or cluster of genes that is found in the region between the small subunit and large subunit of nuclear ribosomal DNA as a target for DNA barcoding [131,132]. This target is commonly used because it is present in all fungus and is highly variable, enabling distinction of different fungal species [133,134]. *Cytochrome C oxidase* (*COX*) genes are found in the mitochondria and commonly used as gene targets. Similar to the *ITS* operon the *COX* gene is commonly used for identification purposes due to it being present in many taxa and at high levels due to multiple copies of mitochondrial DNA [135]. Another gene often used from the mitochondria is *cytochrome B* (*Cyt B*) [136]. The *Cyt B* gene is used for species specific detection and has been used in many phylogenetic studies [137–143]. *SdhB* and *SdhC* are two of the four genes that aid in the succinate dehydrogenase complex [144,145].

For the last 40 years, the most commonly used molecular diagnostic method for fungal pathogens has been Polymerase Chain Reaction (PCR) [146]. PCR detection methods include observing signs and symptoms of the fungal pathogen, taking a sample of the fungus or infected plant tissue and sending it in the mail to the laboratory for identification. In the laboratory the DNA would be extracted, and tests conducted. The type of fungal pathogen and the quantity and quality of the collected material would be dependent on how quickly results could be obtained. In cases where the amount of collected material would not provide adequate DNA yields, methods to grow the pathogen in vitro may be needed, delaying the time to identification. PCR and DNA visualisation use specialised laboratory equipment such as gel electrophoresis and capillary electrophoresis to assess single nucleotide polymorphorisms (SNPs) and single sequence repeat (SSR) markers. For further analysis and confirmation of the pathogen, DNA sequencing and database searches could be used. PCR has been used to aid in the identification of *Alternaria solani* that is a major threat to the potato and tomato industry by targeting the ITS region [147]. Whilst this study was based in the laboratory, strategies to adapt for in-field would be desirable, however, the use of PCR in-field is limited due to need for specialist laboratory equipment, cost of reagents and requirement for trained personal [148]. A variation of PCR called Real Time Polymerase Chain Reaction (real time-PCR) or quantitative Polymerase Chain Reaction (qPCR) couples amplification and detection methods [36]. This assay can quantify the levels of plant pathogen by measuring the time to amplification in real time, essentially giving a precise quantitative relationship between the amount of starting DNA and the quantity of PCR product. The amount of amplification product is visualised using fluorescence and emission is measured during the reaction [36]. Whilst qPCR reduces time, and is suitable for larger-scale diagnostics, its main constraint is the expensive spectrofluorometric thermal cyclers, the reagents and their maintenance [149]. qPCR has been used in plant pathology for the fungal detection in crop biosecurity applications and for the detection of seed borne fungal pathogens such as *Fusarium graminearum* in wheat, *Magnaporthe grisea* in rice and *Phytophthora infestans* in early stage potatoes [74,149–153].

Loop-Mediated Isothermal Amplification or LAMP is a method for rapid and specific amplification of DNA through the use of 4–6 primers making it faster the PCR. A feature of LAMP is the ability to run assays at isothermal or one temperature making it more amenable to field usage. LAMP has been adapted for portable, in-field diagnostics as it can amplify DNA or RNA in short (15–20 min) time-frames due to the formation of loop like DNA structures [33,154]. LAMP assays are more tolerant to crude or dirty DNA making the sample preparation stage easier and quicker [74]. LAMP is measured by its sensitivity or what the least amount of DNA or RNA is required for detection and specificity or how specific is the assay when compared to closely related species.

LAMP can use fluorescence absorbency measured in real time as well as a range of other visualisation methods, including colorimetric LAMP which uses pH-dependent colour change for rapid visualisation of presence/absence of DNA [155]. These approaches are being used for in-field testing of different economically significant pathogenic fungi. LAMP was successfully used to detect *B. cinerea* inoculated rose petals and pelargonium leaf discs in the laboratory [33]. The assay was highly sensitive and no cross reactivity with other fungal pathogens tested such as *Alternaria brassicola*, *Fusarium avenaceum* and *Sclerotinia sclerotiorum* amongst others was observed, however, amplification did occur with the closely related species *B. pelargonii* [33]. Khan et al. (2018) [147] have been able to develop a working LAMP assay in the laboratory for the detection of *A. solani* on potato leaves with a 10-fold greater sensitivity than traditional PCR.

Over time, other isothermal techniques have been developed (Table 1). Currently, many are foccused on the detection of viruses and bacteria for medical applications. There is a growing potential for isothermal techniques such as LAMP to be further developed and optimised in agricultural settings for the rapid detection of fungal pathogens [97]. All techniques have their advantages and disadvantages. The LAMP assay is more sensitive and specific than traditional PCR however it can be easily contaminated [4,73].

Nucleic acid sequence-based amplification (NASBA) is a transcription based method designed to amplify RNA sequences [97]. This method has a decreased chance of contamination however it cannot amplify double stranded DNA [97]. NASBA has been used to identify *Candida* species in human blood using primers targeting the 18S rRNA sequences, with a further 19 different fungi being tested and showing postive detection [156].

Rolling circle ampification (RCA) is a DNA replication method that uses a Padlock probe to identify a single nucleotide polymorphorisms (SNPs) in the genome. A padlock probe uses a phosphorylated enzyme at the 5′ end and a complement 5′-biotinylated enzyme at the 3′-end. This method can be relatively easily expanded to multiple allele or loci identifications using Ligation-RCA, however, due to the high sensitivity requires precautions to avoid contamination and false positives [99,157]. In 2014 a study reliably identified fungal pathogens at the species and subspecies level including but limited to *Aspergillus*, *Scedosporium* spp., *Cyphellophora* spp., *Fusarium* spp. and *Cryptococcus* spp., many of which effect immunocompromised people [99].

Whole genome amplification (WGA) is considered part of the Multiple Displacement Amplification (MDA) methods [97]. WGA is a non-selective amplification method that enables sequencing of DNA from a single cell. This method is easily contaminated by competing DNA contamination or endogenously generated DNA [97]. To date, MDA has been used to amplify less than 10 ng of DNA template routinely creating 10 kb amplicons of *Penicillium paxilli* and the slow growing endophyte *Epichloe festacae* [102,158].

Strand displacement amplification (SDA) is a recently developed technique of WGA that is used to pre-amplify DNA templates before PCR analysis. This method can detect single spores, however, the technique can only amplify DNA not RNA [106]. This method has been tested on single spores of the arbuscular mycorrhizal fungi *Glomus* and *Gigaspora*. The results showed that SDA was able to amplify products 3.8 to 5.4 μg per reaction for the *Glomus* spp. and 5.8 μg for *Gigaspora. rosea* [106].

Helicase-Dependent Amplification (HDA) is similar to PCR but does not require a heat denaturation step, instead uses a single temperature from start to finish [83]. To achieve the separation of single-stranded DNA it uses helicase, polymerases and other enzymes for amplification [97]. HDA was used to successfully detect the fungal pathogen *M. oryzae* in rice seed and when compared with LAMP assays provided similar sensitivity and specificity, showing no amplification from other fungal pathogens [17].

Recombinase Polymerase Amplification (RPA) reaction works by using enzmyes to open strands of dsDNA and amplifies targets with strand substitution activity at a single temperature, usually between 37–42 °C, within 25 min [110]. RPA uses two primers and a low reaction temperature, making it good for in-field diagnostics with the caveat that it requires a complex composition of enzymes and other additives for its reactions [97]. RPA was used in combination with lateral flow strips to detect *Candida albicans* on the human body with a high specificty and limit of detection of 1 CFU/μL proving to be a rapid and sensitive test [110].

5. New Technologies for Rapid, In-Field Diagnostics

Emerging technologies such as lateral flow devices, rapid antigen tests (RATs), biosensors, multiplex assays and point-of-care diagnostics are revolutionizing identification processes by enabling faster, more accurate, and accessible detection of pathogens and other targets. For any new technology to be effectively deployed for the reliable diagnosis or detection of fungal pathogens in planta, it must meet specific criteria: tests should be highly efficient, affordable and user-friendly [159]. Lateral flow devices have been in use for many years, with well-established applications in various fields. Such diagnostic methods have been successfully developed further in human health, notably with the creation of Rapid Antigen Tests (RATs) during the COVID-19 pandemic that began in December 2019 [159]. As COVID-19 spread rapidly worldwide, numerous countries collaborated on the development of diagnostic tools, including RATs, which detect viral proteins from live SARS-CoV-2 virus particles [159]. A lateral flow assay is a paper-based technique used for the detection and quantification of specific antigens or antibodies [160]. It works by binding gold nanoparticles to form an antibody-antigen complex that moves along a test strip, where it binds to additional antibodies at the test line [160]. The accumulation of nanoparticles at the test line produces a visible fluorescent or coloured line [81,160]. The pregnancy test is one of the most common lateral flow assays and has been in use for decades [160]. Despite criticisms regarding RATs' sensitivity and susceptibility to false positives, they remain simple, user-friendly devices applicable across various sectors, including healthcare, agriculture, food safety and environmental monitoring [160]. In the realm of fungal detection, lateral flow devices are primarily used for medically important fungal pathogens, such as the *Aspergillus galactomannan* and *Scedosporium* species [79,81]. Companies such as Pocket Diagnostic (https://www.pocketdiagnostic.com/, accessed 25 October 2024) and AgDia (https://www.agdia.com/, accessed 25 October 2024) have developed lateral flow devices for the detection of the fungal pathogen *Phytophthora*, however, review articles of the success have been unable to be sourced [161,162].

Another area gaining traction in molecular diagnostics are biosensors [163]. A biosensor is a device which analyses and converts a biological response into a processable signal [164]. A biosensor takes samples such as human, food, cell cultures and environmental samples, incorporates transducers such as bioreceptors (enzymes, antibodies and nucleic acids) and combines them with an electrical interface such as nanoparticles or electrodes, which then converts into an electronic system [163]. Most of the work in the area has been for the benefit of human health in the early detection of many ailments including cancer, neurodegenerative disorders and viral infections [165]. There are four classifications of biosensors depending on the signal system or transducer; these are electrochemical, optical, piezoelectric or thermometric [84]. Electrochemical biosensors combined with the use of fluorescence imaging has proven a viable option for fungal detection as was proven by [166] who were able to detect *Yarrowia lipolytica*, a type of yeast in fuel [167,168]. Electrochemical biosensors have the advantage of a low limit of detection, high specificity and ease of use and ability to produce devices for POC field-based analysis [168]. Imaging for fungal pathogens is a non-invasive plant disease detection method and when combined with the low limit of detection of the electrochemical biosensors has the potential to discover the presence of fungal infections before disease symptoms manifest in the host [169]. Currently there is limited information on the combination on electrochemical biosensors and imaging for fungal detection.

The next generation of disease detection will likely involve real-time detection of multiple fungal species simultaneously, referred to as multiplexing [73]. The technology is based on using one reaction with multiple primer pairs to simultaneously amplify multiple pathogens [74]. Visualisation of multiplex PCR currently requires electrophoresis that generate the amplicons by separating the bands in the same column [74]. The current issue with multiplexing PCR is decreased sensitivity [74]. The most common form of multiplexing is through PCR however advancement with LAMP has proven successful in identifying multiple pathogens [170]. The multiplex LAMP (mLAMP) has been used in the detection of *Pyricularia oryzae* and *Triticum* species. These fungal pathogens cause blast diseases in rice and cereals and showed similar results when the mLAMP was compared to the individual LAMP analysis [170]. To utilise LAMP as a multiplexing technique requires the introduction of an endonuclease recognition site to the LAMP primers that generate amplicons that are specific to the target pathogen [170]. Current methods under investigation for mLAMP are approaches to visualise the result. Currently, there are two options: gel electrophoresis and the addition of post-amplification dyes [170]. Further developments in colorimetric LAMP assays is the hue-saturation-value (HSV) colour model. HSV is a variation of the Red, Green and Blue (RGB) colour model, and describes each colour in terms of its shade (hue and saturation) and brightness (value) [155]. This is an image processing quantitative analysis method that can be graphed using the results from the colorimetric assay and a mobile phone [155].

The point of the care diagnostics area has enormous potential to advanced molecular diagnostics. The agriculture industry would benefit from further development in the areas of field-based multiplex assays for the pathogen detection, and even add the potential for fungicide resistance testing [83]. This, coupled with eDNA sampling and swab-based methods where growers can conduct a very quick and cheap on the spot test, would revolutionise the agriculture industries fight against fungal pathogens.

Author Contributions: All authors contributed to writing the manuscript and preparation of the figure and table. A.C. wrote the majority of the manuscript. K.J., P.D. and T.B. performed a detailed review and revision of the manuscript. All authors have read and agreed to the published version of the manuscript.

Funding: This work was supported by the Australian Research Council (ARC) Research Hub for Medicinal Agriculture (IH180100006), with funding provided to KJ and TB. AC is supported by a La Trobe University ARC Research Hub for Medicinal Agriculture Graduate Research Scholarship. Cann Group Limited is a partner organization of the ARC Research Hub for Medicinal Agriculture.

Acknowledgments: We thank Zahir shah safari (Leibniz University Hannover), Marlo Molinaro (La Trobe University), CABI Plantwise Plus website and Schmidt Dávid for the use of their photos in this review. We thank La Trobe University Industry Partner for the photo of a fungal infected Cannabis leaf. Niloofar Vaghefi (University of Melbourne) for supplying the *Pythium* photo. We acknowledge the use of images under a creative commons license (*CC BY-NC-ND 4.0*). Credit is given to the original photographers and sources where applicable.

Conflicts of Interest: The authors declare no conflicts of interest.

References

1. Hawksworth, D.; Lücking, R. Fungal Diversity Revisited: 2.2 to 3.8 Million Species. *Microbiol. Spectr.* **2017**, *5*. [CrossRef] [PubMed]
2. Zhang, Y.J.; Zhang, S.; Liu, X.Z.; Wen, H.A.; Wang, M. A simple method of genomic DNA extraction suitable for analysis of bulk fungal strains. *Lett. Appl. Microbiol.* **2010**, *51*, 114–118. [CrossRef] [PubMed]
3. Debegnach, F.; Patriarca, S.; Brera, C.; Gregori, E.; Sonego, E.; Moracci, G.; De Santis, B. Ergot Alkaloids in Wheat and Rye Derived Products in Italy. *Foods* **2019**, *8*, 150. [CrossRef] [PubMed]
4. Dayarathne, M.; Mridha, A.; Wang, Y. Diagnosis of Fungal Plant Pathogens Using Conventional and Molecular Approaches. In *Diagnostics of Plant Diseases*; IntechOpen: London, UK, 2021.
5. Figueroa, M.; Hammond-Kosack, K.E.; Solomon, P.S. A review of wheat diseases-a field perspective. *Mol. Plant Pathol.* **2018**, *19*, 1523–1536. [CrossRef]
6. Chen, X. Epidemiology and control of stripe rust [*Puccinia striiformis* f. sp *tritici*] on wheat. *Can. J. Plant Pathol.* **2005**, *27*, 314–337. [CrossRef]
7. Addai, D.; Hafi, A.; Randall, L.; Tennant, P.; Arthur, T.; Gomboso, J. *Potential Economic Impacts of the Wheat Stem Rust Strain Ug99 in Australia*; Research Report 18.9; Australian Government: Canberra, Australia, 2018.
8. Huerta-Espino, J.; Singh, R.; German, S.; McCallum, B.; Park, R.; Chen, W.Q.; Bhardwaj, S.; Goyeau, H. Global status of wheat leaf rust caused by *Puccinia triticina*. *Euphytica* **2011**, *179*, 143–160. [CrossRef]
9. Rossman, A.Y. The impact of invasive fungi on agricultural ecosystems in the United States. In *Ecological Impacts of Non-Native Invertebrates and Fungi on Terrestrial Ecosystems*; Langor, D.W., Sweeney, J., Eds.; Springer: Dordrecht, The Netherlands, 2009; pp. 97–107.
10. Benke, K.; Tomkins, B. Future food-production systems: Vertical farming and controlled-environment agriculture. *Sustain. Sci. Pract. Policy* **2017**, *13*, 13–26. [CrossRef]
11. Adenaeuer, L. Up, Up and Away! The Economics of Vertical Farming. *J. Agric. Stud.* **2014**, *2*, 40–60. [CrossRef]
12. Connellan, G. Selection of greenhouse design and technology options for high temperature regions. *Acta Hortic.* **2002**, *578*, 113–117. [CrossRef]
13. Tesoriero, L. *Effective Management of Root Diseases in Hydroponic Lettuce*; Austrlian Governemnt: Sydney, NSW, Australia, 2008.
14. Frew, A.; Antunes, P.M.; Cameron, D.D.; Hartley, S.E.; Johnson, S.N.; Rillig, M.C.; Bennett, A.E. Plant herbivore protection by arbuscular mycorrhizas: A role for fungal diversity? *New Phytol.* **2021**, *233*, 1022–1031. [CrossRef]
15. Rivarez, M.P.S.; Vučurović, A.; Mehle, N.; Ravnikar, M.; Kutnjak, D. Global Advances in Tomato Virome Research: Current Status and the Impact of High-Throughput Sequencing. *Front. Microbiol.* **2021**, *12*, 671925. [CrossRef] [PubMed]
16. Vatistas, C.; Avgoustaki, D.D.; Bartzanas, T. A Systematic Literature Review on Controlled-Environment Agriculture: How Vertical Farms and Greenhouses Can Influence the Sustainability and Footprint of Urban Microclimate with Local Food Production. *Atmosphere* **2022**, *13*, 1258. [CrossRef]
17. Prasannakumar, M.K.; Parivallal, P.B.; Pramesh, D.; Mahesh, H.B.; Raj, E. LAMP-based foldable microdevice platform for the rapid detection of *Magnaporthe oryzae* and *Sarocladium oryzae* in rice seed. *Sci. Rep.* **2021**, *11*, 178. [CrossRef] [PubMed]
18. Devanna, B.N.; Jain, P.; Solanke, A.U.; Das, A.; Thakur, S.; Singh, P.K.; Kumari, M.; Dubey, H.; Jaswal, R.; Pawar, D.; et al. Understanding the Dynamics of Blast Resistance in Rice-Magnaporthe oryzae Interactions. *J. Fungi* **2022**, *8*, 584. [CrossRef]
19. Jerushalmi, S.; Maymon, M.; Dombrovsky, A.; Freeman, S. Fungal Pathogens Affecting the Production and Quality of Medical Cannabis in Israel. *Plants* **2020**, *9*, 882. [CrossRef]
20. Ghosh, S.K.; Banerjee, S.; Pal, S.; Chakraborty, N. Encountering epidemic effects of leaf spot disease (*Alternaria brassicae*) on *Aloe vera* by fungal biocontrol agents in agrifields-An ecofriendly approach. *PLoS ONE* **2018**, *13*, e0193720. [CrossRef]

21. Punja, Z.K.; Collyer, D.; Scott, C.; Lung, S.; Holmes, J.; Sutton, D. Pathogens and Molds Affecting Production and Quality of *Cannabis sativa* L. *Front. Plant Sci.* **2019**, *10*, 1120. [CrossRef]
22. Mekapogu, M.; Jung, J.A.; Kwon, O.K.; Ahn, M.S.; Song, H.Y.; Jang, S. Recent Progress in Enhancing Fungal Disease Resistance in Ornamental Plants. *Int. J. Mol. Sci.* **2021**, *22*, 7956. [CrossRef]
23. Arvanitis, M.; Anagnostou, T.; Fuchs, B.B.; Caliendo, A.M.; Mylonakis, E. Molecular and nonmolecular diagnostic methods for invasive fungal infections. *Clin. Microbiol. Rev.* **2014**, *27*, 490–526. [CrossRef]
24. Vashist, S.K. Point-of-Care Diagnostics: Recent Advances and Trends. *Biosensors* **2017**, *7*, 62. [CrossRef]
25. Luethy, P.M. Point-of-Care Testing for the Diagnosis of Fungal Infections: Current Testing Applications and Potential for the Future. *Clin. Lab. Med.* **2023**, *43*, 209–220. [CrossRef] [PubMed]
26. Punja, Z.K. Emerging diseases of *Cannabis sativa* and sustainable management. *Pest. Manag. Sci.* **2021**, *77*, 3857–3870. [CrossRef] [PubMed]
27. Aranyi, N.R.K.; Varga, I.; Poczai, P.T.; CernΓŸk, I.; Vida, G.; MolnΓŸr-LΓŸng, M.; Hoffmann, B. What types of powdery mildew can infect wheat-barley introgression lines? *Eur. J. Plant Pathol.* **2014**, *139*, 19–25. [CrossRef]
28. Malarczyk, D.; Panek, J.; Frąc, M. Alternative Molecular-Based Diagnostic Methods of Plant Pathogenic Fungi Affecting Berry Crops—A Review. *Molecules* **2019**, *24*, 1200. [CrossRef]
29. Escrivá, L.; Font, G.; Manyes, L. In vivo toxicity studies of fusarium mycotoxins in the last decade: A review. *Food Chem. Toxicol.* **2015**, *78*, 185–206. [CrossRef]
30. Safari, Z.S.; Ding, P.; Nakasha, J.J.; Yusoff, S.F. Controlling Fusarium oxysporum Tomato Fruit Rot under Tropical Condition Using Both Chitosan and Vanillin. *Coatings* **2021**, *11*, 367. [CrossRef]
31. Valero-Jiménez, C.A.; Veloso, J.; Staats, M.; van Kan, J.A.L. Comparative genomics of plant pathogenic Botrytis species with distinct host specificity. *BMC Genom.* **2019**, *20*, 203. [CrossRef]
32. Cheung, N.; Tian, L.; Liu, X.; Li, X. The Destructive Fungal Pathogen Botrytis cinerea—Insights from Genes Studied with Mutant Analysis. *Pathogens* **2020**, *9*, 923. [CrossRef]
33. Tomlinson, J.A.; Dickinson, M.J.; Boonham, N. Detection of Botrytis cinerea by loop-mediated isothermal amplification. *Lett. Appl. Microbiol.* **2010**, *51*, 650–657. [CrossRef]
34. Petrasch, S.; Knapp, S.; Kan, J.; Blanco-Ulate, B. Grey mould of strawberry, a devastating disease caused by the ubiquitous necrotrophic fungal pathogen Botrytis cinerea. *Mol. Plant Pathol.* **2019**, *20*, 877–892. [CrossRef]
35. Mehrotra, R.S.; Aneja, K.R. *An Introduction to Mycology*; New Age International Limited: New Delhi, India, 1990.
36. Suarez, M.B.; Walsh, K.; Boonham, N.; O'Neill, T.; Pearson, S.; Barker, I. Development of real-time PCR (TaqMan) assays for the detection and quantification of *Botrytis cinerea* in planta. *Plant Physiol. Biochem.* **2005**, *43*, 890–899. [CrossRef] [PubMed]
37. McPartland, J.M. *Hemp Diseases and Pests: Management and Biological Control: An Advanced Treatise*; Wallingford: CABI Pub: Wallingford, UK, 2000.
38. Uysal-Morca, A.; Kinay-Teksür, P.; Egercï, Y. Morphological and phylogenetic identification of Botrytis cinerea causing blossom blight and fruit rot of sweet cherries in Aegean region, Turkey. *J. Plant Dis. Prot.* **2021**, *128*, 1051–1060. [CrossRef]
39. van Kan, J.A.; Shaw, M.W.; Grant-Downton, R.T. Botrytis species: Relentless necrotrophic thugs or endophytes gone rogue? *Mol. Plant Pathol.* **2014**, *15*, 957–961. [CrossRef] [PubMed]
40. Kiss, L.; Vaghefi, N.; Bransgrove, K.; Dearnaley, J.D.W.; Takamatsu, S.; Tan, Y.P.; Marston, C.; Liu, S.-Y.; Jin, D.-N.; Adorada, D.L.; et al. Australia: A Continent Without Native Powdery Mildews? The First Comprehensive Catalog Indicates Recent Introductions and Multiple Host Range Expansion Events, and Leads to the Re-discovery of Salmonomyces as a New Lineage of the Erysiphales. *Front. Microbiol.* **2020**, *11*, 1571. [CrossRef]
41. Braun, U. A monograph of the Erysiphales (*Powdery mildews*). *Beih. Zur. Nova Hedwig.* **1987**, *89*, 629–653.
42. Cunnington, J.H.; Lawrie, A.C.; Pascoe, I.G. Genetic characterization of the Golovinomyces cichoracearum complex in Australia. *Plant Pathol.* **2010**, *59*, 158–164. [CrossRef]
43. Wiseman, M.S.; Bates, T.; Garfinkel, A.; Ocamb, C.M.; Gent, D.H. First report of Powdery Mildew Caused by *Golovinomyces ambrosiae* on *Cannabis sativa* in Oregon. *Plant Dis.* **2021**, *105*, 2733. [CrossRef]
44. Cho, S.-E.; Park, M.-J.; Park, J.-H.; Han, K.-S.; Shin, H.-D. First report of *Golovinomyces cichoracearum* associated with powdery mildew on *Helianthus tuberosus* in Korea. *Australas. Plant Dis. Notes* **2012**, *7*, 35–37. [CrossRef]
45. Feechan, A.; Kabbara, S.; Dry, I.B. Mechanisms of powdery mildew resistance in the Vitaceae family. *Mol. Plant Pathol.* **2011**, *12*, 263–274. [CrossRef]
46. Koh, S.; André, A.; Edwards, H.; Ehrhardt, D.; Somerville, S. Arabidopsis thaliana subcellular responses to compatible *Erysiphe cichoracearum* infections. *Plant J.* **2005**, *44*, 516–529. [CrossRef]
47. Benavent-Celma, C.; Puertolas, A.; McLaggan, D.; van West, P.; Woodward, S. Pathogenicity and Host Range of Pythium kashmirense-A Soil-Borne Oomycete Recently Discovered in the UK. *J. Fungi* **2021**, *7*, 479. [CrossRef] [PubMed]
48. Pringsheim, N. *Beiträge zur Morphologie der Meeres-Algen*; Druckerei der Königlichen Akademie der Wissenschaften: Berlin, Germany, 1862.
49. Fukuta, S.; Takahashi, R.; Kuroyanagi, S.; Ishiguro, Y.; Miyake, N.; Nagai, H.; Suzuki, H.; Tsuji, T.; Hashizume, F.; Watanabe, H.; et al. Development of loop-mediated isothermal amplification assay for the detection of *Pythium myriotylum*. *Lett. Appl. Microbiol.* **2014**, *59*, 49–57. [CrossRef] [PubMed]

50. Fukuta, S.; Takahashi, R.; Kuroyanagi, S.; Miyake, N.; Nagai, H.; Suzuki, H.; Hashizume, F.; Tsuji, T.; Taguchi, H.; Watanabe, H.; et al. Detection of *Pythium aphanidermatum* in tomato using loop-mediated isothermal amplification (LAMP) with species-specific primers. *Eur. J. Plant Pathol.* **2013**, *136*, 689–701. [CrossRef]
51. Martin, F.N.; Loper, J.E. Soilborne Plant Diseases Caused by *Pythium* spp.: Ecology, Epidemiology, and Prospects for Biological Control. *Crit. Rev. Plant Sci.* **1999**, *18*, 111–181. [CrossRef]
52. Arora, H.; Sharma, A.; Sharma, S.; Haron, F.F.; Gafur, A.; Sayyed, R.Z.; Datta, R. Pythium Damping-Off and Root Rot of *Capsicum annuum* L.: Impacts, Diagnosis, and Management. *Microorganisms* **2021**, *9*, 823. [CrossRef]
53. Cloyd, R.A. Ecology of Fungus Gnats (*Bradysia* spp.) in Greenhouse Production Systems Associated with Disease-Interactions and Alternative Management Strategies. *Insects* **2015**, *6*, 325–332. [CrossRef]
54. Feng, H.; Ye, W.; Liu, Z.; Wang, Y.; Chen, J.; Wang, Y.; Zheng, X. Development of LAMP Assays Using a Novel Target Gene for Specific Detection of *Pythium terrestris*, *Pythium spinosum*, and 'Candidatus Pythium huanghuaiense'. *Plant Dis.* **2021**, *105*, 2888–2897. [CrossRef]
55. Chun, S.C.; Schneider, R.W. Sites of infection by pythium species in rice seedlings and effects of plant age and water depth on disease development. *Phytopathology* **1998**, *88*, 1255–1261. [CrossRef]
56. Handoko, R.N.S.; Tu, C.K.; Ou, J.H.; Chen, Y.N.; Lee, M.H. First report of Pythium aristosporum causing root rot on rice seedling in Taiwan. *Plant Dis.* **2022**, *107*, 236. [CrossRef]
57. Garfinkel, A. First report of *Sclerotinia sclerotiorum* causing stem canker on *Cannabis sativa* L. in Oregon. *Plant Dis.* **2021**, *105*, 2245. [CrossRef]
58. O'Sullivan, C.A.; Belt, K.; Thatcher, L.F. Tackling Control of a Cosmopolitan Phytopathogen: Sclerotinia. *Front. Plant Sci.* **2021**, *12*, 707509. [CrossRef] [PubMed]
59. Duan, Y.; Ge, C.; Zhang, X.; Wang, J.; Zhou, M. A rapid detection method for the plant pathogen *Sclerotinia sclerotiorum* based on loop-mediated isothermal amplification (LAMP). *Australas. Plant Pathol.* **2014**, *43*, 61–66. [CrossRef]
60. Bolton, M.D.; Thomma, B.P.H.J.; Nelson, B.D. *Sclerotinia sclerotiorum* (Lib.) de Bary: Biology and molecular traits of a cosmopolitan pathogen. *Mol. Plant Pathol.* **2006**, *7*, 1–16. [CrossRef] [PubMed]
61. Hossain, M.M.; Sultana, F.; Li, W.; Tran, L.P.; Mostofa, M.G. *Sclerotinia sclerotiorum* (Lib.) de Bary: Insights into the Pathogenomic Features of a Global Pathogen. *Cells* **2023**, *12*, 1063. [CrossRef]
62. Purdy, L.H. *Sclerotinia sclerotiorum*: History, diseases and symptomatology, host range, geographic distribution, and impact. *Phytopathology* **1979**, *69*, 875–880. [CrossRef]
63. Koledenkova, K.; Esmaeel, Q.; Jacquard, C.; Nowak, J.; Clément, C.; Ait Barka, E. Plasmopara viticola the Causal Agent of Downy Mildew of Grapevine: From Its Taxonomy to Disease Management. *Front. Microbiol.* **2022**, *13*, 889472. [CrossRef]
64. Toffolatti, S.L.; Russo, G.; Campia, P.; Bianco, P.A.; Borsa, P.; Coatti, M.; Torriani, S.F.; Sierotzki, H. A time-course investigation of resistance to the carboxylic acid amide mandipropamid in field populations of *Plasmopara viticola* treated with anti-resistance strategies. *Pest. Manag. Sci.* **2018**, *74*, 2822–2834. [CrossRef]
65. Beccari, G.; Hao, G.; Liu, H. Editorial: *Fusarium pathogenesis*: Infection mechanisms and disease progression in host plants. *Front. Plant Sci.* **2022**, *13*, 1020404. [CrossRef]
66. Arie, T. Fusarium diseases of cultivated plants, control, diagnosis, and molecular and genetic studies. *J. Pestic. Sci.* **2019**, *44*, 275–281. [CrossRef]
67. Diaz, J.; Garcia, J.; Lara, C.; Hutmacher, R.B.; Ulloa, M.; Nichols, R.L.; Ellis, M.L. Characterization of Current *Fusarium oxysporum* f. sp. *vasinfectum* Isolates from Cotton in the San Joaquin Valley of California and Lower Valley El Paso, Texas. *Plant Dis.* **2021**, *105*, 1898–1911. [CrossRef]
68. Wang, L.; Calabria, J.; Chen, H.-W.; Somssich, M. The Arabidopsis thaliana–*Fusarium oxysporum* strain 5176 pathosystem: An overview. *J. Exp. Bot.* **2022**, *73*, 6052–6067. [CrossRef] [PubMed]
69. Szymańska, R.; Gabruk, M.; Kruk, J. Arabidopsis thaliana accessions—A tool for biochemical and phylogentical studies. *Postep. Biochem.* **2015**, *61*, 102–113.
70. Koch, R.I. Die Aetiologie der Tuberculose: Nach einem in der physiologischen Gesellschaft zu Berlin am 24. März cr. gehaltenen Vortrage. *Zentralblatt Bakteriol. Mikrobiol. Hygiene. 1 Abt. Originale. A Med. Mikrobiol. Infekt. Parasitol.* **1982**, *251*, 287–296. [CrossRef]
71. Walker, L.; Levine, H.; Jucker, M. Koch's postulates and infectious proteins. *Acta Neuropathol.* **2006**, *112*, 1–4. [CrossRef] [PubMed]
72. Ainsworth, G.C. Observations, Botanical and Physiological, on the Potato Murrain, Together with Selections from Berkeley's Vegetable Pathology. Made by the Plant Pathology Committee of the British Mycological Society. Phytopathological Classics, No. 8, American Phytopathological Society, East Lansing, Mich. (1948) 104 pp., 6 plates, 1.50 dollars. *Trans. Br. Mycol. Soc.* **1950**, *33*, 192–193. [CrossRef]
73. Bernreiter, A. Molecular diagnostics to identify fungal plant pathogens—a review of current methods. *Ecuad. Es Calid. Rev. Científica Ecuat.* **2017**, *4*, 26–35.
74. Hariharan, G.; Prasannath, K. Recent Advances in Molecular Diagnostics of Fungal Plant Pathogens: A Mini Review. *Front. Cell. Infect. Microbiol.* **2021**, *10*, 600234. [CrossRef]
75. Yoshida, K.; Schuenemann, V.J.; Cano, L.M.; Pais, M.; Mishra, B.; Sharma, R.; Lanz, C.; Martin, F.N.; Kamoun, S.; Krause, J.; et al. The rise and fall of the Phytophthora infestans lineage that triggered the Irish potato famine. *Elife* **2013**, *2*, e00731. [CrossRef]
76. Birch, P.R.; Cooke, D.E. The early days of late blight. *Elife* **2013**, *2*, e00954. [CrossRef]

77. Scholthof, K.B. The disease triangle: Pathogens, the environment and society. *Nat. Rev. Microbiol.* **2007**, *5*, 152–156. [CrossRef]
78. Lücking, R.; Aime, M.C.; Robbertse, B.; Miller, A.N.; Ariyawansa, H.A.; Aoki, T.; Cardinali, G.; Crous, P.W.; Druzhinina, I.S.; Geiser, D.M.; et al. Unambiguous identification of fungi: Where do we stand and how accurate and precise is fungal DNA barcoding? *IMA Fungus* **2020**, *11*, 14. [CrossRef]
79. Davies, G.E.; Thornton, C.R. A Lateral-Flow Device for the Rapid Detection of Scedosporium Species. *Diagnostics* **2024**, *14*, 847. [CrossRef]
80. Manessis, G.; Gelasakis, A.I.; Bossis, I. Point-of-Care Diagnostics for Farm Animal Diseases: From Biosensors to Integrated Lab-on-Chip Devices. *Biosensors* **2022**, *12*, 455. [CrossRef] [PubMed]
81. Jenks, J.D.; Miceli, M.H.; Prattes, J.; Mercier, T.; Hoenigl, M. The Aspergillus Lateral Flow Assay for the Diagnosis of Invasive Aspergillosis: An Update. *Curr. Fungal Infect. Rep.* **2020**, *14*, 378–383. [CrossRef] [PubMed]
82. Heidt, B.; Siqueira, W.F.; Eersels, K.; Diliën, H.; van Grinsven, B.; Fujiwara, R.T.; Cleij, T.J. Point of Care Diagnostics in Resource-Limited Settings: A Review of the Present and Future of PoC in Its Most Needed Environment. *Biosensors* **2020**, *10*, 133. [CrossRef] [PubMed]
83. Lau, H.Y.; Botella, J.R. Advanced DNA-Based Point-of-Care Diagnostic Methods for Plant Diseases Detection. *Front. Plant Sci.* **2017**, *8*, 2016. [CrossRef] [PubMed]
84. Hussain, K.K.; Malavia, D.; Johnson, E.M.; Littlechild, J.; Winlove, C.P.; Vollmer, F.; Gow, N.A.R. Biosensors and Diagnostics for Fungal Detection. *J. Fungi* **2020**, *6*, 349. [CrossRef]
85. Griesche, C.; Baeumner, A.J. Biosensors to support sustainable agriculture and food safety. *TrAC Trends Anal. Chem.* **2020**, *128*, 115906. [CrossRef]
86. John, M.A.; Bankole, I.; Ajayi-Moses, O.; Ijila, T.; Jeje, O.; Lalit, P. Relevance of advanced plant disease detection techniques in disease and Pest Management for Ensuring Food Security and Their Implication: A review. *Am. J. Plant Sci.* **2023**, *14*, 1260–1295. [CrossRef]
87. Venbrux, M.; Crauwels, S.; Rediers, H. Current and emerging trends in techniques for plant pathogen detection. *Front. Plant Sci.* **2023**, *14*, 1120968. [CrossRef]
88. Brown, L.; Premaratna, D.; Segal, Y.; Beddoe, T. Air sampling for detection of infectious laryngotracheitis (ILT) in commercial poultry flocks. *BMC Res. Notes* **2020**, *13*, 556. [CrossRef] [PubMed]
89. Buttner, M.P.; Stetzenbach, L.D. Monitoring airborne fungal spores in an experimental indoor environment to evaluate sampling methods and the effects of human activity on air sampling. *Appl. Env. Microbiol.* **1993**, *59*, 219–226. [CrossRef] [PubMed]
90. Thiessen, L.; Keune, J.; Neill, T.; Turechek, W.; Grove, G.; Mahaffee, W. Development of a grower-conducted inoculum detection assay for management of grape powdery mildew. *Plant Pathol.* **2015**, *65*, 238–249. [CrossRef]
91. Bessey, C.; Neil Jarman, S.; Simpson, T.; Miller, H.; Stewart, T.; Kenneth Keesing, J.; Berry, O. Passive eDNA collection enhances aquatic biodiversity analysis. *Commun. Biol.* **2021**, *4*, 236. [CrossRef] [PubMed]
92. Lynggaard, C.; Bertelsen, M.F.; Jensen, C.V.; Johnson, M.S.; Frøslev, T.G.; Olsen, M.T.; Bohmann, K. Airborne environmental DNA for terrestrial vertebrate community monitoring. *Curr. Biol.* **2022**, *32*, 701–707.e705. [CrossRef]
93. Sivasubramani, S.K.; Niemeier, R.T.; Reponen, T.; Grinshpun, S.A. Fungal spore source strength tester: Laboratory evaluation of a new concept. *Sci. Total Environ.* **2004**, *329*, 75–86. [CrossRef]
94. Fabian, M.P.; Miller, S.L.; Reponen, T.; Hernandez, M.T. Ambient bioaerosol indices for indoor air quality assessments of flood reclamation. *J. Aerosol Sci.* **2005**, *36*, 763–783. [CrossRef]
95. Peng, J.; Zhang, H.; Chen, F.; Zhang, X.; Xie, Y.; Hou, X.; Li, G.; Pu, J. Rapid and quantitative detection of *Fusarium oxysporum* f. sp. cubense race 4 in soil by real-time fluorescence loop-mediated isothermal amplification. *J. Appl. Microbiol.* **2014**, *117*, 1740–1749. [CrossRef]
96. Duan, Y.B.; Ge, C.Y.; Zhang, X.K.; Wang, J.X.; Zhou, M.G. Development and evaluation of a novel and rapid detection assay for *Botrytis cinerea* based on loop-mediated isothermal amplification. *PLoS ONE* **2014**, *9*, e111094. [CrossRef]
97. Zanoli, L.M.; Spoto, G. Isothermal amplification methods for the detection of nucleic acids in microfluidic devices. *Biosensors* **2013**, *3*, 18–43. [CrossRef]
98. Wang, Y.; Yang, M.; Xia, Y.; Yan, J.; Zou, J.; Zhang, D. Application and evaluation of nucleic acid sequence-based amplification, PCR and cryptococcal antigen test for diagnosis of cryptococcosis. *BMC Infect. Dis.* **2021**, *21*, 1020. [CrossRef] [PubMed]
99. Javaheri Tehrani, S.; Aliabadian, M.; Fata, A.; Najafzadeh, M.J. Rolling Circle Amplification (RCA): An approach for quick detection and identification of fungal species. *J. Mycol. Res.* **2014**, *1*, 55–62.
100. Bhat, A.I.; Rao, G.P. Rolling Circle Amplification (RCA). In *Characterization of Plant Viruses: Methods and Protocols*; Springer: New York, NY, USA, 2020; pp. 377–381.
101. Soares, R.R.G.; Madaboosi, N.; Nilsson, M. Rolling Circle Amplification in Integrated Microsystems: An Uncut Gem toward Massively Multiplexed Pathogen Diagnostics and Genotyping. *Acc. Chem. Res.* **2021**, *54*, 3979–3990. [CrossRef] [PubMed]
102. Foster, S.J.; Monahan, B.J. Whole genome amplification from filamentous fungi using Phi29-mediated multiple displacement amplification. *Fungal Genet. Biol.* **2005**, *42*, 367–375. [CrossRef]
103. Ahsanuddin, S.; Afshinnekoo, E.; Gandara, J.; Hakyemezoğlu, M.; Bezdan, D.; Minot, S.; Greenfield, N.; Mason, C.E. Assessment of REPLI-g Multiple Displacement Whole Genome Amplification (WGA) Techniques for Metagenomic Applications. *J. Biomol. Tech.* **2017**, *28*, 46–55. [CrossRef]

104. Sridapan, T.; Krajaejun, T. Nucleic Acid-Based Detection of *Pythium insidiosum*: A Systematic Review. *J. Fungi* **2023**, *9*, 27. [CrossRef]
105. Fernández-Ortuño, D.; Torés, J.A.; de Vicente, A.; Pérez-García, A. Multiple displacement amplification, a powerful tool for molecular genetic analysis of powdery mildew fungi. *Curr. Genet.* **2007**, *51*, 209–219. [CrossRef]
106. Gadkar, V.; Rillig, M.C. Application of Phi29 DNA polymerase mediated whole genome amplification on single spores of arbuscular mycorrhizal (AM) fungi. *FEMS Microbiol. Lett.* **2005**, *242*, 65–71. [CrossRef]
107. Glökler, J.; Lim, T.S.; Ida, J.; Frohme, M. Isothermal amplifications—A comprehensive review on current methods. *Crit. Rev. Biochem. Mol. Biol.* **2021**, *56*, 543–586. [CrossRef]
108. Silva Zatti, M.; Domingos Arantes, T.; Cordeiro Theodoro, R. Isothermal nucleic acid amplification techniques for detection and identification of pathogenic fungi: A review. *Mycoses* **2020**, *63*, 1006–1020. [CrossRef]
109. Vincent, M.; Xu, Y.; Kong, H. Helicase-dependent isothermal DNA amplification. *EMBO Rep.* **2004**, *5*, 795–800. [CrossRef] [PubMed]
110. Wang, F.; Ge, D.; Wang, L.; Li, N.; Chen, H.; Zhang, Z.; Zhu, W.; Wang, S.; Liang, W. Rapid and sensitive recombinase polymerase amplification combined with lateral flow strips for detecting *Candida albicans*. *Anal. Biochem.* **2021**, *633*, 114428. [CrossRef] [PubMed]
111. Zhao, W.; Chi, Y.-K.; Ye, M.-D.; Wang, T.; Xu, A.M.; Qi, R.-D. Development and application of Recombinase Polymerase Amplification assay for detection of *Bipolaris sorokiniana*. *Crop Prot.* **2021**, *145*, 105619. [CrossRef]
112. Postma, J.; Willemsen-de Klein, M.J.E.I.M.; van Elsas, J.D. Effect of the Indigenous Microflora on the Development of Root and Crown Rot Caused by *Pythium aphanidermatum* in Cucumber Grown on Rockwool. *Phytopathology®* **2000**, *90*, 125–133. [CrossRef]
113. Ehi-Eromosele, C. *Integrated Pest Management*; Covenant University: Ota, Nigeria, 2013; pp. 105–115.
114. Zubrod, J.P.; Bundschuh, M.; Arts, G.; Brühl, C.A.; Imfeld, G.; Knäbel, A.; Payraudeau, S.; Rasmussen, J.J.; Rohr, J.; Scharmüller, A.; et al. Fungicides: An Overlooked Pesticide Class? *Environ. Sci. Technol.* **2019**, *53*, 3347–3365. [CrossRef]
115. Dias, M.C. Phytotoxicity: An Overview of the Physiological Responses of Plants Exposed to Fungicides. *J. Bot.* **2012**, *2012*, 135479. [CrossRef]
116. Belpoggi, F.; Soffritti, M.; Guarino, M.; Lambertini, L.; Cevolani, D.; Maltoni, C. Results of Long-Term Experimental Studies on the Carcinogenicity of Ethylene-bis-Dithiocarbamate (Mancozeb) in Rats. *Ann. N. Y. Acad. Sci.* **2002**, *982*, 123–136. [CrossRef]
117. Bolanos-Carriel, C.; Wegulo, S.N.; Baenziger, P.S.; Funnell-Harris, D.; Hallen-Adams, H.E.; Eskridge, K.M. Effects of fungicide chemical class, fungicide application timing, and environment on Fusarium head blight in winter wheat. *Eur. J. Plant Pathol.* **2020**, *158*, 667–679. [CrossRef]
118. Corkley, I.; Fraaije, B.; Hawkins, N. Fungicide resistance management: Maximizing the effective life of plant protection products. *Plant Pathol.* **2022**, *71*, 150–169. [CrossRef]
119. Donley, N. The USA lags behind other agricultural nations in banning harmful pesticides. *Environ. Health* **2019**, *18*, 44. [CrossRef]
120. Krais, A.M.; de Joode, B.v.W.; Liljedahl, E.R.; Blomberg, A.J.; Rönnholm, A.; Bengtsson, M.; Cano, J.C.; Hoppin, J.A.; Littorin, M.; Nielsen, C.; et al. Detection of the fungicide transformation product 4-hydroxychlorothalonil in serum of pregnant women from Sweden and Costa Rica. *J. Expo. Sci. Environ. Epidemiol.* **2024**, *34*, 270–277. [CrossRef] [PubMed]
121. Cocco, P. Time for Re-Evaluating the Human Carcinogenicity of Ethylenedithiocarbamate Fungicides? A Systematic Review. *Int. J. Environ. Res. Public. Health* **2022**, *19*, 2632. [CrossRef] [PubMed]
122. Li, J.; Zhou, X.; Zhang, C.; Zhao, Y.; Zhu, Y.; Zhang, J.; Bai, J.; Xiao, X. The Effects of Carbendazim on Acute Toxicity, Development, and Reproduction in Caenorhabditis elegans. *J. Food Qual.* **2020**, *2020*, 8853537. [CrossRef]
123. Sánchez-Torres, P. Molecular Mechanisms Underlying Fungicide Resistance in Citrus Postharvest Green Mold. *J. Fungi* **2021**, *7*, 783. [CrossRef]
124. Hahn, M. The rising threat of fungicide resistance in plant pathogenic fungi: *Botrytis* as a case study. *J. Chem. Biol.* **2014**, *7*, 133–141. [CrossRef]
125. Cook, N.M.; Chng, S.; Woodman, T.L.; Warren, R.; Oliver, R.P.; Saunders, D.G. High frequency of fungicide resistance-associated mutations in the wheat yellow rust pathogen *Puccinia striiformis* f. sp. *tritici*. *Pest. Manag. Sci.* **2021**, *77*, 3358–3371. [CrossRef]
126. Golding, J. Assessing and Managing Resistance to Postharvest Fungicides: Mar 2018. 2018. Australian Government. Available online: https://www.dpi.nsw.gov.au/agriculture/horticulture/citrus/content/post-harvest/assessing-resistance-to-postharvest-fungicides-aug-2017 (accessed on 15 April 2024).
127. Peng, J.; Sang, H.; Proffer, T.J.; Gleason, J.; Outwater, C.A.; Jung, G.; Sundin, G.W. A Method for the Examination of SDHI Fungicide Resistance Mechanisms in Phytopathogenic Fungi Using a Heterologous Expression System in Sclerotinia sclerotiorum. *Phytopathology* **2021**, *111*, 819–830. [CrossRef]
128. Wang, Q.; Mao, Y.; Li, S.; Li, T.; Wang, J.; Zhou, M.; Duan, Y. Molecular Mechanism of Sclerotinia sclerotiorum Resistance to Succinate Dehydrogenase Inhibitor Fungicides. *J. Agric. Food Chem.* **2022**, *70*, 7039–7048. [CrossRef]
129. Lebeda, A.; Mieslerová, B. Taxonomy, distribution and biology of lettuce powdery mildew (*Golovinomyces cichoracearum* sensu stricto). *Plant Pathol.* **2011**, *60*, 400–415. [CrossRef]
130. Choe, H.; Deirmengian, C.A.; Hickok, N.J.; Morrison, T.N.; Tuan, R.S. Molecular diagnostics. *J. Am. Acad. Orthop. Surg.* **2015**, *23*, S26–S31. [CrossRef]
131. Hidayat, T.; Priyandoko, D.; Wardiny, P.; Islami, D. Molecular Phylogenetic Screening of Withania somnifera Relative From Indonesia Based on Internal Transcribed Spacer Region. *HAYATI J. Biosci.* **2016**, *23*, 92–95. [CrossRef]

132. Lücking, R.; Aime, M.C.; Robbertse, B.; Miller, A.N.; Aoki, T.; Ariyawansa, H.A.; Cardinali, G.; Crous, P.W.; Druzhinina, I.S.; Geiser, D.M.; et al. Fungal taxonomy and sequence-based nomenclature. *Nat. Microbiol.* **2021**, *6*, 540–548. [CrossRef] [PubMed]
133. Bruns, T.D.; White, T.J.; Taylor, J.W. Fungal Molecular Systematics. *Annu. Rev. Ecol. Syst.* **1991**, *22*, 525–564. [CrossRef]
134. Schoch, C.L.; Seifert, K.A.; Huhndorf, S.; Robert, V.; Spouge, J.L.; Levesque, C.A.; Chen, W. Nuclear ribosomal internal transcribed spacer (ITS) region as a universal DNA barcode marker for Fungi. *Proc. Natl. Acad. Sci. USA* **2012**, *109*, 6241–6246. [CrossRef]
135. Deagle, B.E.; Jarman, S.N.; Coissac, E.; Pompanon, F.; Taberlet, P. DNA metabarcoding and the cytochrome c oxidase subunit I marker: Not a perfect match. *Biol. Lett.* **2014**, *10*, 20140562. [CrossRef]
136. Tobe, S.S.; Kitchener, A.; Linacre, A. Cytochrome b or cytochrome c oxidase subunit I for mammalian species identification—An answer to the debate. *Forensic Sci. Int. Genet. Suppl. Ser.* **2009**, *2*, 306–307. [CrossRef]
137. Anucherngchai, S.; Chontananarth, T.; Tejangkura, T.; Chai, J.Y. The study of Cytochrome B (CYTB): Species-specific detection and phylogenetic relationship of *Echinostoma revolutum*, (Froelich, 1802). *J. Parasit. Dis.* **2019**, *43*, 66–74. [CrossRef]
138. Kocher, T.D.; Thomas, W.K.; Meyer, A.; Edwards, S.V.; Pääbo, S.; Villablanca, F.X.; Wilson, A.C. Dynamics of mitochondrial DNA evolution in animals: Amplification and sequencing with conserved primers. *Proc. Natl. Acad. Sci. USA* **1989**, *86*, 6196–6200. [CrossRef]
139. Parson, W.; Pegoraro, K.; Niederstätter, H.; Föger, M.; Steinlechner, M. Species identification by means of the cytochrome b gene. *Int. J. Leg. Med.* **2000**, *114*, 23–28. [CrossRef]
140. Huyse, T.; Oeyen, M.; Larmuseau, M.H.D.; Volckaert, F.A.M. Co-phylogeographic study of the flatworm Gyrodactylus gondae and its goby host *Pomatoschistus minutus*. *Parasitol. Int.* **2017**, *66*, 119–125. [CrossRef]
141. Valadas, S.Y.; da Silva, J.I.; Lopes, E.G.; Keid, L.B.; Zwarg, T.; de Oliveira, A.S.; Sanches, T.C.; Joppert, A.M.; Pena, H.F.; Oliveira, T.M.; et al. Diversity of *Sarcocystis* spp shed by opossums in Brazil inferred with phylogenetic analysis of DNA coding ITS1, cytochrome B, and surface antigens. *Exp. Parasitol.* **2016**, *164*, 71–78. [CrossRef] [PubMed]
142. Smith, M.F.; Patton, J.L. Variation in mitochondrial cytochrome b sequence in natural populations of South American akodontine rodents (Muridae: Sigmodontinae). *Mol. Biol. Evol.* **1991**, *8*, 85–103. [CrossRef]
143. Bartlett, S.E.; Davidson, W.S. Identification of Thunnus Tuna Species by the Polymerase Chain Reaction and Direct Sequence Analysis of their Mitochondrial Cytochrome b Genes. *Can. J. Fish. Aquat. Sci.* **1991**, *48*, 309–317. [CrossRef]
144. Ding, J.L.; Li, X.H.; Lei, J.H.; Feng, M.G.; Ying, S.H. Succinate Dehydrogenase Subunit C Contributes to Mycelial Growth and Development, Stress Response, and Virulence in the Insect Parasitic Fungus Beauveria bassiana. *Microbiol. Spectr.* **2022**, *10*, e0289122. [CrossRef] [PubMed]
145. Carr, S.M.; Marshall, H.D. Detection of Intraspecific.DNA Sequence Variation in the Mitochondrial Cytochrome b Gene of Atlantic Cod (*Gadus morhua*) by the Polymerase Chain Reaction. *Can. J. Fish. Aquat. Sci.* **1991**, *48*, 48–52. [CrossRef]
146. Khot, P.D.; Fredricks, D.N. PCR-based diagnosis of human fungal infections. *Expert. Rev. Anti Infect. Ther.* **2009**, *7*, 1201–1221. [CrossRef]
147. Khan, M.; Wang, R.; Li, B.; Liu, P.; Weng, Q.; Chen, Q. Comparative Evaluation of the LAMP Assay and PCR-Based Assays for the Rapid Detection of *Alternaria solani*. *Front. Microbiol.* **2018**, *9*, 2089. [CrossRef]
148. Nancy, K.; Lillian, W.; Wilkinson, M.; Claire, M.; Luna, K.; Dorcas, W.; Rosemary, G.; Japheth, L.; Christine, I.; Emily, W.; et al. Optimization of a Loop-Mediated Isothermal Amplification Assay as a Point-of-Care Tool for the Detection of *Wuchereria bancrofti* in Human Blood in Tana River Delta, Kenya. *J. Parasitol. Res.* **2021**, *2021*, 6650870. [CrossRef]
149. Schena, L.; Nigro, F.; Ippolito, A.; Gallitelli, D. Real-time quantitative PCR: A new technology to detect and study phytopathogenic and antagonistic fungi. *Eur. J. Plant Pathol.* **2004**, *110*, 893–908. [CrossRef]
150. Schnerr, H.; Niessen, L.; Vogel, R.F. Real time detection of the tri5 gene in Fusarium species by LightCycler™-PCR using SYBR®Green I for continuous fluorescence monitoring. *Int. J. Food Microbiol.* **2001**, *71*, 53–61. [CrossRef]
151. Qi, M.; Yang, Y. Quantification of Magnaporthe grisea During Infection of Rice Plants Using Real-Time Polymerase Chain Reaction and Northern Blot/Phosphoimaging Analyses. *Phytopathology* **2002**, *92*, 870–876. [CrossRef] [PubMed]
152. Avrova, A.O.; Venter, E.; Birch, P.R.; Whisson, S.C. Profiling and quantifying differential gene transcription in Phytophthora infestans prior to and during the early stages of potato infection. *Fungal Genet. Biol.* **2003**, *40*, 4–14. [CrossRef] [PubMed]
153. Fraaije, B.A.; Lovell, D.J.; Coelho, J.M.; Baldwin, S.; Hollomon, D.W. PCR-based Assays to Assess Wheat Varietal Resistance to Blotch (Septoria Tritici and Stagonospora Nodorum) and Rust (Puccinia Striiformis and Puccinia Recondita) Diseases. *Eur. J. Plant Pathol.* **2001**, *107*, 905–917. [CrossRef]
154. Soroka, M.; Wasowicz, B.; Rymaszewska, A. Loop-Mediated Isothermal Amplification (LAMP): The Better Sibling of PCR? *Cells* **2021**, *10*, 1931. [CrossRef]
155. Yu, L.S.; Chou, S.Y.; Wu, H.Y.; Chen, Y.C.; Chen, Y.H. Rapid and semi-quantitative colorimetric loop-mediated isothermal amplification detection of ASFV via HSV color model transformation. *J. Microbiol. Immunol. Infect.* **2021**, *54*, 963–970. [CrossRef]
156. Widjojoatmodjo, M.N.; Borst, A.; Schukkink, R.A.F.; Box, A.T.A.; Tacken, N.M.M.; Van Gemen, B.; Verhoef, J.; Top, B.; Fluit, A.C. Nucleic acid sequence-based amplification (NASBA) detection of medically important Candida species. *J. Microbiol. Methods* **1999**, *38*, 81–90. [CrossRef]
157. Qi, X.; Bakht, S.; Devos, K.M.; Gale, M.D.; Osbourn, A. L-RCA (ligation-rolling circle amplification): A general method for genotyping of single nucleotide polymorphisms (SNPs). *Nucleic Acids Res.* **2001**, *29*, E116. [CrossRef]
158. Wang, X.; Liu, Y.; Liu, H.; Pan, W.; Ren, J.; Zheng, X.; Tan, Y.; Chen, Z.; Deng, Y.; He, N.; et al. Recent advances and application of whole genome amplification in molecular diagnosis and medicine. *MedComm* **2022**, *3*, e116. [CrossRef]

159. Arshadi, M.; Fardsanei, F.; Deihim, B.; Farshadzadeh, Z.; Nikkhahi, F.; Khalili, F.; Sotgiu, G.; Shahidi Bonjar, A.H.; Centis, R.; Migliori, G.B.; et al. Diagnostic Accuracy of Rapid Antigen Tests for COVID-19 Detection: A Systematic Review With Meta-analysis. *Front. Med.* **2022**, *9*, 870738. [CrossRef]
160. Koczula, K.M.; Gallotta, A. Lateral flow assays. *Essays Biochem.* **2016**, *60*, 111–120. [CrossRef]
161. AgDia. Agdia ImmunoStrip® for Phytophthora (Phyt). Available online: https://www.agdia.com/ (accessed on 25 October 2024).
162. Diagnostics, P. *Phytophthora* spp. Lateral Flow Test. Available online: https://www.pocketdiagnostic.com/ (accessed on 25 October 2024).
163. Grieshaber, D.; MacKenzie, R.; Vörös, J.; Reimhult, E. Electrochemical Biosensors—Sensor Principles and Architectures. *Sensors* **2008**, *8*, 1400–1458. [CrossRef] [PubMed]
164. Lowe, C.R. Biosensors. *Trends Biotechnol.* **1984**, *2*, 59–65. [CrossRef]
165. Olejnik, B.; Kozioł, A.; Brzozowska, E.; Ferens-Sieczkowska, M. Application of selected biosensor techniques in clinical diagnostics. *Expert. Rev. Mol. Diagn.* **2021**, *21*, 925–937. [CrossRef] [PubMed]
166. Radwan, O.; Brothers, M.C.; Coyle, V.; Chapleau, M.E.; Chapleau, R.R.; Kim, S.S.; Ruiz, O.N. Electrochemical biosensor for rapid detection of fungal contamination in fuel systems. *Biosens. Bioelectron.* **2022**, *211*, 114374. [CrossRef]
167. Mehrotra, P. Biosensors and their applications—A review. *J. Oral. Biol. Craniofac Res.* **2016**, *6*, 153–159. [CrossRef]
168. Lopes, L.C.; Santos, A.; Bueno, P.R. An outlook on electrochemical approaches for molecular diagnostics assays and discussions on the limitations of miniaturized technologies for point-of-care devices. *Sens. Actuators Rep.* **2022**, *4*, 100087. [CrossRef]
169. Sankaran, S.; Mishra, A.; Ehsani, R.; Davis, C. A review of advanced techniques for detecting plant diseases. *Comput. Electron. Agric.* **2010**, *72*, 1–13. [CrossRef]
170. Gomez-Gutierrez, S.V.; Goodwin, S.B. Loop-Mediated Isothermal Amplification for Detection of Plant Pathogens in Wheat (*Triticum aestivum*). *Front. Plant Sci.* **2022**, *13*, 857673. [CrossRef]

Journal of *Fungi*

MDPI

Communication

First Report of *Diaporthe goulteri* on Soybean in Germany

Behnoush Hosseini, Maximilian Gerhard Gröbner and Tobias Immanuel Link *

Department of Phytopathology, Institute of Phytomedicine, Faculty of Agricultural Sciences, University of Hohenheim, 70599 Stuttgart, Germany; behnoush.hosseini@uni-hohenheim.de (B.H.); maximilian.groebner@uni-hohenheim.de (M.G.G.)
* Correspondence: tobias.link@uni-hohenheim.de

Abstract: *Diaporthe* (anamorph: *Phomopsis*) species are endophytes or fungal pathogens for many different plant species. Soybean (*Glycine max*) can be infected by many different *Diaporthe* species; among them, *D. caulivora* and *D. longicolla* are responsible for the most significant damages. *Diaporthe goulteri* is a species that was only recently described and has so far been found on sunflower (*Helianthus annuus*) in Australia and an unknown host in Thailand. Here, we report isolation of *D. goulteri* from soybean in southern Germany, molecular species identification, and additional morphological description. We also show that *D. goulteri* can infect soybean and describe the symptoms we observed, both on the plant where the isolate came from and following artificial inoculation.

Keywords: isolation; species identification; phylogeny; morphology; pathogenicity

1. Introduction

Because of increased demand for soybean (*Glycine max*) as a source for oil and protein, its cultivation is increasing in Germany like in other countries. As soybean cultivation area is on the rise, it is expected that the incidence of soybean pathogens will increase. Species of the fungal genus *Diaporthe* are important pathogens for soybean in all major soybean cultivation areas in North and South America, Asia, and Europe [1–6]. *Diaporthe longicolla* was identified as a seed decay pathogen on German soybean seeds together with *D. caulivora*, *D. eres*, and *D. novem* [7]. Since globally many more *Diaporthe* spp. are reported as soybean pathogens [8], the occurrence of more *Diaporthe* spp. on soybean in Germany was expected.

Genus *Diaporthe* is large and complex, containing more than one hundred species. The phylogeny of the genus has been revised based on sequence comparisons [9–11] and studies concerned with updating the molecular phylogeny of *Diaporthe* regularly discover new species belonging to the genus. *Diaporthe goulteri* is among those species discovered relatively recently: Thompson et al. [12] isolated the species from a sunflower (*Helianthus annuus*) seed, described it, and named it. Apart from this, there are scant reports about the species; there are some isolates from China, Taiwan, and Thailand with corresponding sequences deposited in GenBank, but we found only one publication [13] that provides additional information about the species: Bundhun et al. [13] reported the isolation of *D. goulteri* from "a dead branch of an unknown host" in Thailand and described its sexual morph.

Here, we report isolation of *D. goulteri* from soybean in southern Germany. We provide evidence that soybean is a host plant for *D. goulteri* and expand on the species description.

2. Materials and Methods

2.1. Sampling and Isolation

Soybean plants showing signs of *Diaporthe* infection, i.e., pycnidia or perithecia showing up as black dots on the stem were collected from fields shortly before harvest. Seeds, pods, and stem pieces were disinfected with 3% sodium hypochlorite solution for 1 min, rinsed three times with sterile water, dried on filter paper, and placed on Petri plates with

Citation: Hosseini, B.; Gröbner, M.G.; Link, T.I. First Report of *Diaporthe goulteri* on Soybean in Germany. *J. Fungi* **2024**, *10*, 803. https://doi.org/10.3390/jof10110803

Academic Editor: Ofir Degani

Received: 11 October 2024
Revised: 12 November 2024
Accepted: 16 November 2024
Published: 20 November 2024

potato dextrose agar (PDA). These plates were incubated at room temperature for one week. After seven days, cultures, which were tentatively identified as *Diaporthe* sp. based on culture morphology, were subcultured by transferring an agar plug with a small piece of the cultures to fresh PDA plates that were also incubated at room temperature.

2.2. Species Identification

2.2.1. Species Identification from Pure Cultures Using Culture Morphology and Microscopy as Well as Sequencing and Phylogenetic Analysis

Approximately 20 or 40 days-old cultures were evaluated morphologically. This way, it was decided for which cultures molecular species identification should be carried out. The data also now contribute to the species description. The cultures on PDA were observed from the top and from the bottom (front and back side of the plates). Conidiomata on the plates and pycnidia on soybean stems were observed using a Stemi 2000 binocular loupe (Carl Zeiss, Oberkochen, Germany) and conidia using a Primo Star microscope (Carl Zeiss). The images were taken using an AxioCam HRC color camera (Carl Zeiss) and evaluated with AxioVision software, Release 4.8.3 Special Edition 1 (Carl Zeiss, Oberkochen, Germany).

For molecular species identification, genomic DNA was extracted from 7 days-old cultures using the protocol published by [14]. ITS, *TEF1*, and *TUB* genes were amplified using the primer pairs ITS1-F/ITS4 [15], EF1-728F/EF1-986R [16], and Bt-2a/Bt-2b [17] under the same conditions as described by [7] and sequenced. The sequences were then searched against NCBI GenBank using BLASTn (https://blast.ncbi.nlm.nih.gov/Blast.cgi?PROGRAM=blastn&PAGE_TYPE=BlastSearch&BLAST_SPEC=&LINK_LOC=blasttab&LAST_PAGE=blastn accessed on 10 October 2024).

For the isolate (DPC_HOH36) with high sequence similarity to *D. goulteri*, we downloaded highly similar sequences from GenBank and performed phylogenetic analysis. For downloading, we chose >90 percent identity as a criterion and cutoff. The sequences were aligned using ClustalW [18] and the alignments were edited using BioEdit version 7.0.5.3 [19]. For the three-gene phylogeny, the alignments were concatenated using a text editor. Maximum likelihood phylogenetic analysis using the Tamura–Nei model [20] was performed using MEGA X version 10.0.5 [21]. Initial trees were obtained automatically by applying the maximum parsimony method.

2.2.2. Direct Species Identification from Infected Plant Parts Using qPCR

For qPCR detection, pieces of leaves, stems, pods, or seeds were surface disinfected as described in Section 2.1 and used for DNA isolation. Samples of roughly 100 mg were put into 2-mL micro screw tubes (Sarstedt, Nümbrecht, Germany) with two steel balls (4.50 mm in diameter, Niro, Sturm Präzision GmbH, Oberndorf am Neckar, Germany). The tubes with the samples were frozen in liquid nitrogen for 3 min and homogenized for 20 s at 4.5 m/s using a FastPrep®-24 homogenizer (MP Biomedicals GmbH, Santa Ana, CA, USA). DNA extraction was performed using the DNeasy Plant Mini Kit (Qiagen, Hilden, Germany) following the manufacturer's instructions.

Field samples were tested for the presence of *D. caulivora*, *D. eres*, *D. longicolla*, or *D. novem*. The reactions for the quadruplex TaqMan qPCR to detect these species were prepared as previously described [22].

The 20-µL reactions used here for detection of *D. goulteri* in plant samples consisted of 10 µL ProbeMasterMix (2×) No-ROX (Genaxxon bioscience GmbH, Ulm, Germany), 4 pmol of each forward and reverse primers (DPCG-F: 5′-cttacactcacaaaactcgc-3′; DPCG-R: 5′-gctcgattcaccgggttg-3′), 1 pmol probe (DPCG-P: 5′-6-FAM-ccagagcaaacaccaccgacgc-BMN-Q535-3′), and 2 µL template DNA. Design and testing of this primer and probe combination was performed analogous to what was described previously [22]. The reaction mixes were incubated for 3 min at 95 °C and then subjected to 40 cycles of 95 °C for 15 s and 60 °C for 45 s. Reactions were run in technical duplicates on a CFX96 Real-Time PCR system (Bio-Rad Laboratories, Hercules, CA, USA) using FrameStar1 96-Well Skirted PCR Plates (4titude, Brooks Automation, Chelmsford, MA, USA).

2.3. Testing for Pathogenicity on Soybean

Inoculations for the pathogenicity test were performed using the toothpick method [23,24]. To prepare toothpicks for inoculation, autoclaved toothpicks were placed on PDA plates together with agar plugs overgrown with mycelium of the *Diaporthe* isolate. The plates were incubated at room temperature for 20 days. Toothpicks overgrown with mycelium were inserted into the stem in the middle of the internode between cotyledons and the first trifoliate of three weeks old healthy soybean plants (cv. Shouna) at an angle of 90 degrees. Control plants were identically treated with sterile toothpicks. Six plants each were used for inoculation and control plants were kept in the greenhouse at 16 h light / 8 h dark (24 °C/22 °C); air humidity was kept high using a vaporizer (Condair, Norderstedt, Germany). Symptom development was observed weekly for up to two months.

The inoculated plants were tested for infection with *D. goulteri* by two different methods. One was re-isolation, and the other was detection using qPCR with a newly designed primer–probe combination for species-specific detection based on the *TEF* sequence of *D. goulteri*, as described in Section 2.2.2.

For re-isolation, the stem of an infected plant was cut close (1 cm removed) to the inoculation site. After surface disinfection (as described in Section 2.1), 2 to 3 mm long and 1 mm thick pieces of the center part of these sections of the stems were placed on PDA plates. The developing fungal colonies were determined as identical to the inoculum based on culture morphology to fulfil Koch's postulate.

For qPCR detection, pieces of the center part of the above sections of the stems of the infected plant close to the inoculation site (treated as for re-isolation) and identical stem pieces from a control plant were used for DNA extraction. DNA extraction and qPCR were performed as described in Section 2.2.2.

3. Results

3.1. D. goulteri Found on a Soybean Plant Collected in a Field Close to Tübingen, Germany

As part of a survey monitoring the incidence of *Diaporthe* spp. in Germany, soybean plants showing typical *Diaporthe* symptoms were collected from fields close to Tübingen on 2 October 2023. Using the qPCR assay developed by [22], plants infected with *D. caulivora*, *D. eres*, *D. longicolla*, or *D. novem* were identified (78% of all sampled plants); the highest incidence was found for *D. caulivora*. From some plants, especially those that tested negative in the qPCR assay, parts were also used for gaining *Diaporthe* spp. isolates (see Section 2.1). This way, we gained five new isolates of *Diaporthe* spp.; the other samples either yielded nothing or fungi from different genera. One isolate (designated DPC_HOH36) recovered from a seed of a plant collected from the following coordinates (48.525801, 9.096385) showed high similarity in all three tested genes to sequences annotated with *D. goulteri* in the GenBank database. This, together with morphological observations, allowed for the identification of isolate DPC_HOH36 as *D. goulteri*.

To refine this finding, similar sequences were downloaded from GenBank. Because of the chosen cutoff, the download was restricted to sequences annotated with either *D. goulteri* or *D. ambigua*. To broaden the phylogenetic analysis we also included sequences representing the four *Diaporthe* spp. reported previously on soybean in Germany (*D. caulivora*, *D. eres*, *D. longicolla*, and *D. novem*) [7] as outgroups (Figure 1).

For all three genes, and also in the analysis combining the three genes, our sequences clustered with the sequences annotated as *D. goulteri* in GenBank. The ITS sequence of DPC_HOH36 is identical to that of strain BRIP 55657a, the isolate corresponding to the first description of *D. goulteri* [12] (Figure 1b). The same is true for the *TUB* sequence (Figure 1c), while the *TEF* sequence of DPC_HOH36 is identical to that of isolate MFLUCC 21-0012 [13] (Figure 1d). Additional sequences annotated with *D. goulteri* were found and are represented in the phylogenies; however, no publications were found related to the GenBank records. Altogether, the analysis of our sequence data clearly allowed for the identification of isolate DPC_HOH36 as *D. goulteri*.

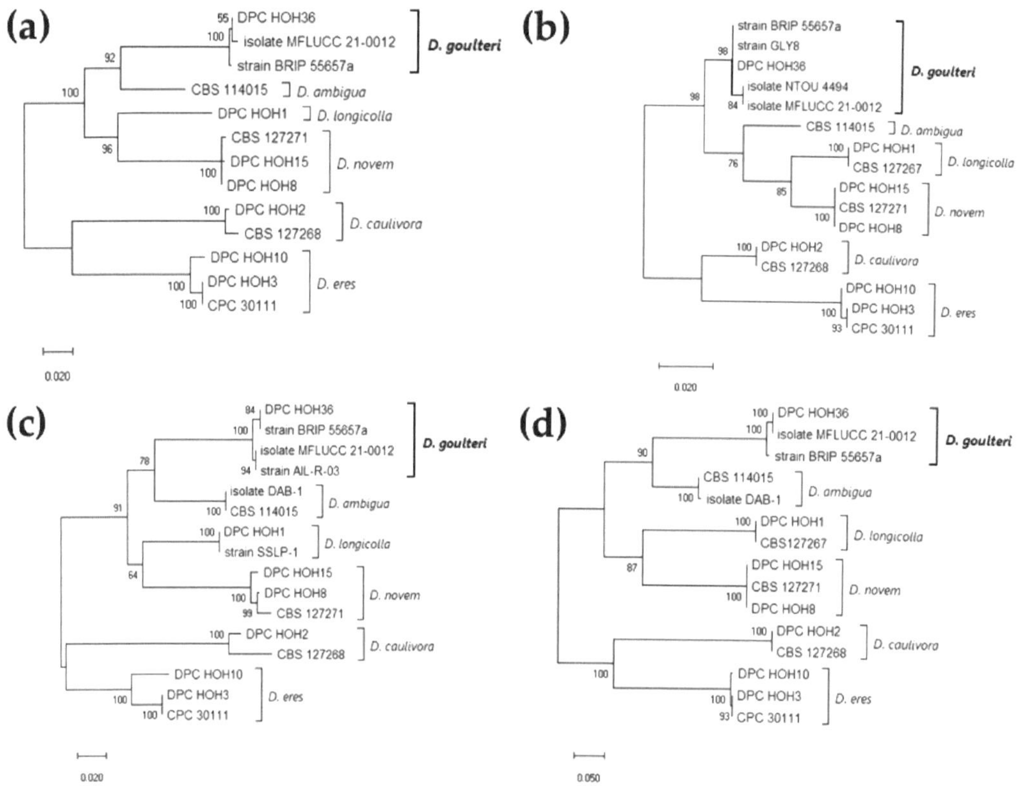

Figure 1. Maximum Likelihood phylogeny identifying isolate DPC_HOH36 as *Diaporthe goulteri*. Trees with the highest log likelihood (lL) are shown. Trees are drawn to scale, with branch lengths indicating the number of substitutions per site. Numbers next to the branches are bootstrapping percentages. (**a**) Combined phylogeny based on 13 concatenated ITS, *TUB*, and *TEF* sequences, 1183 positions in final dataset, lL −4658.69. (**b**) Phylogeny based on 16 ITS sequences, 394 positions in final dataset, lL −1253.80. (**c**) Phylogeny based on 16 *TUB* sequences, 493 positions in final dataset, lL −1581.92. (**d**) Phylogeny based on 15 *TEF* sequences, 324 positions in final dataset, lL −1799.90.

Isolates and corresponding GenBank accession numbers used for the phylogenies are detailed in Table 1.

Table 1. Strains or isolates of *Diaporthe* spp. and GenBank accession numbers of the ITS (internal transcribed spacer), *TUB* (β-tubulin), and *TEF* (translation elongation factor 1-α) sequences used for the phylogenies in Figure 1.

Species	Designation *	Accession Numbers		
		ITS	*TUB*	*TEF*
D. ambigua	CBS 114015	MH862953	KC343978	GQ250299
	Isolate DAB-1		MK463859	MK463861
D. caulivora	DPC_HOH2	MK024677	MK161476	MK099094
	CBS 127268	HM347712	KC344013	HM347691
D. eres	DPC_HOH3	MK024678	MK161477	MK099095
	DPC_HOH10	MK024685	MK161484	MK099102
	CPC 30111	MG281083	MG281256	MG281604

Table 1. *Cont.*

Species	Designation *	Accession Numbers		
		ITS	*TUB*	*TEF*
D. goulteri	DPC_HOH36	PQ008930	PQ014385	PQ014381
	Strain BRIP 55657a	KJ197290	KJ197270	KJ197252
	Isolate MFLUCC 21-0012	MW677456	MW680162	MW680164
	Strain AIL-R-03		ON221693	
	Strain GLY8	MF356582		
	Isolate NTOU 4494	MZ422958		
D. longicolla	DPC_HOH1	MK024676	MK161475	MK099093
	Strain SSLP-1		HQ333510	
	CBS 127267	HM347700		HM347685
D. novem	DPC_HOH8	MK024683	MK161482	MK099100
	DPC_HOH15	MK024690	MK161489	MK099107
	CBS 127271	HM347710	KC344125	HM347695

* Names of strains or isolates for which sequences were deposited in GenBank. Some of these were assigned by the authors who isolated them and deposited the sequences, some represent numbers of culture collections.

3.2. Morphological Description of *D. goulteri* Isolate DPC_HOH36

Growth of *D. goulteri* isolate DPC_HOH36 on PDA was fast compared to other *Diaporthe* spp. (*D. longicolla*, *D. caulivora*, *D. eres*, and *D. novem*) and after 5 days the entire plate was covered with mycelium of white-yellowish-light green color. The back side of the plates was white and after longer incubation (40 days) it became light brown with scattered dark spots (Figure 2b).

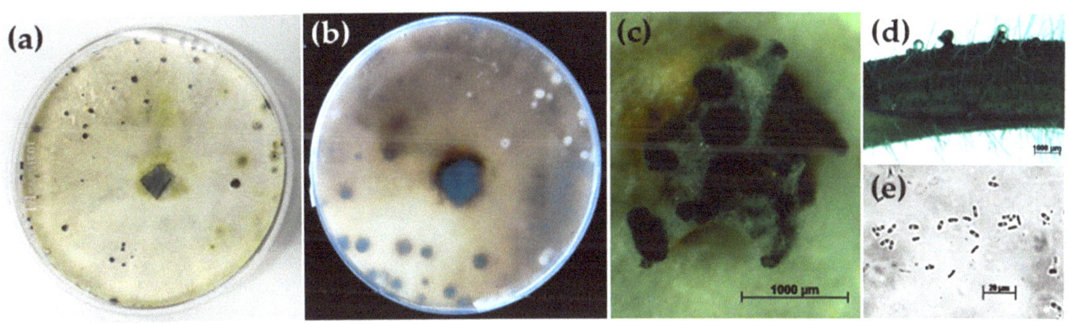

Figure 2. Morphological characteristics of *Diaporthe goulteri* isolate DPC_HOH36. (**a**,**b**) Colony front and back after 40 days on PDA. (**c**) Conidiomata on PDA. (**d**) Pycnidia on soybean stem after three weeks in culture (on water agar). (**e**) α-conidia.

After approximately 40 to 50 days, dark brown pycnidia appeared (Figure 2a,c) that contained numerous α-conidia (Figure 2e). The α-conidia were fusiform to cylinder-shaped, biguttulate, hyaline, and measured 5.19 to 8.06 × 1.92 to 2.88 µm. Perithecia and β-conidia were not observed. The morphological characteristics of isolate DPC_HOH36 were similar to those described for *D. goulteri* strain BRIP 55657a [12]. Thus, the morphology corroborated the identification of isolate DPC_HOH36 as *D. goulteri*.

3.3. *D. goulteri* Can Infect Soybean in the Laboratory Test

We used an established method for isolating fungi growing inside plant tissue to produce our isolates (see Section 2.1). Therefore, our isolate should be either an endophyte or a pathogen. To further establish whether *D. goulteri* can infect soybean, which parts of the plant are affected, and to observe symptoms caused by *D. goulteri*, we performed a pathogenicity test, as described in Section 2.3.

Dark discoloration of the plant tissue extending roughly 1 cm from where the inoculated toothpicks had been inserted was observed at the soybean stem (Figure 3c). When the stems were cracked open, this discoloration, indicating necrosis, could also be found in the center of the stem, where it extended to roughly 2 cm from the inoculation site (Figure 3d). In contrast, control plants that had been stuck with sterile toothpicks showed a tan discoloration that extended less than 2 mm from the insertion site.

Figure 3. *Diaporthe goulteri* on soybean. (**a**) Soybean plant from which *D. goulteri* isolate DPC_HOH36 was recovered (without rootstock). (**b**) Close-up from (**a**) where the stem is broken open. (**c**) Stem of the soybean plant inoculated with *D. goulteri* isolate DPC_HOH36 using the toothpick method 30 days after inoculation. (**d**) Discoloration of central tissue close to the inoculation site two months after the inoculation. (**e**,**f**) Amplification curves of qPCR assays on DNA sample from plant inoculated with DPC_HOH36 and DNA sample from control plant, respectively.

Based on culture morphology, fungal mycelium originating from re-isolation from the discolored plant parts was identified as *D. goulteri*. Using qPCR, we also could detect *D. goulteri* in the stem of inoculated plant close to the infection site (Figure 3e). No *D. goulteri* was detected on any sample taken from control plants (Figure 3f). These observations strongly indicate that *D. goulteri* can infect soybean plants.

4. Discussion

Strain DPC_HOH36 was isolated from a soybean plant in a field close to Tübingen, Germany. BLAST comparisons of the *TUB*, *TEF*, and ITS sequences showed high similarity to published sequences of *D. goulteri*. This similarity can also be seen in phylogenies built based on these sequences. These results strongly suggest that our isolate DPC-HOH36 is *D. goulteri*.

Thompson et al. [12] described *D. goulteri* colonies growing on PDA and OMA (oatmeal agar) with relatively fast-growing mycelium that was first white before dark spots appeared. On PDA, we made the same observation. Droplets containing conidia formed on OMA were quite colorful—pale yellow, orange, or ochre and reddish, or as the authors described it, "sienna colored". It is probably due to the medium that we could not observe these colors on our PDA plates. The form of the conidia they observed was exactly as seen by us, and with 6 to 9 × 2 to 3 µm, the dimensions they observed also encompass what we measured. The conidiomata observed by [12] on OMA and on sterilized wheat straw are slightly different from what we observed on PDA and on sterilized soybean stems. Nevertheless, morphological observations also indicate that isolate DPC_HOH36 is the same species. Our description of the species should be seen as an extension of the existing descriptions.

Thompson et al. [12] isolated *D. goulteri* from a sunflower seed. No further tests regarding the pathogenicity of *D. goulteri* on sunflower or on association to any other plant species were performed. The species from which other researchers isolated *D. goulteri* are either unidentified or not mentioned [13]. Therefore, the host range of *D. goulteri* remains largely unknown. Our pathogenicity test on soybean is the first demonstration of infection of a plant by *D. goulteri*, so it can be stated that *G. max* definitely is a host for *D. goulteri*.

We performed surface disinfection of the plant material before isolation of *Diaporthe* spp. Using this procedure, no growth of fungi only randomly associated (i.e., spores sticking to the surface) with the sampled soybean plants should occur. Therefore, isolating *D. goulteri* from a symptomatic soybean plant means that the fungus was growing inside the plant, which fulfils the first of Koch's postulates and indicates that it is either an endophyte or a pathogen. Together with that, inoculation, infection and detection of the fungus in the inoculated soybean tissue demonstrated that *D. goulteri* can infect and grow in soybean. This constitutes the first demonstration of the ability of *D. goulteri* to infect plants. More findings of *D. goulteri* in the field together with observations of symptoms and damages will be necessary to establish whether *D. goulteri* acts as a pathogen and how severe damages caused by the species may be.

Author Contributions: Conceptualization, B.H. and T.I.L.; methodology, B.H.; validation, B.H. and T.I.L.; formal analysis, T.I.L.; investigation, B.H., M.G.G. and T.I.L.; resources, T.I.L.; writing—original draft preparation, B.H. and T.I.L.; writing—review and editing, B.H. and T.I.L.; visualization, B.H., M.G.G. and T.I.L.; project administration, T.I.L.; funding acquisition, T.I.L. All authors have read and agreed to the published version of the manuscript.

Funding: This research was funded by the German Federal Ministry of Food and Agriculture, grant number 2815EPS082 (SoySound) to T.I.L. The APC was funded by the University of Hohenheim.

Institutional Review Board Statement: Not applicable.

Informed Consent Statement: Not applicable.

Data Availability Statement: Sequences underlying our analyses were submitted to GenBank, accession numbers are given in the manuscripts. All other data are contained within the manuscript.

Acknowledgments: We thank Heike Popovitsch for technical assistance.

Conflicts of Interest: The authors declare no conflicts of interest. The funders had no role in the design of the study; in the collection, analyses, or interpretation of data; in the writing of the manuscript; or in the decision to publish the results.

References

1. Pioli, R.N.; Morandi, E.N.; Bisaro, V. First report of Soybean Stem Canker caused by *Diaporthe phaseolorum* var. *caulivora* in Argentina. *Plant Dis.* **2001**, *85*, 95. [CrossRef] [PubMed]
2. Costamilan, L.M.; Yorinori, J.T.; Almeida, A.M.; Seixas, C.D.S.; Binneck, E.; Araújo, M.R.; Carbonari, J.A. First report of *Diaporthe phaseolorum* var. *caulivora* infecting soybean plants in Brazil. *Trop. Plant Pathol.* **2008**, *33*, 381–385. [CrossRef]
3. Santos, J.M.; Vrandečić, K.; Ćosić, J.; Duvnjak, T.; Phillips, A.J.L. Resolving the *Diaporthe* species occurring on soybean in Croatia. *Persoonia* **2011**, *27*, 9–19. [CrossRef] [PubMed]
4. Zhao, X.; Li, K.; Zheng, S.; Yang, J.; Chen, C.; Zheng, X.; Wang, Y.; Ye, W. *Diaporthe* diversity and pathogenicity revealed from a broad survey of Soybean Stem Blight in China. *Plant Dis.* **2022**, *106*, 2892–2903. [CrossRef] [PubMed]
5. Petrović, K.; Riccioni, L.; Vidić, M.; Đorđević, V.; Balešević-Tubić, S.; Đukić, V.; Miladinov, Z. First report of *Diaporthe novem*, *D. foeniculina*, and *D. rudis* associated with soybean seed decay in Serbia. *Plant Dis.* **2016**, *100*, 2324. [CrossRef]
6. Petrović, K.; Skaltsas, D.; Castlebury, L.A.; Kontz, B.; Allen, T.W.; Chilvers, M.I.; Gregory, N.; Kelly, H.M.; Koehler, A.M.; Kleczewski, N.M.; et al. *Diaporthe* seed decay of soybean [*Glycine max* (L.) Merr.] is endemic in the United States, but new fungi are involved. *Plant Dis.* **2021**, *105*, 1621–1629. [CrossRef]
7. Hosseini, B.; El-Hasan, A.; Link, T.; Voegele, R.T. Analysis of the species spectrum of the *Diaporthe/Phomopsis* complex in European soybean seeds. *Mycol. Prog.* **2020**, *19*, 455–469. [CrossRef]
8. Hosseini, B.; Voegele, R.T.; Link, T.I. Diagnosis of soybean diseases caused by fungal and oomycete pathogens: Existing methods and new developments. *J. Fungi* **2023**, *9*, 587. [CrossRef]
9. Gomes, R.R.; Glienke, C.; Videira, S.I.R.; Lombard, L.; Groenewald, J.Z.; Crous, P.W. *Diaporthe*: A genus of endophytic, saprobic and plant pathogenic fungi. *Persoonia* **2013**, *31*, 1–41. [CrossRef]
10. Udayanga, D.; Castlebury, L.A.; Rossman, A.Y.; Hyde, K.D. Species limits in *Diaporthe*: Molecular re-assessment of *D. citri*, *D. cytosporella*, *D. foeniculina* and *D. rudis*. *Persoonia* **2014**, *32*, 83–101. [CrossRef]
11. Santos, L.; Alves, A.; Alves, R. Evaluating multi-locus phylogenies for species boundaries determination in the genus *Diaporthe*. *PeerJ* **2017**, *5*, e3120. [CrossRef] [PubMed]
12. Thompson, S.; Tan, Y.P.; Shivas, R.G.; Neate, S.; Morin, L.; Bissett, A.; Aitken, E. Green and brown bridges between weeds and crops reveal novel *Diaporthe* species in Australia. *Persoonia* **2015**, *35*, 39–49. [CrossRef] [PubMed]
13. Bundhun, D.; Senanayake, I.; Jayawardena, R.S.; Camporesi, E.; Huang, Y.; Dong, Z.; Hyde, K.D.; To-Anun, C.; Cheewangkoon, R. First reports of the sexual morphs of *Diaporthe forlicesenica* nom. nov. and *Diaporthe goulteri* (Diaporthaceae, Diaporthales) revealed by molecular phylogenetics. *Phytotaxa* **2021**, *516*, 1–27. [CrossRef]
14. Liu, D.; Coloe, S.; Baird, R.; Pedersen, J. Rapid mini-preparation of fungal DNA for PCR. *J. Clin. Microbiol.* **2000**, *38*, 471. [CrossRef]
15. White, T.J.; Bruns, T.; Lee, S.; Taylor, J. Amplification and direct sequencing of fungal ribosomal RNA genes for phylogenetics. In *PCR Protocols: A Guide to Methods and Applications*; Innis, N., Gelfand, D., Sninsky, J., White, T., Eds.; Academic Press Inc.: New York, NY, USA, 1990; pp. 315–322.
16. Carbone, I.; Kohn, L.M. A method for designing primer sets for speciation studies in filamentous ascomycetes. *Mycologia* **1999**, *91*, 553–556. [CrossRef]
17. Glass, N.L.; Donaldson, G.C. Development of primer sets designed for use with the PCR to amplify conserved genes from filamentous ascomycetes. *Appl. Environ. Microbiol.* **1995**, *61*, 1323–1330. [CrossRef]
18. Thompson, J.D.; Higgins, D.G.; Gibson, T.J. CLUSTAL W: Improving the sensitivity of progressive multiple sequence alignment through sequence weighting, position-specific gap penalties and weight matrix choice. *Nucleic Acids Res.* **1994**, *22*, 4673–4680. [CrossRef]
19. Hall, T.A. BioEdit: A user-friendly biological sequence alignment editor and analysis program for Windows 95/98/NT. *Nucleic Acids Symp. Ser.* **1999**, *41*, 95–98.
20. Tamura, K.; Nei, M. Estimation of the number of nucleotide substitutions in the control region of mitochondrial DNA in humans and chimpanzees. *Mol. Biol. Evol.* **1993**, *10*, 512–526. [CrossRef]
21. Kumar, S.; Stecher, G.; Li, M.; Knyaz, C.; Tamura, K. MEGA X: Molecular Evolutionary Genetic Analysis across computing platforms. *Mol. Biol. Evol.* **2018**, *35*, 1547–1549. [CrossRef]
22. Hosseini, B.; Voegele, R.T.; Link, T.I. Establishment of a quadruplex real-time PCR assay to distinguish the fungal pathogens *Diaporthe longicolla*, *D. caulivora*, *D. eres*, and *D. novem* on soybean. *PLoS ONE* **2021**, *16*, e0257225. [CrossRef] [PubMed]
23. Ghimire, K.; Petrović, K.; Kontz, B.J.; Bradley, C.A.; Chilvers, M.I.; Mueller, D.S.; Smith, D.L.; Wise, K.A.; Mathew, F.M. Inoculation method impacts symptom development associated with *Diaporthe aspalathi*, *D. caulivora*, and *D. longicolla* on soybean (*Glycine max*). *Plant Dis.* **2019**, *103*, 677–684. [CrossRef] [PubMed]
24. Campbell, M.A.; Li, Z.; Buck, J.W. Development of southern stem canker disease on soybean seedlings in the greenhouse using a modified toothpick inoculation assay. *Crop Protect.* **2017**, *100*, 57–64. [CrossRef]

Journal of *Fungi*

MDPI

Article

Evaluating Two Fungicides, Prochloraz–Manganese Chloride Complex and Seboctylamine Acetate, to Control Cobweb Disease in White Button Mushroom Caused by *Cladobotryum mycophilum*

Qiqi Chen [1,†], Yazhen Yuan [1,†], Gang Chen [1], Ning Li [1], Xinrong Li [1], Yufei Lan [2] and Hongyan Wang [1,*]

[1] Department of Plant Protection, Shandong Agricultural University, Tai'an 271018, China; qiqichen19970905@163.com (Q.C.); cg15052715639@163.com (G.C.); 13465333875@163.com (N.L.); lxr15550892735@163.com (X.L.)

[2] Tai'an Academy of Agricultural Sciences, Tai'an 271018, China; lanyufei526@163.com

* Correspondence: sdauwhy@163.com or hongyanwang@sdau.edu.cn

† These authors contributed equally to this work.

Abstract: Cobweb disease in white button mushroom (*Agaricus bisporus*) is a newly identified disease caused by *Cladobotryum mycophilum* in China. Currently, there are few highly effective and safe fungicides for controlling this disease in the field. This study assessed the fungicidal effect of prochloraz–manganese chloride complex and seboctylamine acetate against *C. mycophilum*, as well as their ability to control cobweb disease. Additionally, the residues of these fungicides in the mycelium and the mushroom were evaluated. The extent of the fungicidal effect against the pathogen was determined based on the efficiency of crop production. The results revealed that, in addition to the potent inhibitory effect of prochloraz–manganese chloride complex on the hyphae of *C. mycophilum*, the domestically developed seboctylamine acetate exhibited high toxicity, inhibiting both mycelial growth and spore germination of *C. mycophilum*, with EC_{50} values of 0.990 mg/L and 0.652 mg/L, respectively. Furthermore, the application of the two chemical agents had no adverse effects on the mycelial growth and fruiting body growth of *A. bisporus*, and the residual amount of chemical agent was lower than the maximum residue limit standard. The field application results showed that 400 mg/L of prochloraz–manganese chloride complex and 6 mg/L of seboctylamine acetate resulted in 61.38% and 81.17% disease control respectively. This study presents efficient and safe fungicides for controlling cobweb disease in white button mushroom. Additionally, a residue determination analysis of the fungicide seboctylamine acetate in mushroom crops is described.

Keywords: *Agaricus bisporus*; *Cladobotryum mycophilum*; cobweb disease; seboctylamine acetate; prochloraz–manganese chloride complex

Citation: Chen, Q.; Yuan, Y.; Chen, G.; Li, N.; Li, X.; Lan, Y.; Wang, H. Evaluating Two Fungicides, Prochloraz–Manganese Chloride Complex and Seboctylamine Acetate, to Control Cobweb Disease in White Button Mushroom Caused by *Cladobotryum mycophilum*. *J. Fungi* **2024**, *10*, 676. https://doi.org/10.3390/jof10100676

Academic Editor: Ofir Degani

Received: 15 August 2024
Revised: 25 September 2024
Accepted: 25 September 2024
Published: 27 September 2024

1. Introduction

Edible fungi are important in the diets of people all over the world [1]. As the edible fungus industry has flourished and expanded its scale significantly [2], China has become a major producer in this sector [3]. The species *Agaricus bisporus* (*A. bisporus*), also known as the white button mushroom [4], is widely cultivated, has a wide distribution range and high production yield, and is highly favored [5,6]. The *A. bisporus* mushroom has tender flesh, a delicious taste and excellent commodity value due to its high protein content, low fat and low calories [7]. On the other hand, its medicinal value is also very high, and it has a certain preventive effect on viral diseases, stomach diseases, etc. [8].

With the rapid development of the *A. bisporus* planting industry, the occurrence of *A. bisporus* diseases is becoming more common [9], such as ulcer disease, brown spot, soft rot, and cobweb disease. The yield is generally reduced by 20~30% when these diseases occur [10]. Cobweb disease is particularly severely affecting *A. bisporus* in China [11], and is caused by *Cladobotryum mycophilum* (*C. mycophilum*). It can infect various edible fungi such as *Lentinula edodes* [11]. Cobweb disease, together with *Lecanicillium fungicola*, *Trichoderma aggressivum* and *Mycogone perniciosa*, are acknowledged as the four primary diseases of edible fungi attributed to parasitic fungi [5]. In the initial stages of *A. bisporus* cobweb disease, downy mildew-like mycelium adheres to the soil layer and gradually spreads, eventually reaching the tip of the fruiting body [12]. In one study, small white round patches appeared in the casing soil or basidiospore. As the mycelium of the pathogen progressed, it gradually took on a cottony, flocculent appearance and simultaneously produced a copious number of dry spores. These spores were highly susceptible to being dislodged upon physical disturbance, such as watering, whereupon the conidia were disseminated throughout the mushroom facility via air currents, ultimately causing the appearance of brown spots on the fruiting bodies of infected *A. bisporus* mushrooms [13]. Finally, they rotted, and the whole fruiting body shrunk if it was serious. The pathogen can infect *A. bisporus* throughout the entire development period [14]. To prevent the spread of cobweb disease, it is crucial to take immediate action before spatulation sets in. This can be achieved by covering the affected area with a thick, dampened paper sheet to hinder the release of conidia and arrest the further dissemination of the disease [15].

Prevention and control strategies for edible mushroom diseases encompass the breeding of disease-resistant varieties, rigorous strain quality control, meticulous cultivation environment management, the induction and application of disease resistance mechanisms, chemical intervention, biological control methods, and more [16]. However, the breeding of excellent strains takes a long time and is prone to variation [15]. While biological pesticides are indeed safe for humans and the environment, there remains a lack of an effective biological control method specifically tailored to combatting cobweb disease. Furthermore, the exact mechanism of action of these potential biological agents is not yet fully understood. Consequently, in field applications, their control efficacy often falls short of that achieved by chemical fungicides [17]. Therefore, the control of mushroom diseases still depends on the use of chemical fungicides [18]. In China, only seven fungicides, including prochloraz–manganese chloride complex, are allowed to be used on mushrooms, but there is no registered fungicide for the prevention and control of cobweb disease on *A. bisporus*. Prochloraz–manganese salt, also known as prochloraz–manganese chloride complex, is composed of prochloraz and manganese chloride. Its disease prevention performance is very similar to prochloraz. Prochloraz–manganese was found to control green mold on white mushrooms (*A. bisporus*) and increase the yield of *A. bisporus*, and can effectively control wet bubble disease (WBD) caused by *Mycogone rosea* [19]. Chakwiya et al. focused on the sensitivity of *C. mycophilum* to carbendazim and prochloraz–manganese; the prochloraz–manganese ED_{50} values varied from 0.00001 mg/L to 0.55 mg/L, and it is to be used in a disease management strategy [20]. Pathogen sensitivity to fungicides was estimated by probit analyses. Fungicide susceptibility tests showed that *C. mycophilum* strains were highly sensitive both to prochloraz ($ED_{50} < 0.087$ μg mL^{-1}) and the newly introduced metrafenone ($ED_{50} < 0.15$ μg mL^{-1}) [21]. Prochloraz can no longer be used in the European Union (EU) as of June 2023, and over-reliance on metrafenone has resulted in putative resistant pathogenic strains emerging. Prochloraz still showed good control of two different isolates of *C. mycophilum*, with efficacy values consistently reaching 70% [22]. However, some studies found that *C. mycophilum* type II isolate 192B1 was sensitive to prochloraz-Mn, and easily lead to drug resistance, resulting in its inability to prevent the occurrence of cobweb disease symptoms [23]. Consequently, there is an urgent need to identify chemical agents capable of effectively managing this disease.

In recent years, extensive research has highlighted that the presence of chemicals in the human body, even at trace levels, poses a significant risk to health. Therefore, the stringent regulation of pesticide residue levels in agricultural products, along with regular comparison to the Maximum Residue Limits (MRLs), and a concerted effort to promote reduced pesticide usage, are crucial steps towards ensuring the safety of our food supply [24]. Various methods have been developed for detecting residual pesticides over time, including the ultrasound-assisted solvent extraction, solid phase extraction and QuEChERS methods. Analytical techniques such as high-performance liquid chromatography (HPLC) and high-performance liquid chromatography–mass spectrometry (HPLC-MS) have been widely applied in this context [25]. The advancement and implementation of the aforementioned pesticide residue detection technologies have significantly enhanced analysis accuracy, streamlined procedural complexity, and bolstered analysis efficiency [26]. Among these methods, HPLC-MS is notable for its high resolution and ability to perform both qualitative and quantitative analysis [27].

In this study, we conducted an indoor toxicity test to evaluate the antifungal effects of prochloraz–manganese chloride complex and seboctylamine acetate (former name: Xinjuan; the chemical structure is shown in Figure 1), which were selected as the agents with a higher antifungal effect than *A. bisporus* cobweb disease. The control effects of prochloraz–manganese chloride complex and seboctylamine acetate were investigated for their effectiveness against the cobweb disease pathogen in the field. Additionally, the safety of these compounds on the mycelial growth and fruiting body development of *A. bisporus* was assessed. Meanwhile, the method for detecting the residue of seboctylamine acetate in *A. bisporus* was established, and the effect of the level of chemical on yield was thoroughly explored, to provide a basis for the development of application rates and the risk assessment of fungicides in the cultivation of *A. bisporus*.

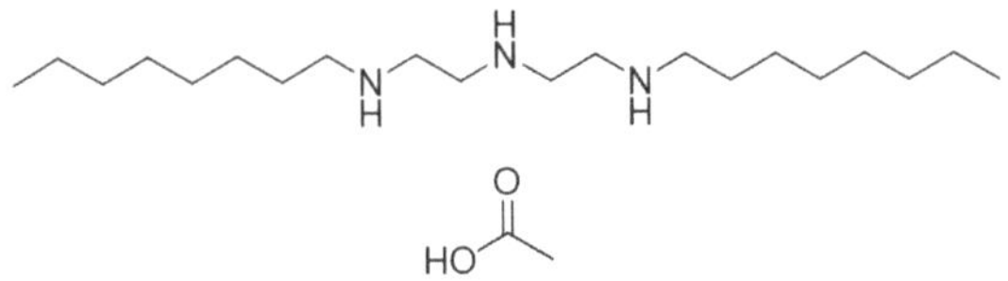

Figure 1. The structure of seboctylamine acetate.

2. Materials and Methods

2.1. Test Strains and Fungicides

The first step was to investigate the strain *C. mycophilum* at the Tai'an Academy of Agricultural Science in Shandong Province. The researchers collected and isolated the fruiting bodies of *A. bisporus* cobweb disease from a mushroom shed in Jining, and purified and identified the strain as *C. mycophilum*. They stored it at 4 °C for later use.

A. bisporus varieties S2796, W192, and W2000 were provided by Tai'an Academy of Agricultural Sciences.

Several fungicides, including 97% quatrimycin TC, purchased by Shandong Union Pesticide Industry Co., Ltd. in Taian, China; 98% pyrimethanil TC, purchased by Shandong Weifang Rainbow Chemical Co., Ltd. in Weifang, China; 97% kresoxim-methyl TC, purchased by Shandong Hailier Chemical Co., Ltd. in Qingdao, China; 97% difenoconazole TC, purchased by Shandong Union Pesticide Industry Co., Ltd. in Tai'an, China; 96% tebuconazole TC, purchased by Shandong Union Pesticide Industry Co., Ltd. in Tai'an, China; 98% prochloraz–manganese chloride complex TC, purchased by in China Shandong Weifang Rainbow Chemical Co., Ltd. in Weifang, China; 97% prothioconazole TC, purchased by Shandong Hailier Chemical Co., Ltd. in Qingdao, China; 34% polyoxin TC, purchased by Shandong Yucheng Biochemical Pesticide Co., Ltd. in Weifang, China; 20% seboctylamine acetate AS, purchased by Weifang Voelsing Biopesticide Co., Ltd. in Weifang, China; 50% prochloraz–manganese chloride complex WP, purchased by FMC Investment Co., Ltd. in Philadelphia, PA, USA; and 1.8% seboctylamine acetate AS, purchased by Shaanxi Xi'an Jiake Agricultural Chemical Co., Ltd. in Xi'an, China.

2.2. Toxicity Measurement

To measure toxicity, the indoor biological activity of fungicides against *C. mycophilum* was determined by the mycelial growth rate method [28]. A small amount of acetone was used to dissolve the following fungicides: 97% kresoxim-methyl, 97% quatrimycin,

97% difenoconazole, 96% tebuconazole, 98% prochloraz–manganese chloride complex, 34% polyoxin TC, 97% prothioconazole, and 98% pyrimethanil. The dissolved fungicides were then diluted with sterile water. Additionally, 20% seboctylamine acetate aqueous solution and 50% prochloraz–manganese chloride complex WP were separately dissolved in sterile water to create a 1×10^4 mg/L mother solution. This mother solution was stored in a refrigerator at 4 °C for a brief period.

To prepare the drug-containing medium, 45 mL of pre-prepared PDA medium was added to a 100 mL pre-sterilized and dry Erlenmeyer flask. Following this, 5 mL aliquots of the drug solution were added in ascending order of concentration, with thorough shaking after each addition. The resulting mixture was then poured into three 9 cm diameter Petri dishes to create a drug-containing plate with the desired test design concentration [29]. The treatment without fungicide was set as a blank control, and each treatment had 3 replicates. The final effective component dosage of each agent is shown in Table 1.

Table 1. Concentration gradient settings for 9 fungicides.

Fungicide	Experimental Chemical Concentration (mg/L)
Quatrimycin	0.075, 0.150, 0.300, 0.600, 1.200, 2.400, 4.800
Prochloraz–manganese chloride complex	0.150, 0.300, 0.600, 1.200, 2.400, 4.800, 9.600
Seboctylamine acetate	0.196, 0.391, 0.781, 1.563, 3.125, 6.250, 12.500
Prothioconazole	0.075, 0.150, 0.300, 0.600, 1.200, 2.400, 4.800
Tebuconazole	0.150, 0.300, 0.600, 1.200, 2.400, 4.800, 9.600
Difenoconazole	0.781, 1.563, 3.125, 6.250, 12.500, 25.000, 50.000
Pyrimethanil	6.250, 12.500, 25.000, 50.000, 100.000, 200.000
Kresoxim-methyl	6.250, 12.500, 25.000, 50.000, 100.000, 200.000
Polyoxin	0.781, 1.563, 3.125, 6.250, 12.500, 25.000, 50.000

The indoor toxicity was assessed using the mycelium growth rate method [30]. The *C. mycophilum* mycelium was collected from the edge of the mycelium using a sterile puncher (5 mm diameter) and inoculated onto a PDA plate containing the drug. The culture dish was placed in a dark environment with a temperature of 25 ± 1 °C for culturing. The growth diameter of the mycelium in each treatment group, along with that of the blank control, was precisely measured when the mycelium of the blank control group had covered three-quarters of the area within the Petri dish. The inhibition rate of the fungicide on mycelial growth was calculated using following formula:

Growth inhibition rate (%) = [(Pathogen diameter in control − Pathogen diameter under fungicidal action)/Pathogen diameter in control] × 100.

The virulence regression equation and EC_{50} value were also calculated.

2.3. Spore Germination Test

In the spore germination test, the indoor biological activity of fungicides against *A. bisporus* cobweb disease was determined using the spore germination method [31]. The final effective component dosage of each agent is shown in Table 2. The spore suspension was prepared using sterile water, and its concentration was adjusted to 10^6 spores per milliliter. Subsequently, 20 microliters of this spore suspension was pipetted onto a concave glass slide. To prepare the fungicide-containing spore suspension, 20 microliters of each respective concentration gradient of the fungicide was then added to the spore suspension on the slide. The control group was administered an equivalent volume of sterile water. Each treatment was replicated three times to ensure the reliability and reproducibility of the results. The concave slides containing the prepared spore suspensions were carefully placed in Petri dishes and then incubated at a constant temperature of 25 °C in a dark, humid environment. After 24 h of incubation, the spore germination was meticulously examined under a microscope to assess the effectiveness of the treatments. Each treatment was randomly observed in three fields, totaling 100 spores. Spore production was assessed using the blood cell counting plate method, and the spore germination rate and drug inhibition rate during spore germination were calculated after 5 days. The relevant calculation formula is as follows:

$$\text{Spore germination rate (\%)} = \text{spore germination number/total number of spores examined} \times 100\%$$

$$\text{Spore germination inhibition rate (\%)} = (\text{spore germination rate of control group} - \text{spore germination rate of treatment group})/\text{spore germination rate of control group} \times 100\%$$

Table 2. Concentration gradient settings for 8 fungicides.

Fungicide	Experimental Chemical Concentration (mg/L)
Quatrimycin	0.150, 0.300, 0.600, 1.200, 2.400, 4.800, 9.600
Prochloraz–manganese chloride complex	2.000, 4.000, 8.000, 16.000, 32.000, 64.000, 128.000
Seboctylamine acetate	0.075, 0.150, 0.300, 0.600, 1.200, 2.400, 4.800
Prothioconazole	2.000, 4.000, 8.000, 16.000, 32.000, 64.000, 128.000
Tebuconazole	0.150, 0.300, 0.600, 1.200, 2.400, 4.800, 9.600
Difenoconazole	2.000, 4.000, 8.000, 16.000, 32.000, 64.000, 128.000
Pyrimethanil	1.000, 2.000, 4.000, 8.000, 16.000, 32.000, 64.000
Kresoxim-methyl	2.000, 4.000, 8.000, 16.000, 32.000, 64.000, 128.000

The calculation method of efficiency improvement is the same as above.

2.4. Effect of Fungicides on Mycelial Growth of *A. bisporus*

The indoor effectiveness of fungicides on a *A. bisporus* was determined by the mycelium growth rate method [32]. A 98% prochloraz–manganese chloride complex and 20% seboctylamine acetate aqueous solution were dissolved and diluted with sterile water. The mother liquor concentration was 1×10^4 mg/L, and it was stored in a refrigerator at 4 °C. The test gradients of prochloraz–manganese chloride complex were 0.94 mg/L, 1.88 mg/L, and 3.76 mg/L, while test gradients of seboctylamine acetate were 0.99 mg/L, 1.98 mg/L, and 3.96 mg/L.

The preparation of drug-containing medium was as follows: A total of 45 mL of pre-prepared PDA medium was dispensed into a 100 mL pre-sterilized Erlenmeyer flask, and the mixture was poured into three 9 cm diameter Petri dishes to create drug-containing plates with the desired test concentrations. The concentration range of fungicides affecting the growth of *A. bisporus* mycelium was designed according to the method of Gao et al. (2017), and the medium plates with three treatment concentrations were prepared for use [30]. The treatment without adding chemicals was set as a blank control, and each treatment was repeated 3 times. The *C. mycophilum* mycelium was obtained from the edge of the mycelium using a sterile puncher (5 mm diameter) and inoculated onto a PDA plate containing the drug. The culture dish was placed in a dark environment at a temperature of 25 ± 1 °C. The growth diameter of each treatment and blank control mycelium was measured when the blank control covered three-quarters of the Petri dish. The inhibition rate of each fungicide during mycelial growth was calculated using the following formula [33]:

$$\text{Growth inhibition rate (\%)} = [(\text{Pathogen diameter in control} - \text{Pathogen diameter under fungicidal action})/\text{Pathogen diameter in control}] \times 100.$$

2.5. Field Safety of Fungicides on *A. bisporus*

According to the requirements of NY/T 1965.1-2010 [34], the fungicide test was conducted once, twice and four times the recommended dosage in the field, using the dosage of active ingredients. Spraying treatment was carried out when the mushroom buds rose from the covering soil, and the test results were investigated after 10 days. The specific fungicide dosage was as follows: 6, 12, and 24 mg/L of 20% aqueous seboctylamine acetate solution; the concentrations of the 98% prochloraz–manganese chloride complex TC were set at 400 mg/L, 800 mg/L, and 1600 mg/L. The water treatment without the agent was used as the blank control, and each treatment was sprayed with 250 mL/m^2 fungicide solution. Each plot was arranged in random blocks, and each treatment was repeated three times. Using the five-point sampling method, 20 fruiting bodies were collected from each cell, and the area of each plot was 2 m^2.

2.6. Field Control Efficacy

The experiment was carried out in the mushroom shed of the mushroom resources and utilization practice base of Shandong Agricultural University in Tai'an City, China, in October 2020, and the experiment was conducted in the mushroom shed for the continuous cultivation of *A. bisporus*. The previous crop was *A. bisporus*, and all the experimental areas conformed to local scientific agricultural practice, and the cultivation conditions were uniform.

The variety of *A. bisporus* was S2796. When the daytime temperature stabilized at 20–24 °C and the culture material remained below 28 °C, the strain was sown. Initially, half the inoculum was spread on the material's surface, and the remaining inoculum was evenly distributed. After about 20 days, the mycelium was ready for soil covering. The prepared spores of *A. bisporus* cobweb disease were covered with soil 5 days later.

The mushroom buds were sprayed at the beginning of 13 October 2020. The experiment was divided into seven treatments: 50% prochloraz–manganese chloride complex WP 400, 800, and 1600 mg/L; 1.8% seboctylamine acetate AS 2, 4, and 6 mg/L; and a blank control. Each treatment was sprayed with 250 mL/m^2 fungicide solution, and each plot was arranged randomly.

Based on the NY/T 1464.10-2007 [35], according to the requirements, the investigation included recording the occurrence of diseased mushrooms in each plot as well as calculating the incidence of *A. bisporus* and the control effect of fungicides. The *A. bisporus* harvest time was approximately 10 days after application, and the investigation results were obtained on 25 October 2021. The efficacy was calculated using the formula $P = (CK - PT)/CK \times 100\%$, where P represents the control effect percentage (%); CK represents the number of diseased mushrooms (fungi) in the blank control area; and PT represents the number of diseased mushrooms (fungi) in the treatment area.

2.7. Effect of Fungicides on the Yield of A. bisporus

As for the effect of fungicides on the yield of *A. bisporus*, the fungicide test was conducted strictly in accordance with the requirements of the NY/T 1965.1-2010 [34]. The fungicide test was 1 time, 2 times, and 4 times the recommended dosage in the field, all at the active ingredient dosage. Spraying treatment took place when the mushroom buds rose from the covering soil, and the investigation was conducted after 12 days. The concentration of 1.8% seboctylamine acetate aqueous solution was set at 6, 12, and 24 mg/L, while the concentrations of prochloraz–manganese chloride complex wettable powder were set at 400, 800, and 1600 mg/L. A water treatment without chemical agents was used as the blank control. Each treatment was sprayed with 250 mL/m^2 fungicide solution. The experiment design included random blocks for each plot, with each treatment repeated three times and an area of 2 m^2. The five-point sampling method was used to collect 20 subentities in each cell.

2.8. Drug Residue Test

The drug residue test was conducted using QuECHERS to detect the residues of prochloraz–manganese salt and seboctylamine acetate in *A. bisporus* [36]. *A. bisporus* was planted in the mushroom shed of Shandong Agricultural University's fungal resources and utilization training practice base. The test involved three treatments: 50% prochloraz–manganese chloride complex WP, active ingredient dosage, 400 mg/L; 1.8% seboctylamine acetate AS, active ingredient dosage, 6 mg/L; and a blank control. After 3 days of covering soil and fruiting, the test was divided. Each treatment also involved spraying 250 mL/m^2 fungicide solution, and the experimental design was the same as that for determining the effect of fungicide on the yield of *A. bisprus*. Fresh *A. bisporus* samples were collected at 2 hours, 1 day, 3 day, 5 day, 7 day, and 9 day after application. From each plot, at least 1 kg of fresh *A. bisporus* was sampled by chopping. Two portions of 150 g samples were collected and stored at a low temperature in the refrigerator. The residual dynamics and final residues of prochloraz–manganese chloride complex and seboctylamine acetate in *A. bisporus* were calculated [25].

The QuECHERS method was used to prepare nutrient solutions and plant samples [36].

Chromatographic conditions: We used a poroshell 120 EC-C_{18}; the chromatographic column temperature was 40 °C, with an injection volume of 5.0 µL and a flow rate of 0.8 mL/min. Mobile phase A was 0.1% formic acid in water, pH ≈ 3.5, while mobile phase B was acetonitrile.

The parameters of the mass spectrometer were as follows: The ion source was an Electrospray Ionization (ESI) source, with positive ion scanning mode and Multiple Reaction Monitoring (MRM) as the mass spectrometry scanning method. The ion source temperature (TEM) was set at 600 °C. The ion spray voltage (IS) was 4500 V. The curtain gas (CUR) flow rate was 30 psi, the nebulizer gas (Gas1) flow rate was 55 psi, the auxiliary gas (Gas2) flow rate was 60 psi, and the collision gas (CAD) pressure was 7 psi. The other mass spectrometry parameters are shown in Table 3.

Table 3. Residual test-related parameters.

Fungicides	Qualitative Ion Pair (*m/z*)	Quantitative Ion Pair (*m/z*)	DP (V)	CE (eV)	Retention Time (min)
Seboctylamine acetate	328.5/199.2	328.5/156.2	166	25/35	3.080
Prochloraz–manganese chloride complex	376.0/308.0	376.0/266.0	42	8/16	3.820

2.9. Statistical Analysis

SPSS 25.0 software was used to calculate the biological activity regression equation, EC_{50} value, and 95% confidence limit of the agent against the mycelial growth of the pathogen of *A. bisporus*. A safety evaluation and a field efficacy test using Ducan's method were performed to compare the differences between different treatments in each test. A value of $p < 0.05$ was considered statistically significant, and the data were expressed as mean ± standard deviation.

3. Results

3.1. Toxicity Measurement

To measure the inhibitory effects of nine fungicides on the mycelial growth of *C. mycophilum*, we used the mycelial growth rate method (Table 4). The EC_{50} values of quatrimycin, prochloraz–manganese chloride complex, seboctylamine acetate, prothioconazole, tebuconazole, difenoconazole, and polyoxin were 0.269, 0.939, 0.991, 0.240, 1.435, 6.589, and 4.396 mg/L, respectively. These fungicides exhibited high toxicity against the mycelial growth of cobweb disease in *A. bisporus*. Pyrimethanil and kresoxim-methyl had low toxicity against the mycelial growth of *A. bisporus*, with EC_{50} values of 34.584 and 31.271 mg/L, respectively (Figure 2).

Table 4. Bioassay testing of 9 fungicides against *A. bisporus* cobweb disease.

Fungicides	Regression Equation	Coefficient of Determination	$EC_{50}/(mg \cdot L^{-1})$	95% Confidence Interval
Quatrimycin	y = 0.803 + 1.409x	0.927	0.269	0.095~0.582
Prochloraz–manganese chloride complex	y = 0.027 + 1.002x	0.932	0.939	0.171~3.093
Seboctylamine acetate	y = 0.006 + 1.645x	0.968	0.991	0.367~2.696
Prothioconazole	y = 0.820 + 1.324x	0.809	0.240	0.001~0.574
Tebuconazole	y = −0.150 + 0.956x	0.864	1.435	0.426~12.328
Difenoconazole	y = −0.585 + 0.714x	0.892	6.589	-
Pyrimethanil	y = −2.077 + 1.350x	0.663	34.584	15.469~324.450
Kresoxim-methyl	y = −1.843 + 1.233x	0.773	31.271	13.563~340.752
Polyoxin	y = −1.044 + 1.623x	0.623	4.396	1.129~19.953

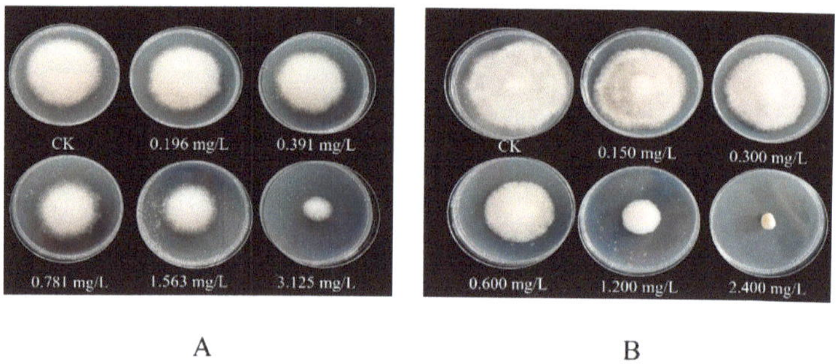

Figure 2. Bioassay testing of prochloraz–manganese chloride complex and seboctylamine acetate to *C. mycophilum*. (**A**) Seboctylamine acetate; (**B**) Prochloraz–manganese chloride complex.

3.2. Spore Germination Test

The results of the inhibition of eight fungicides on the spore germination of *A. bisporus* are shown in Table 5. The EC_{50} values of these eight fungicides on spore germination range from 0.338 mg/L to 80.588 mg/L. Among them, seboctylamine acetate, tebuconazole, and quatrimycin showed strong inhibitory activity against spore germination; the EC_{50} values were 0.652, 0.338, and 1.365 mg/L, respectively. On the other hand, the inhibitory activities of prochloraz–manganese chloride complex, prothioconazole, pyrimethanil, difenoconazole, and kresoxim-methyl against the spore germination of *C. mycophilum* were found to be weak, with EC_{50} values of 80.588, 43.766, 13.393, 48.227, and 57.117 mg/L, respectively. These results indicate that both prochloraz–manganese chloride complex and seboctylamine acetate have good indoor activity against *C. mycophilum*.

The above results indicate that both prochloraz–manganese chloride complex and seboctylamine acetate exhibited effective inhibition against the mycelial growth and spore germination of the pathogenic fungus.

Table 5. Toxicity of 8 fungicides against spores of *C. mycophilum*.

Fungicides	Regression Equation	Coefficient of Determination	EC_{50}/(mg·L^{-1})	95% Confidence Interval
Prochloraz–manganese chloride complex	y = −1.958 + 1.027x	0.677	80.588	55.482~125.775
Seboctylamine acetate	y = 0.264 + 1.417x	0.845	0.652	0.509~0.824
Prothioconazole	y = −3.636 + 2.216x	0.38	43.766	24.923~91.635
Pyrimethanil	y = −1.564 + 1.388x	0.506	13.393	7.598~21.968
Difenoconazole	y = −2.025 + 1.203x	0.824	48.227	32.280~85.313
Kresoxim-methyl	y = −3.191 + 1.816x	0.5	57.117	31.357~276.730
Tebuconazole	y = −1.346 + 2.86x	0.912	0.338	0.295~0.387
Quatrimycin	y = −1.563 + 1.970x	0.933	1.365	0.897~3.652

3.3. Safety Test

Following the results of the indoor toxicity test, prochloraz–manganese chloride complex and seboctylamine acetate were chosen for further testing. The results of the indoor safety test results are presented in Table 6. The data indicate that at three different concentrations of 0.99, 1.98, and 3.96 mg/L of seboctylamine acetate and three concentrations of 0.94, 1.88, and 3.76 mg/L of prochloraz–manganese chloride complex, there were no significant differences in the inhibition of mycelial growth across the three varieties of *A. bisporus* (*A. bisporus* W2000, *A. bisporus* W192, and *A. bisporus* S2796 (Figure 3). This demonstrates that the use of these two agents on *A. bisporus* will not affect growth.

Table 6. Indoor safety test of prochloraz–manganese chloride complex and seboctylamine acetate on *A. bisporus*.

Fungicides	Effective Concentration/ (mg·L^{-1})	*A. bisporus* W2000		*A. bisporus* W192		*A. bisporus* S2796	
		Colony Diameter/(cm)	Inhibition Ratio	Colony Diameter /(cm)	Inhibition Ratio	Colony Diameter/(cm)	Inhibition Ratio
CK	-	6.43 ± 0.17 a	-	6.52 ± 0.18 a	-	6.51 ± 0.16 a	-
Prochloraz–manganese chloride complex	0.94	6.28 ± 0.05 ab	0.02	6.41 ± 0.06 a	0.02	6.23 ± 0.05 bcd	0.04
	1.88	6.25 ± 0.03 ab	0.03	6.35 ± 0.06 a	0.03	6.17 ± 0.06 cd	0.05
	3.76	6.13 ± 0.04 b	0.05	6.33 ± 0.06 a	0.03	6.14 ± 0.06 d	0.06
Seboctylamine acetate	0.99	6.32 ± 0.05 ab	0.02	6.39 ± 0.07 a	0.02	6.39 ± 0.05 abcd	0.02
	1.98	6.26 ± 0.07 ab	0.03	6.37 ± 0.07 a	0.02	6.26 ± 0.10 abcd	0.04
	3.96	6.22 ± 0.07 ab	0.03	6.29 ± 0.10 a	0.04	6.25 ± 0.09 abcd	0.04

Note: Different lowercase letters in the same column in the table indicate significant differences at the 0.05 level between different treatments.

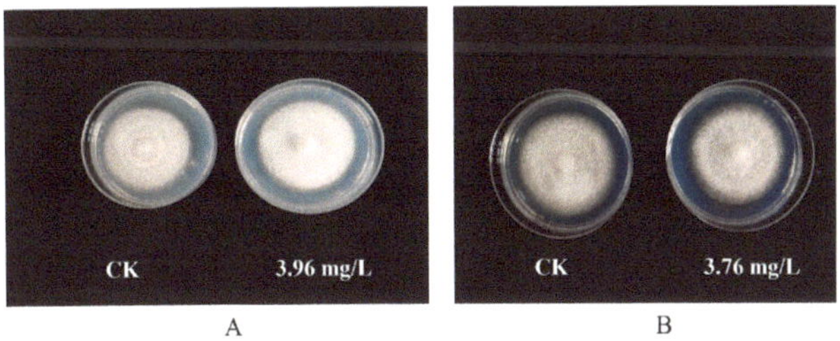

Figure 3. Safety of prochloraz–manganese chloride complex and seboctylamine acetate against *A. bisporus* mycelium. (**A**) Seboctylamine acetate; (**B**) prochloraz–manganese chloride complex.

Based on indoor toxicity determination, field-recommended doses of prochloraz–manganese chloride complex and seboctylamine acetate were selected for follow-up tests. Table 7 clearly demonstrates that the application of prochloraz–manganese chloride complex at concentrations of 400 mg/L, 800 mg/L, and 1600 mg/L, as well as seboctylamine acetate at concentrations of 6 mg/L, 12 mg/L, and 24 mg/L, did not have any negative impact on the yield of *A. bisporus* fruiting bodies. Furthermore, there was no significant difference observed between the treated groups and the control group in terms of yield. Importantly, no signs of necrosis, deformity, or other adverse symptoms were noted, indicating that these two agents can be safely used on *A. bisporus*.

Table 7. Field safety test of prochloraz–manganese chloride complex and seboctylamine acetate on *A. bisporus* (effect on yield).

Fungicides	Effective Concentration/(mg·L^{-1})	Yield/(g)
CK	-	529.3 ± 21.7 a
Prochloraz–manganese chloride complex	400	519.9 ± 3.4 a
	800	502.1 ± 10.4 a
	1600	492.6 ± 3.6 a
Seboctylamine acetate	6	493.0 ± 3.9 a
	12	527.5 ± 10.0 a
	24	530.0 ± 16.6 a

Note: Yield is the fresh weight of *A. bisporus* per 20 fruited bodies of different treatments (unit: g). Different lowercase letters in the same column in the table indicate significant differences at the 0.05 level between different treatments.

The indoor and field safety tests indicated that prochloraz–manganese chloride complex and seboctylamine acetate were not only highly toxic to *C. mycophilum*, but also safe for the growth of *A. bisporus*.

3.4. Control Effect of Prochloraz–Manganese Chloride Complex and Seboctylamine Acetate Against Cobweb Disease in A. bisporus

Table 8 summarizes the results of the efficacy test, revealing that both 50% prochloraz–manganese chloride complex WP and 1.8% seboctylamine acetate AS exhibited excellent control over the cobweb disease caused by *C. mycophilum* in *A. bisporus*. Spraying treatment was carried out when the mushroom buds appeared 10 days after soil covering. The test results after 10 days clearly indicate that 800 mg/L and 1600 mg/L concentrations of 50% prochloraz–manganese chloride complex WP were effective in controlling cobweb disease in *A. bisporus*, with control effects of 54.11% and 61.38%, respectively. It is evident that the higher the concentration, the better the control effect. Similarly, for 1.8% seboctylamine acetate AS, concentrations of 2 mg/L, 4 mg/L, and 6 mg/L were all found to be beneficial for preventing and treating cobweb disease in *A. bisporus*, with control effects of 71.76%, 78.82%, and 81.17%, respectively. Notably, the 6 mg/L concentration of 1.8% seboctylamine acetate aqueous solution achieved a control effect of 81.17%, suggested that this concentration offers the most optimal control among the tested doses.

Table 8. Field efficacy test of prochloraz–manganese chloride complex and seboctylamine acetate on cobweb disease in *A. bisporus*.

Fungicides	Experimental Chemical Concentration/(mg·L^{-1})	Incidence/(%)	Control Effect/(%)
CK	-	61.0	-
Prochloraz–manganese chloride complex	400	29.4	49.41 ± 6.35 d
	800	26.6	54.11 ± 5.69 d
	1600	22.4	61.38 ± 4.63 c
Seboctylamine acetate	2	16.7	71.76 ± 3.65 b
	4	12.4	78.82 ± 2.89 a
	6	10.9	81.17 ± 6.51 a

Note: Different lowercase letters in the same column in the table indicate significant differences at the 0.05 level between different treatments.

3.5. Effects of Prochloraz–Manganese Chloride Complex and Seboctylamine Acetate Against Yield in A. bisporus

The results presented in Table 7 and Figure 4 indicate that the application of 50% prochloraz–manganese chloride complex WP and 1.8% seboctylamine acetate AS as chemical treatments did not significantly impact the yield of *A. bisporus* fruiting bodies compared to the untreated control group. There were no observed instances of necrosis or deformity in the mushrooms treated with these agents, further validating their safety for use on *A. bisporus*. Consequently, it can be confidently stated that 50% prochloraz–manganese chloride complex WP and 1.8% seboctylamine acetate AS are suitable for application in the management of cobweb disease in *A. bisporus* cultivation, ensuring both disease control and the preservation of mushroom quality and yield.

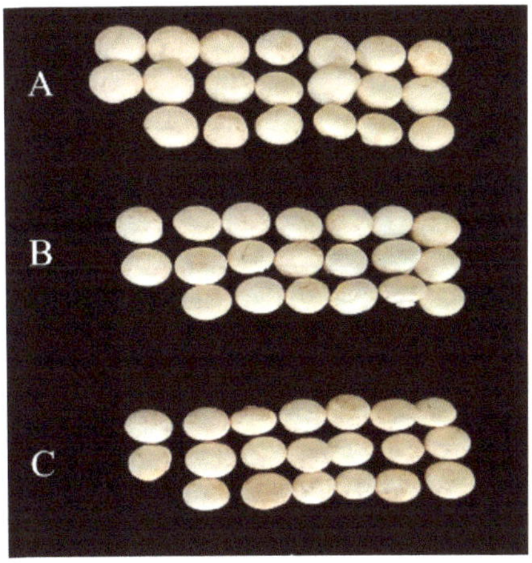

Figure 4. The effects of the highest concentrations of two fungicides on the yield of *A. bisporus*. (**A**) CK; (**B**) 24 mg/L 1.8% seboctylamine acetate; (**C**) 1600 mg/L 50% prochloraz–manganese chloride complex.

3.6. Effects of Prochloraz–Manganese Chloride Complex and Seboctylamine Acetate on Pesticide Residues in A. bisporus

The results regarding pesticide residue indicate that prochloraz–manganese chloride complex and seboctylamine acetate decompose relatively quickly. Specifically, when applied at concentrations of 1.8% for seboctylamine acetate aqueous solution (AS) and 50% for prochloraz–manganese chloride complex wettable powder (WP), with effective ingredient levels of 6 mg/L and 400 mg/L, respectively, the residual amounts after nine days of application at a rate of 250 mL per square meter were notably low. The final residue of seboctylamine acetate AS was measured at 0.116 mg/kg, while that of prochloraz–manganese chloride complex WP was 1.325 mg/kg. The maximum residue limit (MRL) for prochloraz–manganese chloride complex in edible fungi is 2 mg/kg, and our experimental results fell below this threshold, adhering to the national standard. For seboctylamine acetate, while a specific MRL for mushrooms is not established, the MRL for vegetables is 0.5 mg/kg. Notably, our experimental findings revealed residues of seboctylamine acetate to be less than 0.5 mg/kg, significantly lower than the established limit for vegetables (Figure 5).

As a result, the application of 1.8% seboctylamine acetate AS and 50% prochloraz–manganese chloride complex WP during the fruiting bud stage of *A. bisporus* is deemed suitable. This practice effectively mitigates *A. bisporus* cobweb disease. The recommended dosages of active ingredients in the field are 1600 mg/L and 2–6 mg/L, respectively.

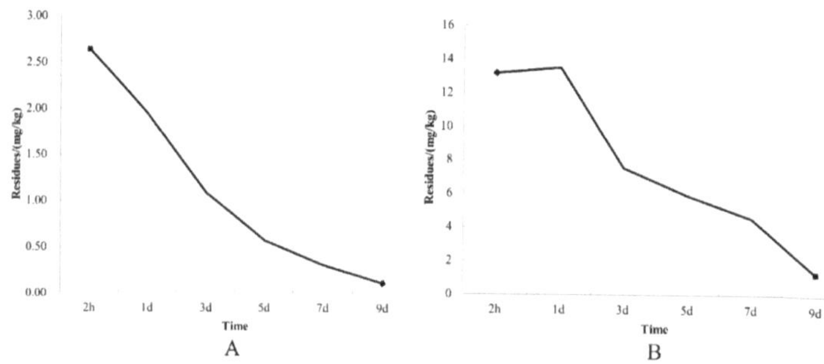

Figure 5. Degradation curves of chemical agents in *A. bisporus* fruiting body. (**A**) Degradation curve of seboctylamine acetate in sub-body of *A. bisporus* at 6 mg/L; (**B**) degradation curve of prochloraz–manganese chloride complex in fruiting body of *A. bisporus* at 400 mg/L.

4. Discussion

As the *A. bisporus* planting industry rapidly develops, diseases affecting *A. bisporus* are becoming more common. Pathogens such as fungi, bacteria, and viruses pose a significant threat to *A. bisporus* production [9], with cobweb disease pathogens being especially prevalent. In China, cobweb disease in *A. bisporus* is caused by *C. mycophilum*, a pathogenic microorganism that infects mushrooms through their stomata, leading to the rapid decay and widespread contamination of surrounding mushrooms, resulting in substantial economic losses [11,37,38]. At present, chemical control measures are currently employed to manage *A.bisporus* cobweb disease, but the choice and number fungicides is limited, and there is a lack of field evaluation regarding the safety of these chemical agents [21]. Prochloraz-Mn and prochloraz have been utilized effectively to control mushroom diseases. Previous studies have demonstrated the inhibitory effects of prochloraz–manganese chloride complex and 2–benzo imidazole methyl carbamate on cobweb disease in edible mushroom [39], with Prochloraz–manganese chloride complex displaying effective antifungal properties. Additionally, research has shown the effectiveness of metrafenone (EC_{50} = 0.025 mg/L) and prochloraz-Mn (EC_{50} = 0.045 mg/L) as alternative fungicides for disease control [39]. Furthermore, prochloraz–manganese has been found to control green mold on white mushroom (*A. bisporus*) and increase the yield of *A. bisporus*, and can effectively control wet bubble (WBD) disease caused by *Mycogone rosea* while also increasing the yield of *A.bisporus* [19]. Chakwiya et al. focused on the sensitivity of *C. mycophilum* to carbendazim and prochloraz–manganese; prochloraz–manganese ED_{50} values varied from 0.00001 mg/L to 0.55 mg/L, and it is to be used in a disease management strategy [20]. Pathogen sensitivity to fungicides was estimated by probit analyses. Fungicide susceptibility tests showed that *C. mycophilum* strains were highly sensitive both to prochloraz (ED_{50} < 0.087 μg mL^{-1}) and the newly introduced metrafenone (ED_{50} < 0.15 μg mL^{-1}) [21]. Prochloraz can no longer be used in the European Union (EU) as of June 2023, and over-reliance on metrafenone has resulted in putative resistant pathogenic strains emerging. Prochloraz still showed good control of two different isolates of *C. mycophilum*, with efficacy values consistently reaching 70% [22]. This supports the findings of Stanojević et al. who also found that prochloraz performed better than the tested BCA in green mold and dry bubble disease trials in vivo [40]. However, the long-term use of prochloraz–manganese chloride complex can lead to resistance in mushroom disease, affecting its efficacy. For example, *C. mycophilum* type II isolate 192B1 has been found to develop resistance to prochloraz-Mn, rendering it ineffective in preventing cobweb disease symptoms [23]. In this study, we selected prochloraz–manganese chloride complex, due to its commonly used status in edible fungi, and found that it exhibited higher inhibitory toxicity towards the mycelial growth of *C. mycophilum*, and the EC_{50} was 0.939 mg/L. Discrepancies in the results between this study and others may be attributed to variations in pathogen virulence among different regions, which is also an important reason for this situation. In an effort to mitigate the development of resistance to a single fungicide, this study screened eight additional chemical fungicides, including seboctylamine acetate, a product currently under research and development in China. Seboctylamine acetate, a broad-spectrum fungicide with potential for controlling diseases in fruits and vegetables, demonstrated potent inhibitory effects against the mycelial growth and spore germination of *C. mycophilum*, with EC_{50} values of 0.990 mg/L and 0.652 mg/L, respectively. This finding presents a promising alternative for the prevention and treatment of *A. bisporus* cobweb disease in the future.

The application of fungicides for controlling mushroom crop diseases necessitates a thorough consideration of diverse factors, including the intricate interplay between edible fungi and pathogens. This interaction encompasses not only the modes of pathogen infection, but also the defensive mechanisms of mushrooms and sensitivity to fungicides. There is an urgent need for fungicides that have a good inhibitory effect on pathogens and are harmless to the host [41]. Hence, when employing chemical agents to combat pathogens, it is indispensable to assess their potential harm to *A. bisporus* strains, aiming to minimize or eliminate any adverse effects. Bhat et al. demonstrated that carbendazim is both highly efficacious and cost-effective in inhibiting the mycelial growth and conidial germination of pathogenic fungi, while posing no threat to the spores of edible fungi [42]. Liu et al. found that *Bacillus subtilis* B154 has antagonistic activity against *Alternaria alternata*, but does not affect the growth of *A. bisporus* [43]. Our results are similar to the results of Bhat et al. and Liu et al. [42,43] and show that the application of prochloraz–manganese chloride complex and seboctylamine acetate had no detrimental impact on the regular growth of *A. bisporus*. The field experiment results further revealed that at a concentration of 400 mg/L, prochloraz–manganese chloride complex achieved control efficacy of 61.38%. When the concentration of seboctylamine acetate was 6 mg/L, the control effect was 81.17%. The application of two chemical agents could prevent and control cobweb disease in *A. bisporus*, and does not affect the yield of *A. bisporus*.

Bioactive compounds extracted from plants or microorganisms can be used to prevent and treat diseases in edible fungi. These compounds have strong antifungal properties, especially in the treatment of spider web disease [41]. Clarke et al. also found that 48 h of *C. mycophilum* culture was exposed to 25% v/v, and 96 h of *B. velezensis* culture to a filtrate, and the biomass ($p < 0.0002$) was reduced by 57% [44]. Additionally, the essential oil extracted from cinnamon, geranium, and mint showed a good inhibitory effect on *C. mycophilum*, but caused damage to mushroom mycelium [17]. Furthermore, an EEO-TG emulsion prepared by Han et al. was able to inhibit a pathogen of *A. bisporus* from changing its membrane permeability [7]. However, the mechanism of action underlying biological pesticides is not well understood, potentially affecting their effectiveness compared to chemical fungicides in the field. Nevertheless, the burgeoning trend toward green agriculture presents a promising outlook for the development of biological pesticides in the realm of edible fungi [45].

Both chemical agents and biological pesticides are used to prevent and manage diseases, with a focus on minimizing adverse impacts on human health and the environment. It is crucial to safeguard crops from these diseases [41,46]. Pesticide residues can pose a significant risk to food safety, and the excessive use of pesticides during pest management can threaten human health [47]. Research has shown high occurrences of pesticide residues in edible fungi, with carbendazim, acephate, diuron, prochloraz, and dichlorvos being the predominant pesticides found. Wang et al. discovered that *Flammulina velutipes*, *Lentinus edodes*, and *Pleurotus ostreatus* exhibited a high occurrence of pesticide residues [38]. Chemical agents are commonly employed in the control of edible fungi. However, post-treatment, the residues that persist in agricultural products may exceed their permissible safe limits, thereby posing potential health hazards to humans [48]. To date, Han et al. have utilized an enhanced QuEChERS method coupled with ultrahigh-performance liquid chromatography–tandem mass spectrometry to accurately quantify xinjuan (seboctylamine) pesticide residues in plant-derived food products [49]. In addition, a method for the determination of seboctylamine acetate was established. The pesticide residue in *A. bisporus* is lower than the maximum limit standard of pesticide residue, so the modified agent is suitable for use in *A. bisporus*. In this study, a method for the determination of seboctylamine acetate residue was established, the results showed that the final residues of seboctylamine acetate and prochloraz–manganese chloride complex at 250 mL/m^2 for 9 d were 0.116 and 1.325 mg/kg, respectively, which were lower than the maximum residue limit standard and could be used in *A. bisporus*. In a separate study that evaluated the residue levels of prochloraz–manganese chloride complex in grapes and soil, it was found that the minimum detection limit was 1.0×10^{-10} g, and the minimum detectable concentration was 0.04 mg/kg in grapes and 0.05 mg/kg in soil [50]. The findings presented in this study align with the aforementioned results, suggesting that the application of prochloraz–manganese chloride complex and seboctylamine acetate on *A. bisporus* is both safe and effective.

5. Conclusions

In conclusion, this study presented effective and safe fungicides to control cobweb disease in *A. bisporus*. It provides a detailed description of the residue determination analysis of the fungicide seboctylamine acetate in mushroom crop. This study proved that prochloraz–manganese chloride complex and seboctylamine acetate had a good inhibitory effect on the pathogen *C. mycophilum*. Moreover, the application of the two agents had no adverse effect on the mycelial growth and fruiting body growth of *A. bisporus*, and the residual amount of the agents was lower than the maximum residue limit standard. Field applications showed that 400 mg/L prochloraz–manganese chloride complex and 6 mg/L seboctylamine acetate gave 61.38% and 81.17% disease control, respectively.

Author Contributions: Q.C.: Investigation, Methodology, Writing—Original Draft, Formal Analysis; Y.Y.: Investigation, Formal Analysis; G.C.: Investigation, Formal Analysis; N.L.: Investigation; X.L.: Investigation, Formal Analysis; Y.L.: Funding Acquisition, Resources; H.W.: Supervision, Conceptualization, Resources, Writing—Original Draft, Writing–Review and Editing, Funding Acquisition. All authors have read and agreed to the published version of the manuscript.

Funding: This research was supported by the National Key R&D Program of China (2023YFD1401200), the Modern Agricultural Technology Industry System of Shandong Province (SDAIT-07-04), and the National Natural Science Foundation of China (32102259).

Institutional Review Board Statement: No applicable.

Informed Consent Statement: No applicable.

Data Availability Statement: The data that support the findings of this study will be available from the corresponding author upon reasonable request.

Conflicts of Interest: The authors declare that they have no known competing interests or personal relationships that could have appeared to influence the work reported in this paper.

References

1. Kumar, H.; Bhardwaj, K.; Kuča, K.; Sharifi Rad, J.; Verma, R.; Machado, M.; Kumar, D.; Cruz Martins, N. Edible mushrooms' enrichment in food and feed: A mini review. *Int. Food Sci. Technol.* **2022**, *53*, 1386–1398. [CrossRef]
2. Hou, X.; Luo, C.; Chen, S.; Jiang, J.; Yang, Z.; Wang, F.; Xie, X. Progress in research on diseases of edible fungi and their detection methods: A review. *Crop Prot.* **2023**, *174*, 106420. [CrossRef]
3. Bellettini, M.; Bellettini, S.; Fiorda, F.; Pedro, A.; Bach, F.; Morón, M.; Ribani, R. Diseases and pests noxious to *Pleurotus* spp. mushroom crops. *Rev. Argent. Microbiol.* **2018**, *50*, 216–226. [CrossRef] [PubMed]
4. Chen, M.; Liao, J.; Li, H.; Cai, Z.; Guo, Z.; Wach, M.; Wang, Z. iTRAQ-MS/MS proteomic analysis reveals differentially expressed proteins during post-harvest maturation of the white button mushroom *Agaricus bisporus*. *Curr. Microbiol.* **2017**, *74*, 641–649. [CrossRef] [PubMed]
5. Oke, F.; Aslim, B. Protective effect of two edible mushrooms against oxidative cell damage and their phenolic composition. *Food Chem.* **2011**, *128*, 613–619. [CrossRef]
6. Wu, J.; Wang, R.; Liu, X.; Ni, Y.; Sun, H.; Deng, X.; Wan, L.; Liu, F.; Tang, J.; Yu, J.; et al. Calcium dynamics during the growth of *Agaricus bisporus*: Implications for mushroom development and nutrition. *Chem. Biol. Technol. Agric.* **2023**, *10*, 99. [CrossRef]
7. Han, J.; Jiang, X.; Feng, L.; Wang, J.; Wang, X.; Zhou, Q.; Kitazawa, H.; Guo, Y.; Li, L. Identification of *Trichderma harzianum* in postharvest *Agaricus bisporus* and a novel control approach using eucalyptus essential oil emulsion. *Sci. Hortic.* **2024**, *329*, 113029. [CrossRef]

8. Xue, Y.; Xie, J.; Xu, X.; Yong, L.; Hu, B.; Liang, J.; Li, X.; Qing, L. UPLC-QqQ/MS combined with similarity assessment of 17 nucleic acid constituents in 147 edible fungi from Sichuan Basin, China. *Food Res. Int.* **2019**, *120*, 577–585. [CrossRef]
9. Zhang, G.; Tang, C. First report of *Acrostalagmus* luteo-albus causing red rust of needle mushroom (*Flammulina velutipes*) in China. *Plant Dis.* **2015**, *99*, 158. [CrossRef]
10. Umar, M.; Van Griensven, L. Studies on the morphogenesis of *Agaricus bisporus*: The dilemma of normal versus abnormal fruit body development. *Mycol. Res.* **1999**, *103*, 1235–1244. [CrossRef]
11. Lan, Y.; An, X.; Wang, Q.; Tang, L. Study on biological characteristics of *A. bisporus* cobweb disease. *Edible Fungi China* **2017**, *36*, 62–65.
12. Dhawan, A.; Chakraborty, S. Pulsed light treatment of whole white button mushroom (*Agaricus bisporus*): Kinetics and mechanism of microbial inactivation and storage study. *J. Food Sci.* **2024**, *89*, 5319–5334. [CrossRef] [PubMed]
13. Navarro, J.; Carrasco, J.; González, A.; Suz, L. First report of cobweb on white button mushroom (*Agaricus bisporus*) in Spain caused by *Cladobotryum mycophilum*. *Plant Dis.* **2012**, *96*, 1067. [PubMed]
14. Gea, F.; Navarro, M.; Suz, L.M. First report of cobweb disease caused by *Cladobotryum dendroides* on shiitake mushroom (*Lentinula edodes*) in Spain. *Plant Dis.* **2018**, *102*, 1030–1031. [CrossRef]
15. Sossah, F.; Liu, Z.; Yang, C.; Okorley, B.; Sun, L.; Fu, Y.; Li, Y. Genome sequencing of *Cladobotryum protrusum* provides insights into the evolution and pathogenic mechanisms of the cobweb disease pathogen on cultivated mushroom. *Genes* **2019**, *10*, 124. [CrossRef] [PubMed]
16. Bian, Y.; Xiao, Y.; Guo, M.P. Research progress on prevention and control of edible fungi diseases. *Acta Edulis Fungi* **2021**, *28*, 121–131.
17. Lan, Y.; Wang, Q.; Yu, C.; Cong, Q.; Tang, L.; An, X.; Kong, F.; Li, X. First Report of *Cladobotryum mycophilum* Causing Cobweb Disease on *Agaricus bisporus* in China. *Plant Dis.* **2016**, *100*, 2334. [CrossRef]
18. Shi, N.; Ruan, H.; Jie, Y.; Chen, F.; Du, Y. Sensitivity and efficacy of fungicides against wet bubble disease of *Agaricus bisporus* caused by *Mycogone perniciosa*. *Eur. J. Plant Pathol.* **2020**, *157*, 873–885. [CrossRef]
19. Du, Y.; Shi, N.; Ruan, H.; Miao, J.; Yan, H.; Shi, C.; Chen, F.; Liu, X. Analysis of the prochloraz-Mn resistance risk and its molecular basis in Mycogone rosea from *Agaricus bisporus*. *Pest Manag. Sci.* **2021**, *77*, 4680–4690. [CrossRef]
20. Chakwiya, A.; van der Linde, E.J.; Korsten, L. In vitro sensitivity testing of *Cladobotryum mycophilum* to carbendazim and prochloraz manganese. *S. Afr. J. Sci.* **2015**, *111*, 1–7. [CrossRef]
21. Luković, J.; Potočnik, I.; Šantrić, L.; Radivojević, L.; Todorović, B.; Milijašević-Marčić, S.; Szűcs, A.; Kredics, L.; Vágvölgyi, C.; Hatvani, L. Response of the mushroom pathogen *Cladobotryum mycophilum* to prochloraz and metrafenone fungicides and *Streptomyces flavovirens* actinobacteria. *J. Environ. Sci. Health B* **2022**, *57*, 636–643. [CrossRef] [PubMed]
22. Clarke, J.; McGuinness, B.; Fitzpatrick, D.; Kavanagh, K.; Grogan, H. Response of the mushroom pathogen *Cladobotryum mycophilum* to the fungicides prochloraz and metrafenone and two *Bacillus*-based biological control agents in mushroom crop trials. *Crop Prot.* **2024**, *177*, 106530. [CrossRef]
23. Grogan, H. Fungicide control of mushroom cobweb disease caused by *Cladobotryum* strains with different benzimidazole resistance profiles. *Pest Manag. Sci.* **2006**, *2*, 153–161. [CrossRef] [PubMed]
24. Qin, G.; Chen, Y.; He, F.; Yang, B.; Zou, K.; Shen, N.; Zuo, B.; Liu, R.; Zhang, W.; Li, Y. Risk assessment of fungicide pesticide residues in vegetables and fruits in the mid-western region of China. *J. Food Compos. Anal.* **2021**, *95*, 103663. [CrossRef]
25. Marquette, C.; Blum, L.J. Applications of the luminol chemiluminescent reaction inanalytical chemistry. *Anal. Bioanal. Chem.* **2006**, *385*, 546–554. [CrossRef] [PubMed]
26. Guo, S.J.; Zhang, C.C.; Li, B.; Zhang, F.; Han, J.; Chen, X.; Su, X.F.; Sun, J.H.; Feng, J.L. Simultaneous determination of quaternary phosphonium compounds and phosphine oxides in environmental water and solid samples by ultrahigh performance liquid chromatography–tandem mass spectrometry. *J. Chromatogr. A* **2024**, *1733*, 465280. [CrossRef]
27. AI-Amoud, A.; Clark, B.; Chrystyn, H. Determinationof gentamicin in urine samples after inhalation by reversed-phase high-performance liquid chromatography using pre-column derivatisation with o-phthalaldehyde. *J. Chromatogr. B* **2002**, *769*, 89–95. [CrossRef]
28. Chen, Q.; Qiu, Y.; Yuan, Y.; Wang, K.; Wang, H. Biocontrol activity and action mechanism of *Bacillus velezensis* strain SDTB038 against Fusarium crown and root rot of tomato. *Front. Microbiol.* **2022**, *13*, 994716. [CrossRef]
29. Qiu, Y.; Yan, H.; Sun, S.; Wang, Y.; Zhao, X.; Wang, H. Use of *Bacillus velezensis* SDTB022 against tobacco black shank (TBS) and the biochemical mechanism involved. *Biol. Control* **2022**, *165*, 104785. [CrossRef]
30. Gao, Y.; He, L.; Li, B.; Mu, W.; Lin, J.; Liu, F. Sensitivity of *Colletotrichum acutatum* to six fungicides and reduction in incidence and severity of chili anthracnose using pyraclostrobin. *Australas. Plant Pathol.* **2017**, *46*, 521–528. [CrossRef]
31. Yang, X.; Deng, P.; Liu, Q.; Meng, Y.; Dong, P.; Xu, L.; Huang, L. Exploring the efficacy of carvacrol as a biocontrol agent against pear Valsa canker. *Pestic. Biochem. Phys.* **2023**, *196*, 105641. [CrossRef] [PubMed]
32. Chen, Q.; Gao, J.; Yang, X.; Qiu, X.; Wang, Y.; Wang, H. Synergistic effects of *Bacillus velezensis* SDTB038 and phenamacril on Fusarium crown and root rot of tomato. *Plant Pathol.* **2023**, *72*, 1453–1462. [CrossRef]
33. Fang, Q.; Yao, G.; Shi, Y.; Ding, C.; Wang, Y.; Wu, X.; Hua, R.; Cao, H. Residue dynamics and risk assessment of prochloraz and its metabolite 2,4,6-trichlorophenol in apple. *Molecules* **2017**, *22*, 1780. [CrossRef] [PubMed]
34. *NY/T 1965.1-2010*; Guidelines for Crop Safety Evaluation of Pesticides. Part 1: Laboratory Test for Crop Safety Evaluation of Fungicides and Insecticides. Ministry of Agriculture, Pesticide Testing Center: Beijing, China, 2010.

35. *NY/T 1464.10-2007*; Guidelines on Efficacy Evaluation of Pesticides. Part 10: Fungicides against wet Bubble Disease of Mushrooms. Ministry of Agriculture, Pesticide Testing Center: Beijing, China, 2007.
36. Santana-Mayor, A.; Rodríguez-Ramos, R.; Herrera-Herrera, A.; Socas-Rodríguez, B.; Rodríguez-Delgadoet, M. Updated overview of QuEChERS applications in food, environmental and biological analysis (2020–2023). *TrAC Trend Anal. Chem.* **2023**, *169*, 117375. [CrossRef]
37. Qin, Y.; Liu, D.; Wu, Y.; Yuan, M.; Li, L.; Yang, J. Effect of PLA/PCL/cinnamaldehyde antimicrobial packaging on physicochemical and microbial quality of button mushroom (*Agaricus bisporus*). *Postharvest Biol. Technol.* **2015**, *99*, 73–79. [CrossRef]
38. Wang, T.; Yun, J.; Zhang, Y.; Bi, Y.; Zhao, F.; Niu, Y. Effects of ozone fumigation combined with nano-film packaging on the postharvest storage quality and antioxidant capacity of button mushrooms (*Agaricus bisporus*). *Postharvest Biol. Technol.* **2021**, *176*, 111501. [CrossRef]
39. Carrasco, J.; Navarro, M.; Santos, M.; Gea, F. Effect of five fungicides with different modes of action on cobweb disease (*Cladobotryum mycophilum*) and mushroom yield. *Ann. Appl. Biol.* **2017**, *171*, 62–69. [CrossRef]
40. Stanojević, O.; Berić, T.; Potočnik, I.; Rekanović, E.; Stanković, S.; Milijašević-Marčić, S. Biological control of green mould and dry bubble diseases of cultivated mushroom (*Agaricus bisporus* L.) by *Bacillus* spp. *Crop Protect.* **2019**, *126*, 104944. [CrossRef]
41. Gea, F.J.; Navarro, M.J.; Santos, M.; Diánez, F.; Carrasco, J. Control of fungal diseases in mushroom crops while dealing with fungicide resistance: A Review. *Microorganisms* **2021**, *9*, 585. [CrossRef]
42. Bhat, A.; Bhat, Z.; Khan, N.; Mohiddin, F.; Mir, G. Response of fungicides against *Verticillium fungicola* and the host fungus *Agaricus bisporus* under in vitro conditions. *Pestic. Res. J.* **2017**, *29*, 6–11.
43. Liu, C.; Sheng, J.; Chen, L.; Zheng, Y.; David, Y.; Yang, Y.; Xu, M.; Shen, L. Biocontrol activity of *Bacillus subtilis* isolated from *Agaricus bisporus* mushroom compost against pathogenic fungi. *J. Agric. Food Chem.* **2015**, *63*, 6009–6018. [CrossRef] [PubMed]
44. Clarke, J.; Grogan, H.; Fitzpatrik, D.; Kavanagh, K. Analysis of the effect of Bacillus velezensis culture filtrate on the growth and proteome of *Cladobotryum mycophilum*. *Fungal Biol.* **2022**, *126*, 11–19. [CrossRef] [PubMed]
45. Largeteau, M.; Savoie, J. Microbially induced diseases of *Agaricus bisporus*: Biochemical mechanisms and impact on commercial mushroom production. *Appl. Microbiol. Biot.* **2010**, *86*, 63–73. [CrossRef] [PubMed]
46. Diamantopoulou, P.; Philippoussis, A.; Kastanias, M.; Flouri, F.; Tokousbalides, M. Effect of famoxadone, tebuconazole and trifloxystrobin on *Agaricus bisporus* productivity and quality. *Sci. Hortic.* **2006**, *109*, 190–195. [CrossRef]
47. Schusterova, D.; Mraz, P.; Uttl, L.; Drabova, L.; Kocourek, V.; Hajslova, J. Pesticide residues in fresh and processed edible mushrooms from Czech markets. *Food Addit. Contam. B.* **2023**, *16*, 384–392. [CrossRef]
48. Maron, D.; Smith, T.; Nachman, K. Restrictions on antimicrobial use in food animal production: An international regulatory and economic survey. *Glob. Health* **2013**, *9*, 1–11. [CrossRef]
49. Han, C.; Hu, B.; Chen, S.; Wang, N.; Hou, J.; Jin, N.; Shen, Y. Determination of Xinjunan pesticide residue in foodstuffs of plant origin by a modified QuEChERS method and ultra performance liquid chromatography-tandem mass spectrometry. *LWT* **2021**, *151*, 112101. [CrossRef]
50. Zheng, X.; Xu, J.; Sun, H. Study on the analysis method of prochloraz-manganese chloride complex 50% wettable powder in grape and soil by high performance liquid chromatography. *Pestic. Sci. Adm.* **2014**, *35*, 38–41.

Journal of *Fungi*

MDPI

Article

Integrated Transcriptomics–Proteomics Analysis Reveals the Response Mechanism of *Morchella sextelata* to *Pseudodiploöspora longispora* Infection

Shurong Wang [1,2,†], Jingyi Wang [1,†], Tengyun Wang [1], Tonglou Li [1], Lijing Xu [1,3], Yanfen Cheng [1,3], Mingchang Chang [1,2], Junlong Meng [1,2,*] and Ludan Hou [1,3,*]

1 College of Food Science and Engineering, Shanxi Agricultural University, Taigu, Jinzhong 030801, China; wzlj2005@163.com (S.W.); wangjingyi0304@163.com (J.W.); wangtengyun2021@163.com (T.W.); 15582690226@163.com (T.L.); xulijingsx@hotmail.com (L.X.); cyf2341986@163.com (Y.C.); sxndcmc@163.com (M.C.)

2 Shanxi Research Center for Engineering Technology of Edible Fungi, Taigu, Jinzhong 030801, China

3 Shanxi Key Laboratory of Edible Fungi for Loess Plateau, Taigu, Jinzhong 030801, China

* Correspondence: mengjunlongseth@hotmail.com (J.M.); houludan@126.com (L.H.)

† These authors contributed equally to this work and share first authorship.

Abstract: Morels (*Morchella* spp.) are valuable and rare edible mushrooms with unique flavors and high nutritional value. White mold disease occurring during cultivation has seriously affected the quality and yield of morels in China. In this study, the fungus causing white mold disease in morels was isolated, purified, and identified as *Pseudodiploöspora longispora* by morphology and molecular biology. In addition, research has shown that *P. longispora* infection causes wrinkled and rupturing asci, loosened cell walls, and obvious membrane breakage accompanied by severe cytoplasmic leakage in *M. sextelata*. Interestingly, research has shown that infection with *P. longispora* can induce the production of an unknown substance in the cells of *M. sextelata*, which accumulates on the cell membrane, leading to membrane breakage. Furthermore, integrated transcriptomics–proteomics analysis revealed the response mechanism of *M. sextelata* to *P. longispora* infection. The results indicate that DEGs and DEPs can be significantly enriched in pathways involved in oxidoreductase activity; peroxisomes, lipid transport, and metabolism; cell wall assembly; and integral components of membranes. Further electron microscopy analysis clarified the important role of changes in the cell membrane and cell wall in the response of mycelia to biological stress. This study clarified the response mechanism of *M. sextelata* to *P. longispora*, laying a foundation for further clarifying the infection mechanism of *P. longispora*.

Keywords: *Morchella sextelata*; *Pseudodiploöspora longispora*; infestation process; response mechanism; ultrastructure

Citation: Wang, S.; Wang, J.; Wang, T.; Li, T.; Xu, L.; Cheng, Y.; Chang, M.; Meng, J.; Hou, L. Integrated Transcriptomics–Proteomics Analysis Reveals the Response Mechanism of *Morchella sextelata* to *Pseudodiploöspora longispora* Infection. *J. Fungi* **2024**, *10*, 604. https://doi.org/10.3390/jof10090604

Academic Editor: Ofir Degani

Received: 31 July 2024
Revised: 23 August 2024
Accepted: 23 August 2024
Published: 26 August 2024

1. Introduction

Morels (*Morchella* spp.), named for their sheep belly-like shape formed by the uneven pileus [1], are famous and rare fungi for both food and medicine worldwide. Morels are rich in polysaccharides, phenolic compounds, amino acids, carotenoids, and other active ingredients. In addition to lowering blood glucose and blood lipids, the beneficial biological properties of morels include immunomodulatory, antitumor, anti-inflammatory, and antioxidant properties, and gut flora regulation [2]. Indoor cultivation has drawn much attention globally because of its tremendous economic potential. The success of indoor cultivation was first reported in 1982 by Ower [3]. After more than a century of research, field morels have been successfully cultivated for commercial purposes [4]. The scale of cultivation of morel mushrooms in China has increased annually. The *Morchella* species currently cultivated in China include *M. sextelata*, *M. importuna*, and *M. septimelata* [5]. Since 2012, the Chinese morel mushroom cultivation industry, exemplified by Sichuan Province, has been developing rapidly [6]. Unfortunately, with the development of the cultivation industry, fungal infections have become the main factor affecting the yield and quality of morels. Currently, various fungal diseases have been reported during the cultivation of morels, such as white mold disease (caused by *Pseudodiploöspora longispora* or *Paecilomyces penicillatus*) [7,8], stipe rot disease (*Fusarium incarnatum, Fusarium equiseti*) [9], cobweb disease (*Cladobotryum mycophilum, Cladobotryum protrusum*) [10,11], white rot disease (*Aspergillus* sp.) [12], stalk rot (*Fusarium nematophilum*) [13], and rot disease (*Clonostachys rosea, Lecanicillium aphanocladii, Trichoderma atroviride*) [14–16].

White mold disease commonly occurs in the main cultivation areas of morel mushrooms in China. It has become one of the most important diseases in the cultivation and production of morels, with the highest incidence in cultivation [17]. The disease is characterized by white, velvety hyphae on infected fruiting bodies, which expand rapidly at high temperatures (>25 °C) and high humidity (>90%), leading to wilting, rotting, deformities, and severe damage, which may result in the death of morel mushrooms. The first finding of disease caused by *P. longispora*, which mainly infects the pileus of fruiting bodies, was reported in Chongqing, China [7]. The annual incidence of white mold can reach 80% annually due to the rapid spread of large numbers of conidia around the cultivation area [17]. This is a potential risk to food safety and has been one of the major issues threatening the sustainable cultivation of morel mushrooms. Therefore, effective prevention and control methods for white mold disease are urgently needed.

Many environmental conditions are important for the prevention and control of disease infections in edible mushrooms. However, common strategies are not very effective in controlling the white mold of cultivated morels. Therefore, clarifying disease-causing mechanisms and cultivating disease-resistant varieties are effective ways to control diseases. The main mechanisms by which pathogenic fungi invade edible mushrooms include the mechanical action of mycelia, the toxic or catabolic action of secondary metabolites (toxins, cell wall degrading enzymes), and the action of immune substances that detoxify edible mushrooms. Secondary metabolites are important determinants of the pathogenicity of pathogenic fungi [18]. Peptidaibols are virulence factors that contribute to *P. longispora* invasion of *Morchella* [19]. Genomic characterization and comparative genomic analyses have been performed to reveal the evolution and related pathogenic genes of *P. longispora* through genomic characterization and comparative genomic analyses, which have provided valuable insights into the *P. longispora* genome, contributing to pathogenicity studies and drug development [20].

In recent years, the genomes of many *Morchella* species, including *M. sextelata, M. importuna*, and *M. septimedata*, have been published, laying the foundation for basic research on *Morchella* [21–23]. Current research indicates that the pathogenic mechanism of pathogenic fungi is gradually becoming clearer [24–26]. However, there have been no relevant reports on the response mechanism of *M. sextelata* to pathogenic fungi. This study first purified and identified the pathogen causing white mold disease. On this basis, the response pathways of *M. sextelata* to pathogenic fungi were further analyzed, providing theoretical support for the development of effective white mold control strategies.

2. Materials and Methods

2.1. Collection of Diseased White Mold M. sextelata and Isolation of the Pathogenic Fungus

White mold disease samples were collected for pathogen isolation from an *M. sextelata* cultivation greenhouse in Taigu County, Jinzhong City, Shanxi Province, China. Specifically, the surface of the fruiting bodies was rinsed with sterile water, and then the white mycelia at the diseased spots were placed on potato dextrose agar (PDA) medium and cultivated at 25 °C in the dark. Then, the pathogen was purified and cultured several times by removing the edge of the mycelia, and the purified pathogenic strains were stored at 4 °C. The remaining samples were stored at −80 °C.

2.2. Pathogenicity Test and Microscopic Observation

According to previous research methods, the isolated and purified pathogenic fungi were inoculated on PDA at 25 °C and cultured for 14 d until spore production, after which a spore suspension (1.0×10^6 spores mL^{-1}) was prepared [27]. The surface of the *M. sextelata* was sterilized with 75% ethyl alcohol, a wound of approximately 5 mm was cut on the cap of the *M. sextelata* with a sterile scalpel, and 200 μL of spore suspension was added to the wound. Equal amounts of sterile water were added to the scratches as a blank control. The *M. sextelata* treated were covered with plastic bags for humidity preservation. The onset of disease was observed and recorded every day. After 5 days, the pathogen was re-isolated from infected *M. sextelata* and characterized by morphological and phylogenetic analyses to test Koch's postulates [7].

The samples (1–2 mm^3) were completely submerged in 2.5% glutaraldehyde solution and stored at 4 °C overnight. The samples were rinsed with 0.1 M, pH 7.0 phosphate buffer solution (PBS) after the 2.5% glutaraldehyde fixative was removed. The samples were then dehydrated, critical point-dried, and coated, as previously reported [28]. The treated samples were observed via scanning electron microscopy (SEM) (Regulus 8100, Hitachi, Tokyo, Japan). The samples (1–2 mm^3) were placed into 2.5% glutaraldehyde and fixed at 4 °C for storage. The glutaraldehyde solution was removed, and the samples were rinsed three times with PBS for 15 min each time. The samples were fixed with 1% osmium for 1–2 h, and then rinsed with PBS three times. The samples were then dehydrated with ethanol solution and finally treated with pure acetone for 20 min. The samples were treated with a mixture of embedding agents and acetone. Pure embedding agent was used to treat the samples overnight, and the osmotically treated samples were embedded and heated at 70 °C overnight to obtain the embedded samples [29]. The samples were sectioned in an ultrathin microtome (EMUC7, Leica, Wetzlar, Germany) to obtain 70–90 nm sections, which were stained with lead citrate solution and uranyl acetate 50% ethanol saturated solution for 5–10 min, and then air-dried for visualization via transmission electron microscopy (TEM) (JEM1400, Hitachi, Tokyo, Japan) for observation.

2.3. Identification of Pathogens Causing White Mold Disease in M. sextelata

For morphological identification, the pathogen isolates were cultured on PDA at 25 °C for 10 days. Microstructures, including mycelia, conidia, conidiophores, and chlamydospores, including conidiophores and conidial size, were studied using optical microscopy (Eclipse Ni-U, with a Ri2 camera, Nikon, Tokyo, Japan) [30]. The genomic DNA of pathogen isolates was extracted using the E.Z.N.A.® Forensic DNA Extraction Kit (Omega Bio-Tek, Norcross, GA, USA). The primer pair ITS4 (5′-TCCTCCGCTTATTGATATGC-3′)/ITS5 (5′-GGAAGTAAAAGTCGTAACAAGG-3′) was used for amplifying the internal transcribed sequence (ITS) region [31]. The PCR mixture contained 13 μL of 2 × Easy Taq PCR SuperMix (+dye), 1 μL of each primer, 2 μL of genomic DNA, and ddH_2O for a total reaction volume of 25 μL. PCR analysis was conducted in a Bio-Rad T100 Thermal Cycler (Bio-Rad Laboratories Inc., Hercules, CA, USA). The amplification conditions were as follows: 1 cycle at 95 °C for 3 min; 34 cycles at 94 °C for 30 s, 53 °C for 30 s, and 72 °C for 30 s; and 1 cycle at 72 °C for 10 min. Amplification products were detected using gel electrophoresis on 1% agarose gels stained with Gel Green. The successfully amplified products were cleaned using an E.Z.N.A.® Cycle-Pure Kit (Omega Bio-Tek) and sequenced at Tsingke Biotechnology Co., Ltd. (Xi'an, China). The sequencing results were assembled with DNAMAN 9.0 software (https://www.lynnon.com) (accessed on 12 October 2023), and the assembled results were subsequently aligned by BLAST (Basic Local Alignment Search Tool) on the NCBI website (https://blast.ncbi.nlm.nih.gov/Blast.cgi) (accessed on 25 October 2023) to construct a phylogenetic tree using MEGA X [32,33].

2.4. Inhibitory Effect of the Pathogenic Fungus on M. sextelata on Plates

The inhibitory effect of water-soluble substances from pathogenic fungi on the growth of the mycelia of *M. sextelata* was detected using the plate standoff method [34] and fermentation liquid inhibition tests. *M. sextelata* cakes and pathogenic fungal cakes (diameter = 6 mm) were placed on each side of the PDA plate. Two pathogenic fungal cakes (diameter = 6 mm) were placed on each side of plates covered with *M. sextelata* mycelia. All of the plates were incubated at 25 °C in the dark under static and constant humidity conditions for 5 days of cultivation, and the mycelial growth of the two fungal strains in each treatment was observed. Pathogenic fungal mycelia 6 mm in diameter were inserted into a 250 mL conical flask containing 100 mL of potato dextrose broth (PDB) and incubated at 25 °C and 150 rpm/min for 6 days of oscillatory cultivation. The mycelia were filtered to obtain fermentation broth, which was subsequently filtered through a sterile 0.22 μm filter to obtain a sterile fermentation broth. After the PDA solid medium was autoclaved, it was cooled to approximately 55 °C, and fermentation broth was added, 5 mL of filtrate per 100 mL of PDA medium, mixed well, and poured into the plates to produce PDA plates containing 5% filtrate. In the control group, distilled water was used instead of fermentation broth to make the culture medium. The activated 6 mm diameter mycelia cakes of *M. sextelata* were placed on PDA supplemented with fermentation broth and incubated at 18 °C in the dark to observe the growth of *M. sextelata*. Based on the observed phenotypic changes, mycelial samples from the fermentation broth-treated (MT) and control groups (CK) were scraped and analyzed via transcriptomics and proteomics.

2.5. Effect of Nutrients from M. sextelata on Pathogenic Fungi

To explore whether the nutrients of *M. sextelata* can promote the growth of pathogenic fungi, dried *M. sextelata* were crushed with a pulverizer and sieved through a 100-mesh sieve. The powder of *M. sextelata* was added to PDA to final concentrations of 1% and 2%, with PDA medium used as a control group. Activated pathogen cakes (6 mm diameter) were placed on the three kinds of plates. After 10 days of incubation at 25 °C in the dark, the growth of the pathogen was observed, and the colony diameter was measured.

2.6. Transcriptome Sequencing

Mycelial samples of *M. sextelata* scraped from the MT group and CK group were sent to Majorbio Biopharm Technology Co., Ltd. (Shanghai, China) and sequenced on the Nova Seq6000 platform (Illumina, San Diego, CA, USA). Total RNA was extracted from the tissue using TRIzol® Reagent (Majorbio Biomedical Technology Co., Ltd., Shanghai, China) according the manufacturer's instructions. The differentially expressed genes (DEGs) were analyzed based on a $|\log2FC| \geq 1$ and p-adjusted < 0.05. Gene ontology (GO) enrichment and Kyoto Encyclopedia of Genes and Genomes (KEGG) enrichment analyses were performed by Goatools and R scripts, respectively, and the Majorbio Cloud Platform was utilized to analyze the data (Majorbio Biomedical Technology Co., Ltd., Shanghai, China, www.majorbio.com) (accessed on 20 January 2024).

2.7. Proteomic Sequencing

The same samples used for transcriptome sequencing were used. Proteomic analysis was performed at Majorbio Biopharm Biotechnology Co., Ltd. (Shanghai, China), on an Orbitrap Astral high-resolution mass spectrometer. The differentially expressed proteins (DEPs) were analyzed based on a $|\log2FC| >= 1$ and p-adjusted < 0.05.

2.8. Quantitative PCR (qPCR) Analysis

The qPCR analysis was performed to confirm the validity of the RNA-seq results. The qPCR amplification program was as follows: amplification at 95 °C for 3 min, amplification at 95 °C for 3 s, amplification at 60 °C for 32 s, 40 cycles, and amplification at 72 °C for 3 s. Supplementary Table S1 lists the primers used for qPCR. The relative expression levels of all genes were ascertained according to the $2^{-\Delta\Delta CT}$ method.

2.9. Statistical Analysis and Data Availability

Statistical analysis was performed using SPSS 25 software. The independent samples t-test was used to examine the data, which are presented as the means ± SEs. A p-value less than 0.05 was considered to indicate statistical significance. For figure analysis, GraphPad Prism 8 and Photoshop 2023 were utilized. The reference genome of *M. sextelata* can be downloaded from https://www.ncbi.nlm.nih.gov/genome/86229?genome_assembly_id=748597 (accessed on 18 January 2024).

3. Results

3.1. Pathogen Isolation and Identification

White mold disease spreads rapidly and lacks an effective means of control, which can result in a loss of quality and quantity and lead to economic losses in commercial morel cultivation (Figure 1A–D). White mold disease occurs mainly on the cap of fruiting bodies. In this study, rotting of the pileus of *M. sextelata* was observed (Figure 1E,F). After 8 days of cultivation on PDA plates, the colony diameter of the pathogenic fungi was 3–4 cm. The colonies were projected and white, with smooth edges and sparse to medium aerial mycelia (Figure 1G). The conidia were colorless and transparent, ellipsoidal or citriform, 4.5–5.5 × 15–18 μm, and had 1–3 septates (predominantly 1 septate) (Figure 1H–J). Chlamydospores in terminal or intercalary clusters of 2–10 cells are produced from vegetative hyphal cells (Figure 1K). With a diameter ranging from 6.3 to 11.5 μm and a smooth or verrucose surface, they were spherical, usually released in clusters, short or long chains, and occasionally in pairs or individually. The hyphae had a diameter of 2–4 μm and were smooth, septate, branched, and hyaline (Figure 1L).

The internal transcribed spacer (ITS) region of ribosomal DNA (rDNA-ITS) was sequenced and analyzed. The ITS fragments of strains BZ04, BZ07, and BZ08 were 511 bp, 516 bp, and 491 bp (GenBank accession numbers PP935354, PP935355, and PP935354), respectively. After searching and comparing the sequences with the NCBI database using BLAST, it was discovered that they had more than 99.80% similarity with those of *Pseudodiploospora longispora* and *Paecilomyces penicillatus*. The phylogenetic tree showed that the sequences of strains BZ04, BZ07, and BZ08 were in the same branch as those of *P. longispora* and *P. penicillatus* (Figure 2). The conidia of *P. longispora* have septates, whereas those of *Paecilomyces* spp. have no septates [35]. The conidia of *P. penicillatus* have no septates, whereas the pathogenic isolates of this study had 1–3 septates [7]. Therefore, based on the morphological traits and rDNA-ITS sequencing analysis results, *P. longispora* was finally determined to be the fungus causing pileus rot on cultivated *M. sextelata*.

3.2. Pathogenicity Testing and Effect of *P. longispora* Infestation on the Micromorphology of *M. sextelata*

In the early stage of the disease, irregular lesions and white fluffy mycelia appeared on the cap. With the continued infection of the pathogen, the mold layer continued to expand, leading to shrinkage and drying of the infected area. The cap undergoes decay and even perforation, which severely affects the growth and development of fruiting bodies and ultimately leads to wilting and death. A spore suspension of the isolated strain was made for pathogenicity testing. Pathogenicity results showed that 3 days after inoculation with the spore suspension, typical dense, white, thick mycelia were visible at the inoculation site (Figure 3B). After 6 days, the lesions expanded to form a large layer of white mold (Figure 3C), while the lesions in the control group did not significantly change (Figure 3A). Symptoms from artificial inoculation resembled those seen in the wild. To satisfy Koch's postulates, re-isolation from infected fruiting bodies was carried out, and the pathogen was confirmed to be in alignment with the inoculated isolate based on morphological characteristics.

Figure 1. Occurrences of fungal disease in *M. sextelata* ascocarps cultivated outdoors and the morphological characteristics of *M. sextelata* fungal disease. (**A–D**) Disease-infected young fruiting bodies (red arrow) of *M. sextelata*. (**E,F**) Wilting and shriveling of naturally infested *M. sextelata*. (**G**) Colony of the pathogenic isolate grown on PDA at 25 °C for 14 days and on the reverse side. (**H–J**) Conidia of the pathogenic isolate. (**K**) Chlamydospores of the pathogenic isolate. (**L**) Hyphae of the pathogenic isolate. Scale bars: (**H–L**) = 10 μm.

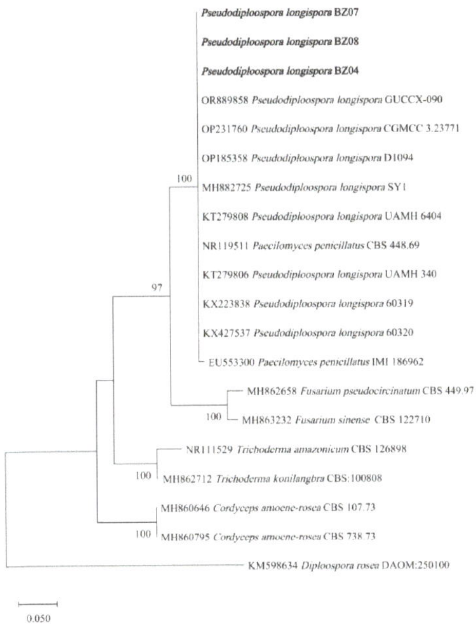

Figure 2. ML phylogenetic tree based on the ITS sequences. The bolded parts are the sequences of the isolates in this study.

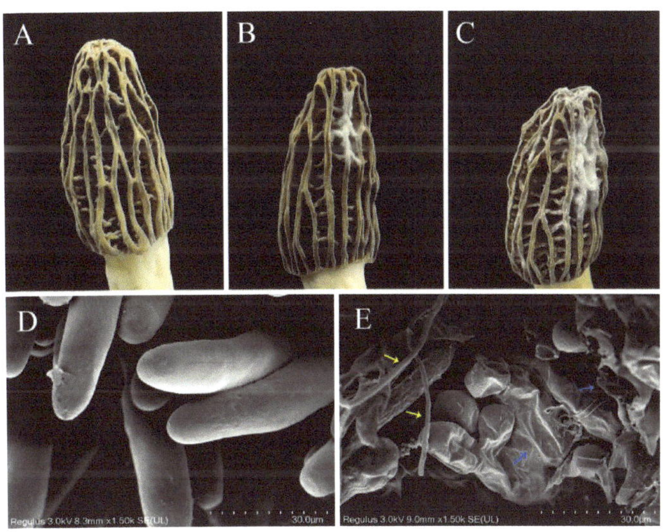

Figure 3. Pathogenicity testing and SEM observations of the fruiting body of *M. sextelata*. (**A**) Mycelial tissue treated with sterile water showed no symptoms. (**B**) Symptoms observed 3 days after wound inoculation of healthy *M. sextelata* with fungal suspensions. (**C**) Symptoms observed 6 days after wound inoculation of healthy *M. sextelata* with fungal suspensions. (**D**) Morphology of the asci of healthy *M. sextelata* via SEM. (**E**) Morphology of the asci of the disease-infected *M. sextelata*; hyphae of *P. longispora* (yellow arrow) crossing the fruiting body of *M. sextelata*, and the asci were crumpled and cracked (blue arrow).

To determine the effect of pathogenic fungal infection on the morphology and ultrastructure of *M. sextelata*, SEM was used. Healthy *M. sextelata* asci were cylindrical, full, and neatly arranged (Figure 3D). However, the mycelia of *P. longispora* invaded the *M. sextelata* along the interstitial space, causing the wall of the asci to shrink and ultimately leading to its rupture (Figure 3E).

3.3. The Response of M. sextelata to P. longispora Infection

Compared with those of healthy cells, the ultrastructures of the cells of *M. sextelata* with different degrees of infestation changed significantly (Figure 4A). The healthy cells had an orderly distribution of internal organelles and a small number of punctate substances. In the early infection stage, the number of punctate substances increased, and the substances were evenly dispersed in the *M. sextelata* cells. In the middle stage, the cell contents gradually flowed out, and the punctate substances were concentrated on the membrane. In the late stage, the cell wall became loose, the cell membrane was severely disrupted, and the number of punctate substances attached to the membrane obviously decreased. This dot-like substance exists in small quantities and is distributed irregularly in the healthy state but increases significantly after infestation by the pathogenic fungus and gradually concentrates on the membrane inside.

This study showed that *P. longispora* infestation leads to the rupture of *M. sextelata* cell membranes and the outflow of contents. To investigate whether *M. sextelata* can promote the growth of *P. longispora*, a dry powder of *M. sextelata* was added to the culture medium of *P. longispora*. The results demonstrated that the mycelial growth rate reached a maximum on medium supplemented with 2% *M. sextelata* powder and reached a minimum on the control PDA medium (Figure 4B,C). This indicates that *P. longispora* can utilize the nutrients of *M. sextelata* to promote its own growth.

A clear inhibition cycle between *P. longispora* and *M. sextelata* was observed (Figure 4D). The isolated strains significantly inhibited the mycelial growth of *M. sextelata* without touching *M. sextelata*. *P. longispora* can significantly inhibit the growth of *M. sextelata* by producing soluble substances. A significant brownish color at the antagonist line was observed on the back of the plate (Figure 4E). The *P. longispora* were inoculated onto plates covered with *M. sextelata* mycelia, and a distinct brown color also appeared around the inoculation site (Figure 4F,G). At the junction of the two strains, obvious brown pigment deposition can be observed. In the inhibition test of the *P. longispora* fermentation solution, the *M. sextelata* cultivated on the fermentation solution media were obviously inhibited compared with those cultivated on the control PDA medium (Figure 4H,I). Moreover, the mycelia on the medium containing fermentation broth were thinner and finer.

3.4. Functional Enrichment Analysis of DEGs

With a $|\log2FC| \geq 2$ and p-adjusted < 0.05, a total of 491 DEGs were identified in diseased *M. sextelata*, including 134 upregulated genes and 357 downregulated genes, compared to healthy controls. To clarify the functional enrichment of DEGs in *M. sextelata*, GO category analysis was applied. A total of 122 GO terms were enriched in the DEGs. Among them, transmembrane transporter activity (GO:0022857), transition metal ion binding (GO:0046914), oxidoreductase activity (GO:0016491), transporter activity (GO:0005215), monooxygenase activity (GO:0004497), and iron ion binding (GO:0005506) were in the molecular function category; sulfate assimilation (GO:0000103), monocarboxylic acid catabolic process (GO:0072329), and carbohydrate catabolic process (GO:0016052) were in the biological processes category; and intrinsic component of membrane (GO:0031224) and integral component of membrane (GO:0016021) were in the cellular component category among the top 20 GO terms (Figure 5A). DEGs related to oxidoreductase activity were screened, including mitochondrial peroxiredoxin, manganese superoxide dismutase, and catalase B. The highest numbers of DEGs enriched in the intrinsic component of membrane (GO:0031224) and integral component of membrane (GO:0016021) terms were 151 and 150, respectively. Several DEGs involved in integral components of the membrane, such as iron transport multicopper oxidase and peroxisomal membrane protein 2, were screened for analysis.

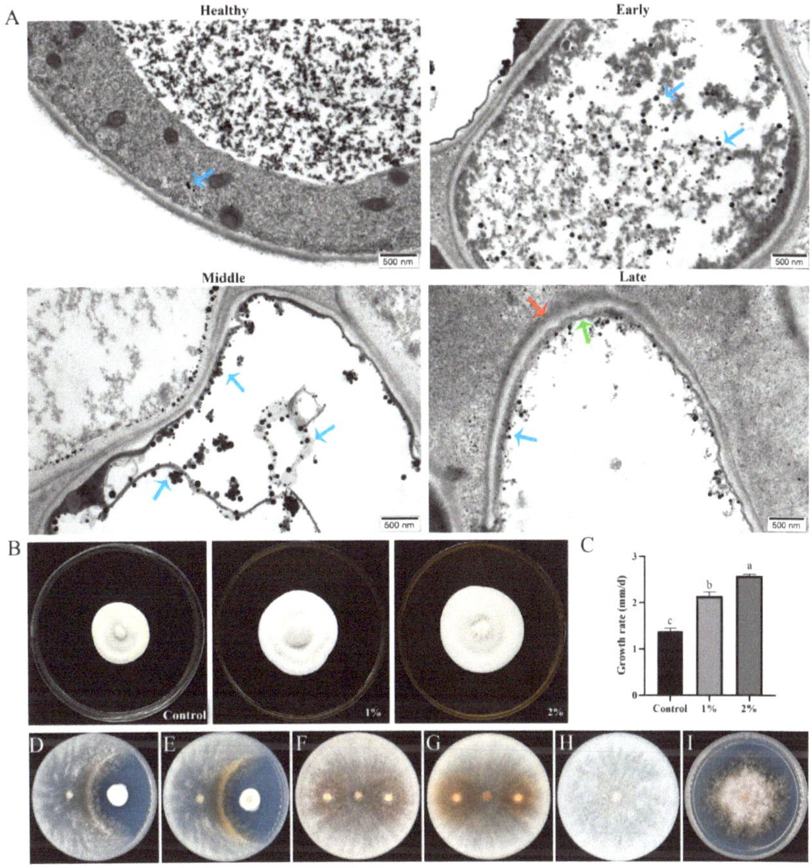

Figure 4. The process by which *P. longispora* infests the cells of *M. sextelata* and *M. sextelata* promotes the growth of *P. longispora*. (**A**) Process by which *P. longispora* infests *M. sextelata* cells. Complex of osmium tetroxide with a substance in the cell (blue arrow), cell wall (red arrow), and cell membrane (green arrow). (**B**) Colony morphology of *P. longispora* on media supplemented with different concentrations of *M. sextelata* powder for 10 days. (**C**) Growth rates of *P. longispora* on media supplemented with different concentrations of *M. sextelata* powder. Different letters indicate significant differences for the comparison of samples ($p < 0.05$ according to Duncan's test). (**D**,**E**) Confrontation cultures of *P. longispora* and *M. sextelata* on PDA media. (**F**,**G**) *P. longispora* cultured on mycelium of *M. sextelata*. (**H**) *M. sextelata* cultured on PDA for 9 days. (**I**) *M. sextelata* cultured on fermentation broth for 9 days.

KEGG pathway enrichment analysis was conducted in detail. The DEGs were mapped to 64 KEGG pathways, including 11 significantly enriched pathways: sulfur metabolism (ko00920), methane metabolism (ko00680), peroxisome (ko04146), glyoxylate and dicarboxylate metabolism (ko00630), citrate cycle (TCA cycle) (ko00020), galactose metabolism (ko00052), starch and sucrose metabolism (ko00500), tyrosine metabolism (ko00350), cyanoamino acid metabolism (ko00460), ascorbate and aldarate metabolism (ko00053), and fatty acid degradation (ko00071) (Figure 5B). Furthermore, peroxisomes (ko04146) were enriched with the greatest number of DEGs. Catalase B was significantly downregulated, and manganese superoxide dismutase was significantly upregulated by pathogen infection.

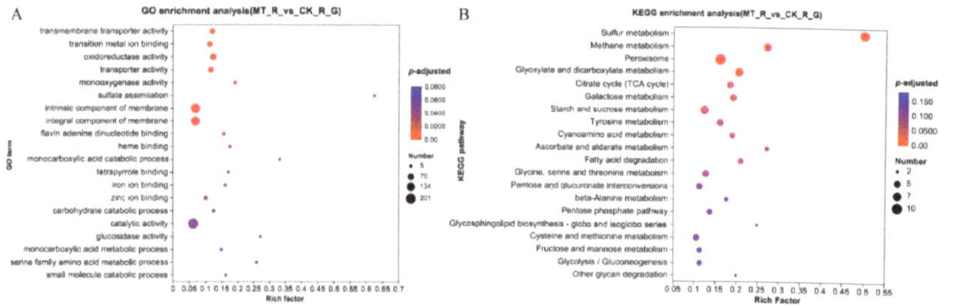

Figure 5. Functional enrichment analyses of DEGs in *M. sextelata*. (**A**) GO enrichment analyses of DEGs. GO term enrichment *p*-values are indicated on the *x*-axis. BP, biological process; MF, molecular function; CC, cellular component. (**B**) KEGG enrichment analyses of DEGs. The x-axis represents the enrichment factor. The size of the dot indicates the number of DEGs involved in the pathway. The color bars on the right represent the *p*-values of the KEGG pathway.

3.5. Functional Annotation and Enrichment Analysis of DEPs

For the identification of differentially expressed proteins (DEPs), FC > 2 or <0.5 and *p*-value < 0.05 were used as criteria. In total, we identified 447 DEPs, including 150 significantly upregulated proteins and 297 significantly downregulated proteins. A total of 88 GO terms were enriched in the DEGs. GO enrichment analyses revealed significant enrichment of the top 20 GO terms (Figure 6A), including catalytic activity (GO:0003824), oxidoreductase activity (GO:0016491), three hydrolase activities (GO:0004553, GO:0016787, GO:0016798), heme binding (GO:0020037), tetrapyrrole binding (GO:0046906), and flavin adenine dinucleotide binding (GO:0050660) in the category of molecular function; amide transport (GO:0042886), lipid catabolic process (GO:0016042), cellular lipid catabolic process (GO:0044242), fatty acid metabolic process (GO:0006631), lipid oxidation (GO:0034440), fatty acid beta-oxidation (GO:0006635), fatty acid oxidation (GO:0019395), triglyceride metabolic process (GO:0006641), and acylglycerol metabolic process (GO:0006639) in the biological processes category; and intracellular lipid transport (GO:0032365), intrinsic component of membrane (GO:0031224), and integral component of membrane (GO:0016021) in the cellular component category. KEGG pathway analysis revealed two important enrichment pathways: peroxisome (ko04146) and pentose and glucuronate interconversions (ko00040) (Figure 6B). Consistent with the RNA-seq results, the peroxisome (ko04146) pathway was the most enriched pathway among the DEPs. Heatmap analysis revealed that all 13 DEPs enriched in the peroxisome pathway were downregulated (Supplementary Figure S3).

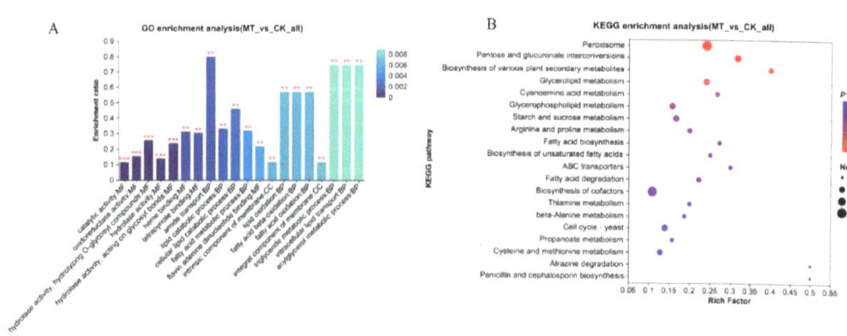

Figure 6. Functional enrichment analyses of DEPs in *M. sextelata*. (**A**) GO enrichment analyses of DEPs. Column color gradients indicate the significance of enrichment; *p* or FDR < 0.001 is labeled ***, and *p* or FDR < 0.01 is labeled **. (**B**) KEGG enrichment analyses of DEPs.

3.6. Combined Transcriptomics and Proteomics Analysis

Venn analysis revealed that the proteins encoded by the 99 DEGs were DEPs (Figure 7A). Among them, 97 DEGs, including 22 upregulated and 75 downregulated genes, showed the same change patterns as their encoded proteins (Figure 7B, Supplementary Tables S2 and S3). The changes in the expression of these two genes were opposite to those in the expression of their encoded proteins. GO functional enrichment analysis of 99 DEGs revealed that fatty acid catabolic process (GO:0006631), fatty acid beta-oxidation (GO:0006635), polysaccharide metabolic process (GO:0005976), fatty acid oxidation (GO:0019395), lipid oxidation (GO:0034440), polysaccharide catabolic process (GO:0000272), and fatty acid metabolic process (GO:0006631) were significantly enriched at both the mRNA and protein levels (Figure 7C). The expressions of two genes' (carboxylesterase and glycerol-1-phosphate phosphohydrolase 2) encoded proteins were jointly upregulated, and the expressions of five genes (peroxisomal hydratase-dehydrogenase-epimerase, mevalonyl-coenzyme A hydratase, sterol carrier protein 2, acyl-CoA ligase, and probable acyl-CoA dehydrogenase), whose encoded proteins are involved in lipid transport and metabolism, were jointly downregulated. The 22 upregulated DEGs coregulated with encoded proteins were significantly enriched in 36 GO terms, with 7 terms associated with the cell wall (Supplementary Figure S1). GO enrichment analyses were performed on 75 jointly downregulated genes (Supplementary Figure S2). Among them, 1,3-β-glucanosyltransferase was jointly upregulated in cell wall assembly, and endopolygalacturonase B was likely to jointly downregulate cell wall organization.

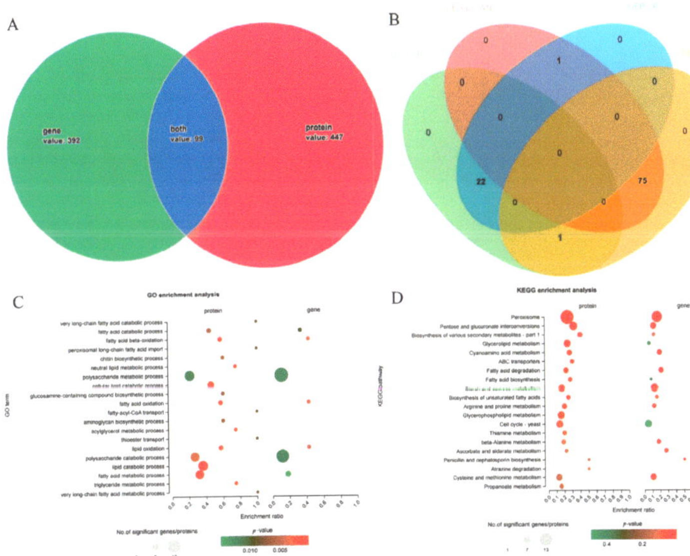

Figure 7. Combined transcriptomic and proteomic analysis. (**A**) Joint Wayne analysis of total DEGs/DEPs. (**B**) Joint Wayne analysis of shared factors. (**C**) GO enrichment analyses. (**D**) KEGG enrichment analyses.

According to the results of the combined KEGG pathway enrichment (Figure 7D), DEGs and DEPs were mainly enriched in peroxisomes (ko04146), pentose and glucuronate interconversions (ko00040), biosynthesis of various secondary metabolites (ko00999), and fatty acid degradation (ko00071). KEGG pathway enrichment analysis of the 22 DEGs whose expression was upregulated revealed that genes involved in beta-alanine metabolism (ko00410) and arginine and proline metabolism (ko00330) were significantly enriched. The 75 downregulated genes that exhibited the same change patterns as their encoded proteins were significantly enriched in pentose and glucuronate interconversions (ko00040), peroxisomes (ko04146), beta-alanine metabolism (ko00410), fatty acid degradation (ko00071), and penicillin and cephalosporin biosynthesis (ko00311).

3.7. Validation of Candidate Genes by qPCR Analysis

To verify the reliability of the RNA-seq results, 9 key DEGs were selected from three important terms and pathways, oxidoreductase activity, peroxisomes, cell wall and integral components of membrane, and subjected to qPCR analysis (Table 1). The results demonstrated that the relative expressions of these genes matched the RNA-seq data (Figure 8). This result confirmed the validity of the RNA-seq data.

Table 1. Information on key genes for qPCR analysis.

Gene ID	Description	Gene Name	Function	Organism
gene-H6S33_012182	Mitochondrial peroxiredoxin	*PRX1*	reduces hydrogen peroxide and alkyl hydroperoxides	*Aspergillus fumigatus* [36]
gene-H6S33_004573	Manganese superoxide dismutase	*SOD2*	neutralizes superoxide anion radicals and protects cells from oxidative stress	*Agaricus bisporus* [37]
gene-H6S33_004442	Catalase B	*catB*	serves to protect cells from the toxic effects of hydrogen peroxide	
gene-H6S33_008738	1,3-beta-glucanosyltransferase	*GAS4*	involved with Gas2p in spore wall assembly	*Saccharomyces cerevisiae* [38]
gene-H6S33_005983	Iron transport multicopper oxidase	*FET3*	multicopper oxidase that oxidizes ferrous (Fe^{2+}) to ferric iron (Fe^{3+}) for subsequent cellular uptake by transmembrane permease Ftr1p	*Fusarium graminearum* [39]
gene-H6S33_000069	Probable peptide transporter	*ptr2*	transport nitrogen-containing substrates	*Fusarium graminearum* [40]
gene-H6S33_000010	Cytochrome P450 monooxygenase	*ORF9*	involved in the metabolism of various endogenous substrates, including fatty acids, steroid hormones, and vitamins	*Penicillium roqueforti* [41]
gene-H6S33_005547	Peroxisomal membrane protein 2	*PXMP2*	involved in pore-forming activity and may contribute to the unspecific permeability of the peroxisomal membrane	NO
gene-H6S33_006216	Hexose transporter	*Hxt1*	monosaccharide transporter and sensor	*Ustilago maydis* [42]

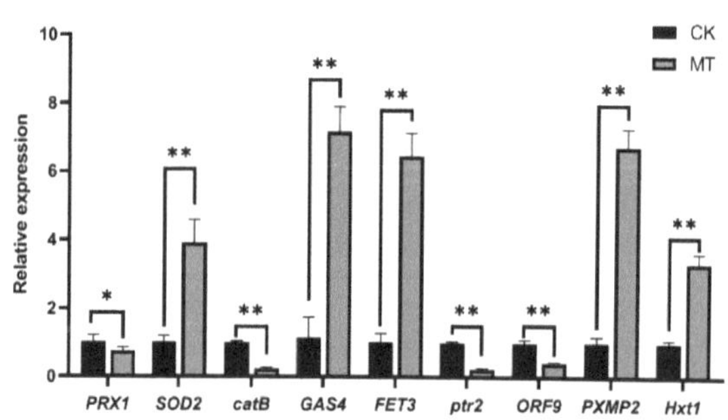

Figure 8. Validation of candidate genes via qPCR. CK: control group; MT: fermentation broth-treated group. Asterisk denotes statistically significant differences, * $p < 0.05$; ** $p < 0.01$.

In summary, integrated transcriptomics–proteomics analysis revealed that metabolic pathways related to the cell wall and cell membrane of *M. sextelata* play important roles in the response to *P. longispora*.

3.8. Changes in the Cell Membrane and Wall of *M. sextelata* Are Markers in Response to *P. longispora* Infection

Furthermore, electron microscopy was used to observe the effects of *P. longispora* on the cell membrane and cell wall of *M. sextelata* mycelia. The results showed that there were a large number of spherical attachments on the surface of the *M. sextelata* mycelia in the MT group (Figure 9B), whereas none were found on the surface of the *M. sextelata* mycelia in the control group (Figure 9A). Combined with the morphology of mycelia on plates, it was speculated that the spherical attachments might be caused by the rapid senescence of the mycelia. TEM observations of mycelial cells showed that the *P. longispora* fermentation filtrate resulted in loosening of the mycelial cell wall and rupture of the cell membrane (Figure 9D).

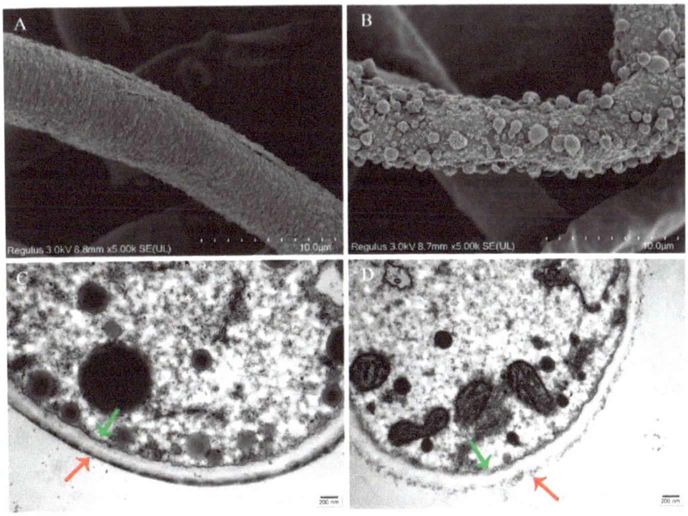

Figure 9. Effects of *P. longispora* metabolites on *M. sextelata* mycelial morphology. (**A**) *M. sextelata* mycelia surface of CK group. (**B**) *M. sextelata* mycelia surface of MT group. (**C**) *M. sextelata* mycelial cells of CK group. (**D**) *M. sextelata* mycelial cells of MT group. Cell wall (red arrow) and membrane (green arrow). CK: control group; MT: fermentation broth-treated group.

4. Discussion

White mold is one of the most prevalent diseases of morel mushrooms and is widespread throughout China. The production of morels in China is seriously threatened by large-scale and extremely invasive pathogenic fungi; however, little research has been conducted on the pathogenic mechanisms of this fungus. In this study, white mold samples were collected from the cultivation area of *M. sextelata* in Taigu County, Shanxi Province. The pathogen was identified as *P. longispora* by morphology, molecular biology, and pathogenicity testing. White mold disease in cultivated *M. importuna* caused by *P. longispora* (previously known as *Diploospora longispora*) was first reported in Chongqing, China [7]. In this study, when performing the dural culture method, a significant brownish color at the antagonist line was observed on the back of the plate. A similar pigment deposition phenomenon was also reported in *Lentinus edodes*, which is thought to be a defense measure [43]. Therefore, brown pigmentation may also be a defense measure of *M. sextelata* against *P. longispora*.

The cell wall is an important physical defense barrier for plants and fungi and plays an important role in resisting infection by pathogenic fungi [44]. The fungal cell wall mainly consists of an outer layer of glycoproteins and an inner skeletal layer of glucans and chitin [45]. In plants, maintaining cell wall integrity (CWI) is crucial for initiating and tracking defense responses [46]. Upon invasion by pathogens, the CWI system perceives changes in cell wall status to activate defense responses [47]. The modification of the chemical composition or structure of the cell wall is an active defense response against pathogen invasion [48]. In this study, cytological analysis revealed that after infection with *P. longispora*, the mycelial cell wall of *M. sextelata* was loose. In addition, chitin synthase I and 1,3-β-glucanosyltransferase activity were upregulated in *M. sextelata*, suggesting that these genes may be involved in cell wall remodeling [13]. Previous studies have also shown that pathogen infection can damage the cell walls of plants or fungi. Infection by *Magnaporthe oryzae* can cause damage to the cell wall of rice leaves, triggering a series of defense responses [49]. *Lycium barbarum* responds to *Fusarium solani* infection by regulating the composition of the cell wall, the stability of the cytoskeleton, and the function of the cell membrane [50]. The infection of *Pseudomonas* sp. can lead to the breakdown of the *Pleurotus eryngii* cell wall, resulting in the leakage of soluble sugars from the contents of *P. eryngii* [51]. Therefore, the CWI plays a crucial role in the stress response. However, it is still unknown how the integrity of edible mushroom cell walls perceives stress and activates defense responses.

The cell membrane plays an important role in material transport, energy conversion, and information transfer between the cell and the external environment [52]. In this study, pathogen infection induced the production of an unknown substance in *M. sextelata* cells, which accumulated on the cell membrane, leading to membrane breakage accompanied by severe cytoplasmic leakage. Previous studies have shown that *Alternaria alternate* toxins can cause changes in the membrane permeability of *Citrus reticulata* cells, leading to electrolyte permeation and affecting their normal metabolism by causing changes in membrane potential [53], which is similar to the results of this study. Lipids are the main components of cell membranes and play an indispensable role in maintaining the structural integrity of the cell; excessive oxidation of lipids changes the physical properties of cell membranes and leads to destruction of the cell membrane structure [54]. DEGs and DEPs were significantly enriched in the lipid oxidation process (Figure 6). Among them, *FET3*, encoding a multicopper oxidase with ferroxidase activity, is upregulated in infected *M. sextelata*. Early research revealed that peptaibols containing Aib residues can form helical structures that allow host cell lipid bilayer membrane iron channels to extend, causing cell death and internal leakage [55]. It is speculated that the high expression of *FET3* in *M. sextelata* may be involved in the process of cell membrane rupture. In addition, transmembrane transporter activity (GO:0022857) was the most enriched GO term in the infected *M. sextelata*. In this regard, the expressions of two oligopeptide transporter genes significantly decreased during the response of *M. sextelata* to disease. Research has shown that the proteins of the oligopeptide transport family can transport their reaction substrates from the external environment or organelles to the cytosol [56]. Therefore, it can be inferred that a decrease in the expression level of oligopeptide transporter genes is a marker of cell membrane damage.

Moreover, one of the first reactions observed when cells are damaged is a shift in the reduction oxidation condition [57]. In this study, DEGs and DEPs were significantly enriched in oxidoreductase activity (GO:0016491) and peroxisomes (ko04146). *SOD2* was significantly upregulated in *M. sextelata* after stimulation by *P. longispora*. Unexpectedly, *CAT* was significantly downregulated. Previous studies have shown that *SOD* and *CAT* are significantly upregulated in *Agaricus bisporus* infected with *Lecanicillium fungicola* [37]. It can be inferred that the antioxidant system also plays an important role in the response of *M. sextelata* to *P. longispora* infection.

5. Conclusions

In summary, white mold samples were collected from a cultivation greenhouse of *M. sextelata* in Taigu County, Shanxi Province, China. The pathogen was identified as *P. longispora* by morphology, molecular biology, and pathogenicity testing. This study clarified the response of *M. sextelata* cells to *P. longispora*. Analyses at the cytological level demonstrated that infection with *P. longispora* can induce the production of an unknown substance in the mycelia of *M. sextelata*, which accumulates on the cell membrane, leading to membrane breakage (Figure 10). Furthermore, integrated transcriptomics–proteomics analysis revealed the response mechanism of *M. sextelata* to *P. longispora* infection. A total of 491 DEGs and 447 DEPs were identified in diseased *M. sextelata*. Most of the DEGs and DEPs were related to oxidoreductase activity, peroxisomes, the cell wall, and integral components of membrane and lipid transport and metabolism processes, such as *PRX1*, *SOD2*, *catB*, *GAS4*, *FET3*, *ptr2*, *ORF9*, *PXMP2*, and *Hxt1*, which play important roles in the response to disease in *M. sextelata*. This study elucidated the mechanism underlying the response of *M. sextelata* to *P. longispora* and provided a foundation for elucidating the infection mechanism of *P. longispora*. In addition, the DEGs and DEPs screened in this study provide targets for the study of the disease resistance mechanism of *M. sextelata*.

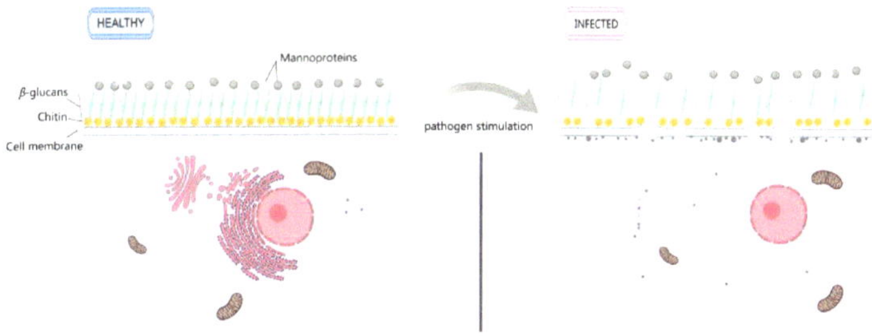

Figure 10. Comparison of healthy and infested *M. sextelata* cells.

Supplementary Materials: The following supporting information can be downloaded at: https://www.mdpi.com/article/10.3390/jof10090604/s1. Figure S1: GO enrichment analysis of jointly upregulated genes; Figure S2: GO enrichment analysis of jointly downregulated genes; Figure S3: Heatmap of the DEPs involved in the peroxisome pathway; Table S1: The primer for qPCR used in this study. Table S2: Joint upregulation of expressed gene information. Table S3: Joint downregulation of expressed gene information.

Author Contributions: L.H., S.W. and J.M.: Conceptualization, methodology, funding acquisition. J.W., T.W. and T.L.: Software, validation. Y.C. and M.C.: Data curation. S.W.: Writing—original draft preparation. J.W. and L.X.: Writing—review and editing. All authors have read and agreed to the published version of the manuscript.

Funding: This research was funded by the Shanxi Province Major Science and Technology Special Project "Open Recruitment and Leadership" Program (202301140601015), Shanxi Provincial Basic Research and Development Project (202103021223159), Shanxi Province Work Award Fund Research Project (SXBYK2022029), Doctoral Science Foundation of Shanxi Agricultural University (2021BQ90), and Special and Excellent Agricultural High Quality Development Science and Technology Support Project in 2024, Shanxi Agricultural University (TYGG24-40).

Institutional Review Board Statement: Not applicable.

Informed Consent Statement: Not applicable.

Data Availability Statement: All data generated or analyzed during this study are included in this published article and the supplementary information files.

Acknowledgments: We thank all members of the research team for their help.

Conflicts of Interest: The authors declare that they have no known competing financial interests or personal relationships.

References

1. Du, X.H.; Wang, H.; Sun, J.; Xiong, L.; Yu, J. Hybridization, Characterization and Transferability of SSRs in the Genus *Morchella*. *Fungal Biol.* **2019**, *123*, 528–538. [CrossRef] [PubMed]

2. Li, Y.; Chen, H.; Zhang, X. Cultivation, Nutritional Value, Bioactive Compounds of Morels, and Their Health Benefits: A Systematic Review. *Front. Nutr.* **2023**, *10*, 1159029. [CrossRef]
3. Ower, R. Notes on the Development of the Morel Ascocarp: *Morchella esculenta*. *Mycologia* **1982**, *74*, 142–144. [CrossRef]
4. Masaphy, S. Biotechnology of Morel Mushrooms: Successful Fruiting Body Formation and Development in A Soilless System. *Biotechnol. Lett.* **2010**, *32*, 1523–1527. [CrossRef]
5. Liu, Q.; Ma, H.; Zhang, Y.; Dong, C. Artificial Cultivation of True Morels: Current State, Issues and Perspectives. *Crit. Rev. Biotechnol.* **2017**, *38*, 259–271. [CrossRef] [PubMed]
6. Peng, W.; Tang, J.; He, X.; Chen, Y.; Tan, H. Status Analysis of Morel Artificial Cultivation in Sichuan. *Edible Med. Mushrooms* **2016**, *24*, 145–150.
7. He, P.; Li, C.; Cai, Y.; Zhang, Y.; Bian, Y.; Liu, W. First Report of Pileus Rot Disease on Cultivated *Morchella importuna* Caused by *Diploöspora longispora* in China. *J. Gen. Plant Pathol.* **2017**, *84*, 65–69. [CrossRef]
8. He, X.L.; Peng, W.H.; Miao, R.Y.; Tang, J.; Chen, Y.; Liu, L.X.; Wang, D.; Gan, B.C. White Mold on Cultivated Morels Caused by *Paecilomyces penicillatus*. *FEMS Microbiol Lett.* **2017**, *364*, fnx037. [CrossRef] [PubMed]
9. Guo, M.P.; Chen, K.; Wang, G.Z.; Bian, Y.B. First report of stipe rot disease on *Morchella importuna* caused by *Fusarium incarnatum—F. equiseti* species complex in China. *Plant Dis.* **2016**, *100*, 2530. [CrossRef]
10. Lan, Y.F.; Cong, Q.Q.; Wang, Q.W.; Tang, L.N.; Li, X.M.; Yu, Q.W.; Cui, X.; An, X.R.; Yu, C.X.; Kong, F.H. First Report of *Cladobotryum protrusum* Causing Cobweb Disease on Cultivated *Morchella importuna*. *Plant Dis.* **2019**, *104*, 977. [CrossRef]
11. Liu, Z.H.; Cong, Y.L.; Sossah, F.L.; Lu, Y.Z.; Kang, J.C.; Li, Y. Characterization and Genome Analysis of *Cladobotryum mycophilum*, the Causal Agent of Cobweb Disease of *Morchella sextelata* in China. *J. Fungi* **2023**, *9*, 411. [CrossRef]
12. Yu, M.; Yin, Q.; He, P.X. Isolation and Identification of Pathogen of Morel White Rot. *N. Hortic.* **2020**, *7*, 142–145. [CrossRef]
13. Yu, Y.; Tan, H.; Liu, T.; Liu, L.; Tang, J.; Peng, W. Dual RNA-Seq Analysis of The Interaction Between Edible Fungus *Morchella sextelata* and Its Pathogenic Fungus *Paecilomyces penicillatus* Uncovers the Candidate Defense and Pathogenic Factors. *Front. Microbiol.* **2021**, *12*, 760444. [CrossRef]
14. Fu, Y.; Xu, X.; Wu, H.; Li, L.; Wang, J.; Sun, Y.; Gu, L.; Yu, Q. First Report of Clonostachys Rosea Causing Rot of *Morchella sextelata* in Anhui Province, China. *Plant Dis.* **2023**, *107*, 1623. [CrossRef]
15. Lv, B.; Yu, S.; Chen, Y.; Yu, H.; Mo, Q. First Report of *Lecanicillium aphanocladii* Causing Rot of *Morchella sextelata* in China. *Plant Dis.* **2022**, *106*, 3202. [CrossRef] [PubMed]
16. Fu, Y.; Wu, H.; Wang, S.; Yu, Q.; Tian, D.; Xu, X. First Report of *Trichoderma atroviride* Causing Rot of *Morchella sextelata* in Anhui Province, China. *Crop Prot.* **2023**, *168*, 106206. [CrossRef]
17. Shi, X.F.; Liu, D.; He, X.; Liu, W.; Yu, F. Epidemic Identification of Fungal Diseases in *Morchella* Cultivation Across China. *J. Fungi* **2022**, *8*, 1107. [CrossRef]
18. Garajova, S.; Mathieu, Y.; Beccia, M.R.; Bennati-Granier, C.; Biaso, F.; Fanuel, M.; Ropartz, D.; Guigliarelli, B.; Record, E.; Rogniaux, H.; et al. Single-domain Flavoenzymes Trigger Lytic Polysaccharide Monooxygenases for 0xidative Degradation of Cellulose. *Sci. Rep.* **2016**, *6*, 28276. [CrossRef]
19. Dong, W.; Chen, B.S.; Zhang, R.; Dai, H.Q.; Han, J.; Lu, Y.; Zhao, Q.; Liu, X.; Liu, H.; Sun, J. Identification and Characterization of Peptaibols As the Causing Agents of *Pseudodiploöspora longispora* Infecting the Edible Mushroom *Morchella*. *J. Agric. Food Chem.* **2023**, *71*, 18385–18394. [CrossRef]
20. Xie, J.T.; Liu, X.; Qin, Z.L.; Mei, S.H.; Tarafder, E.; Li, C.; Zeng, X.Y.; Tian, F.H. Evolution and Related Pathogenic Genes of *Pseudodiploöspora longispora* on *Morchella* Based on Genomic Characterization and Comparative Genomic Analysis. *Sci. Rep.* **2024**, *14*, 18588. [CrossRef]
21. Han, M.; Wang, Q.S.; Baiyintala; Wuhanqimuge. The Whole-genome Sequence Analysis of *Morchella sextelata*. *Sci. Rep.* **2019**, *9*, 15376. [CrossRef]
22. Tan, H.; Kohler, A.; Miao, R.; Liu, T.; Zhang, Q.; Zhang, B.; Jiang, L.; Wang, Y.; Xie, L.; Tang, J.; et al. Multi-omic Analyses of Exogenous Nutrient Bag Decomposition by the Black Morel *Morchella importuna* Reveal Sustained Carbon Acquisition and Transferring. *Environ. Microbiol.* **2019**, *21*, 3909–3926. [CrossRef] [PubMed]
23. Wang, X.; Peng, J.; Sun, L.; Bonito, G.; Guo, Y.; Li, Y.; Fu, Y. Genome Sequencing of *Paecilomyces penicillatus* Provides Insights into Its Phylogenetic Placement and Mycoparasitism Mechanisms on Morel Mushrooms. *Pathogens* **2020**, *9*, 834. [CrossRef]
24. Jia, L.J.; Tang, H.Y.; Wang, W.Q.; Yuan, T.L.; Wei, W.Q.; Pang, B.; Gong, X.M.; Wang, S.F.; Li, Y.J.; Zhang, D.; et al. A Linear Nonribosomal Octapeptide from *Fusarium graminearum* Facilitates cell-to-cell Invasion of Wheat. *Nat. Commun.* **2019**, *10*, 922. [CrossRef] [PubMed]
25. Du, F.Y.; Li, X.M.; Sun, Z.C.; Meng, L.H.; Wang, B.G. Secondary Metabolites with Agricultural Antagonistic Potentials from *Beauveria felina*, a Marine-Derived Entomopathogenic Fungus. *J. Agric. Food Chem.* **2020**, *68*, 14824–14831. [CrossRef]
26. Li, B.; Yuan, B.; Duan, J.; Qin, Y.; Shen, H.; Ren, J.; Francis, F.; Chen, M.; Li, G. Identification of Fcl-29 as an Effective Antifungal Natural Product against *Fusarium graminearum* and Combinatorial Engineering Strategy for Improving Its Yield. *J. Agric. Food Chem.* **2023**, *71*, 5554–5564. [CrossRef]
27. Ni, Y.Y.; Mu, J.Y.; Sang, W.J. Identification of Aniseed Leaf Spot Pathogen with Toxicity Evaluation of Five Fungicides. *J. Mt. Agric. Biol.* **2010**, *29*, 360–363. [CrossRef]
28. Tietel, Z.; Masaphy, S. True Morels (*Morchella*)—Nutritional and Phytochemical Composition, Health Benefits and Flavor: A Review. *Crit. Rev. Food Sci. Nutr.* **2017**, *58*, 1888–1901. [CrossRef]

29. Zheng, Z.K.; Kong, L.; Dai, M.; Chen, Y.D.; Chen, Y.H. ADSC-Exos Outperform BMSC-Exos in Alleviating Hydrostatic Pressure-Induced Injury to Retinal Ganglion Cells by Upregulating Nerve Growth Factors. *World J. Stem Cells* **2023**, *15*, 1077–1092. [CrossRef]
30. Huang, H.; Zhang, X.Y.; Zheng, H.; Cai, Q.Y.; Luo, J.J.; Luo, K.; Li, S.J.; Yang, Y.L. Identification and Cultural Characterization of *Diplospora Longispora* Associated with Pileus Rot Disease on Cultivated Morel. *Plant Prot.* **2022**, *48*, 66–72.
31. White, T.J.; Bruns, T.; Lee, S.; Taylor, J.W. Amplification and Direct Sequencing of Fungal Ribosomal RNA Genes for Phylogenetics. In *PCR Protocols: A Guide to Methods and Applications*; Innis, M.A., Gelfand, D.H., Sninsky, J.J., White, T.J., Eds.; Academic Press: New York, NY, USA, 1990; pp. 315–322.
32. Zhao, D.H.; Zhuang, F.Y.; Ou, C.G.; Yao, X.C.; Liang, C.; Zhang, Y.F.; Liu, X. First Report of Black Rot of Carrot Caused by *Alternaria carotiincultae* in China. *Plant Dis.* **2023**, *108*, 223. [CrossRef]
33. Kumar, S.; Stecher, G.; Li, M.; Knyaz, C.; Tamura, K. MEGA X: Molecular evolutionary genetics analysis across computing platforms. *Mol. Biol. Evol.* **2018**, *35*, 1547–1549. [CrossRef] [PubMed]
34. Zhu, X.T.; Ma, K.L.; Sun, M.Y.; Zhang, J.M.; Liu, L.J.; Niu, S.Q. Isolation and Ddentification of Pathogens of *Morchella sextelata* Bacterial Disease. *Front. Microbiol.* **2023**, *14*, 1231353. [CrossRef]
35. Samson, R.A. Paecilomyces and Some Allied Hyphomycetes. *Stud. Mycol.* **1974**, *6*, 1–119.
36. Rocha, M.C.; de Godoy, K.F.; Bannitz-Fernandes, R.; Fabri, J.H.T.M.; Barbosa, M.M.F.; de Castro, P.A.; Almeida, F.; Goldman, G.H.; da Cunha, A.F.; Netto, L.E.S.; et al. Analyses of the three 1-Cys Peroxiredoxins from Aspergillus fumigatus reveal that cytosolic *Prx1* is central to H_2O_2 metabolism and virulence. *Sci. Rep.* **2018**, *8*, 12314. [CrossRef] [PubMed]
37. Quiroz, L.F.; Ciosek, T.; Grogan, H.; McKeown, P.C.; Spillane, C.; Brychkova, G. Unravelling the Transcriptional Response of *Agaricus Bisporus* under *Lecanicillium Fungicola* Infection. *Int. J. Mol. Sci.* **2024**, *25*, 1283. [CrossRef] [PubMed]
38. Ragni, E.; Coluccio, A.; Rolli, E.; Rodriguez-Peña, M.J.; Colasante, G.; Arroyo, J.; Neiman, A.M.; Popolo, L. GAS2 and GAS4, A Pair of Developmentally Regulated Genes Required for Spore Wall Assembly in *Saccharomyces cerevisiae*. *Eukaryot. Cell* **2007**, *6*, 302–316. [CrossRef]
39. Zheng, Z.T.; Liu, H.Q.; Luo, X.; Liu, R.Z.; Joe, D.A.; Li, H.L.; Sun, H.Y.; Lin, Y.L.; Li, Y.Z.; Wang, Y.P. Comparative Transcriptome Analysis Provides Insights into the Resistance Regulation Mechanism and Inhibitory Effect of Fungicide Phenamacril in *Fusarium Asiaticum*. *Pestic. Biochem. Physiol.* **2024**, *201*, 105848. [CrossRef]
40. Droce, A.; Sørensen, J.L.; Sondergaard, T.E.; Rasmussen, J.J.; Lysøe, E.; Giese, H. PTR2 Peptide Transporters in *Fusarium Graminearum* Influence Secondary Metabolite Production and Sexual Development. *Fungal. Biol.* **2017**, *121*, 515–527. [CrossRef]
41. Coton, E.; Coton, M.; Hymery, N.; Mounier, J.; Jany, J.L. Penicillium Roqueforti: An Overview of Its Genetics, Physiology, Metabolism and Biotechnological Applications. *Fungal Biol. Rev.* **2020**, *34*, 59–73. [CrossRef]
42. Schuler, D.; Wahl, R.; Wippel, K.; Vranes, M.; Münsterkötter, M.; Sauer, N.; Kämper, J. *Hxt1*, A Monosaccharide Transporter and Sensor Required for Virulence of the Maize Pathogen Ustilago Maydis. *New Phytol.* **2015**, *206*, 1086–1100. [CrossRef]
43. Li, G.L.; Xu, Y.; Xing, P.J.; Gao, T.T.; Ji, R.Q. Pathogenicity of *Trichoderma* spp. to Lentinula Edodes and the Inhibitory Effects of Biocontrol Agents. *Mol. Plant Breed.* **2019**, *17*, 6530–6534. [CrossRef]
44. Cantu, D.; Vicente, A.R.; Labavitch, J.M.; Bennett, A.B.; Powell, A.L. Strangers in the Matrix: Plant Cell Walls and Pathogen Susceptibility. *Trends Plant Sci.* **2008**, *13*, 610–617. [CrossRef] [PubMed]
45. Arana, D.M.; Prieto, D.; Roman, E.; Nombela, C.; Alonso-Monge, R.; Pla, J. The Role of the Cell Wall in Fungal Pathogenesis. *Microb. Biotechnol.* **2009**, *2*, 308–320. [CrossRef] [PubMed]
46. Vaahtera, L.; Schulz, J.; Hamann, T. Cell Wall Integrity Maintenance During Plant Development and Interaction with the Environment. *Nat. Plants.* **2019**, *5*, 924–932. [CrossRef] [PubMed]
47. Gigli-Bisceglia, N.; Engelsdorf, T.; Hamann, T. Plant Cell Wall Integrity Maintenance in Model Plants and Crop Species-Relevant Cell Wall Components and Underlying Guiding Principles. *Cell. Mol. Life Sci.* **2020**, *77*, 2049–2077. [CrossRef]
48. Shi, Y.Y.; Li, D.Y.; Zhang, H.J.; Song, F.M. Cell Wall-Mediated Disease Resistance and its Molecular Mechanism in Plants. *Plant Physiol. J.* **2011**, *47*, 661–668. [CrossRef]
49. Liang, Y.L.; Zhao, J.; Liu, L.; Yang, J.; Li, C.Y. The Role of the Plant Cell Wall in Plant Pathogen Interactions. *Mol. Plant Breed.* **2016**, *14*, 1255–1261. [CrossRef]
50. Chi, L.X.; Xie, Z.P.; Feng, R.; Du, X.Q.; Lu, X.Y.; Fan, W.Q.; Gao, H.; Yue, S.J.; Zheng, R. Comparative Transcriptome Analysis of *Lycium Barbarum* Response to *Fusarium Solani* Stress. *Mol. Plant Breed.* 2024, *in press*.
51. Hu, J.J. Study on the Identification and Infection Mechanism of *Pleurotus eryngii* Pathogenic Bacteria. Master's Thesis, South China Agricultural University, Guangzhou, China, 2018.
52. Liu, J.B.; Wang, L.H.; Zuo, X.L. Cell Membranes Functionalization Based on DNA. *Prog. Chem.* **2019**, *31*, 1067–1074. [CrossRef]
53. Kang, Z.T.; Jiang, L.M.; Luo, Y.Y.; Liu, C.J.; Li, X.R. The Research Advances of Mechanism of Pathogenicity of Alternaria Phytopathogenic Fungi. *Chin. Bull. Life Sci.* **2013**, *25*, 908–914. [CrossRef]
54. Wiczew, D.; Szulc, N.; Tarek, M. Molecular Dynamics Simulations of the Effects of Lipid Oxidation on the Permeability of Cell Membranes. *Bioelectrochemistry* **2021**, *141*, 107869. [CrossRef] [PubMed]
55. Nagaoka, Y.; Iida, A.; Kambara, T.; Asami, K.; Tachikawa, E.; Fujita, T. Role of Proline Residue in the Channel-Forming and Catecholamine-Releasing Activities of the Peptaibol, Trichosporin-B-VIa. *Biochim. Biophys. Act.* **1996**, *1*, 31–36. [CrossRef] [PubMed]

56. Lubkowitz, M. The Oligopeptide Transporters: A Small Gene Family with A Diverse Group of Substrates and Functions? *Mol. Plant* **2011**, *4*, 407–415. [CrossRef]
57. Frederickson Matika, D.E.; Loake, G.J. Redox Regulation in Plant Immune Function. *Free Radic. Biol. Med.* **2014**, *21*, 1373–1388. [CrossRef]

MDPI AG
Grosspeteranlage 5
4052 Basel
Switzerland
Tel.: +41 61 683 77 34

Journal of Fungi Editorial Office
E-mail: jof@mdpi.com
www.mdpi.com/journal/jof

www.ingramcontent.com/pod-product-compliance
Lightning Source LLC
Chambersburg PA
CBHW041409210926
51850CB00013B/3218
* 9 7 8 3 7 2 5 8 3 9 9 4 0 *